Zambian Plants
used as
Traditional
Fever Cures

To my dear wife Ena, for her loving support

By the same author:

A Dictionary of Ila Usage, 1860–1960
The Ila Speaking: records of a lost world
Zambian Plants: their vernacular names and uses

Zambian Plants
used as
Traditional
Fever Cures

Dennis G. Fowler

Kew Publishing
Royal Botanic Gardens, Kew

ROYAL BOTANIC GARDENS

First published in 2011 by
Royal Botanic Gardens, Kew
Richmond, Surrey, TW9 3AB, UK
www.kew.org

ISBN 978 1 84246 460 1

British Library Cataloguing in Publication Data
A catalogue record for this book is available from the British Library

Cover design by Publishing, Design & Photography, Royal Botanic Gardens, Kew.

Front Cover Illustration: *Senna occidentalis* – a noxious invasive weed hated by farmers which happens to be also the most widely used of all the fever remedies in this book.

Back Cover Illustration: *Artemisia afra* – a a graceful shrub smelling strongly of wormwood. The most widely used of all plants by traditional practitioners in east Africa to treat malaria.

Printed in the United Kingdom, Australia and the U.S.A. by Lightning Source

For information or to purchase all Kew titles please visit
www.kewbooks.com or email publishing@kew.org

Kew's mission is to inspire and deliver science-based plant conservation worldwide, enhancing the quality of life.

Kew receives half of its running costs from Government through the Department for Environment, Food and Rural Affairs (Defra). All other funding needed to support Kew's vital work comes from members, foundations, donors and commercial activities including book sales.

CONTENTS

FOREWORD

I am delighted to have the opportunity to write this foreword for Dennis Fowler's book *Zambian Plants Used as Traditional Fever Cures*. We live at a time when one in five of the world's plant species is threatened with extinction, and I suspect that if we had the figures they would indicate that loss of knowledge about the usefulness of plants follows a similar trend. Both are vital to our continued ability to adapt and survive on this planet. Traditional knowledge has led to the development of many of our most effective and important medicines, particularly in the treatment of malaria and fever – from quinine to the *Artemisia*-based drugs of today. If we forget what we have learnt over tens of thousands of years, and we allow plants that have evolved over millions of years to become extinct, then we will have failed in our role as custodians for future generations.

The traditional cures that are captured in this book add significantly to our common knowledge of plants and medicine and, regardless of immediate or future application, this publication ensures that the knowledge is not lost. I would like to pay tribute to the author's dedication to this cause. Much of this kind of information resides in the grey literature or in rough notes in someone's attic. Dennis Fowler has meticulously been through both the raw data and the literature, and synthesised it into a format that is readily accessible to all.

Whether you are a medical researcher, a botanist, an entomologist, an anthropologist or an interested reader, I can guarantee that you will find this book useful and entertaining.

Paul Smith
January 2011

ACKNOWLEDGEMENTS

My thanks are due to Dr Olwen Grace, of Kew and Copenhagen University, to Dr Merlin Willcox, of the Research Initiative on Traditional Antimalarial Methods, and to Dr Paul Smith, of the Royal Botanic Gardens, Kew, without whose help, advice, and encouragement this book would never have been written; and to the many pioneer missionaries, travellers and District Officers whose records of Zambian culture of a century ago furnished much of the material for it.

INTRODUCTION

The plants listed below are all found in Zambia, but the uses described are from many countries. Much of the traditional medical lore in Zambia has died with the old men who were its repositories and guardians. At the same time, the shortage of many western drugs in hospitals, together with the high cost of those which are available, has led to a growing health crisis.

It is in this context that the following lists are presented. Many of the sources are over fifty years old, some more than a century. More than half of them do not specify malaria, but refer to "fever". However, the vast majority of "fevers" in Zambia are in fact malarial.

No doubt, most of the remedies listed below were effective only in relieving symptoms, and were not able to destroy the malaria parasites in their victims' bodies. Nevertheless, plant-derived compounds hold promise for innovations in modern medicine, and it is possible that one or more may be found to be the basis of the next generation of anti-malarial drugs.

How to use this book

For each plant, the following information is given:
a) The NAME(S) of the plant, its family, and the botanist(s) who described it are presented in the conventional scientific shorthand. Obselete names are given in square brackets.

b) A simple DESCRIPTION of the plant and its typical locations

c) VERNACULAR NAMES. These have to be presented in a very compressed fashion. More than seventy languages are used in Zambia, and to try to record all the languages having the same name for a particular plant would have rendered the list too unwieldy for practical use.They have therefore been grouped according to regions, as follows:
B: Bemba/ Ambo/ Ausi/ Bisa/ Cisinga/ Kabende/ Kambonsenga/ Kawendi/ Lala/ Mukulu/ Ngumbo/ Syila/ Swaka/ Taabwa/ Unga
I: Ila
K: Kaonde
Ku: Kunda
Lb: Lamba/ Lima/ Seba
Le: Lenje/ Soli
Ln: Luba/ Lunda/ Ndembo
Lo: Lozi/ Kwangwa/ Simaa/ Totelai
Lv: Luvale/ Cokwe/ Lucazi/ Bunda
M: Mambwe/ Fipa/ Inamwanga/ Iwa/ Lungu/ Nkhonde
N: Nkoya/ Lukolwe/ Lusyange
Nm: Nyamwezi/ Sukuma
Ny: Nyanja/ Cewa/Ngoni/ Nsenga/ Tonga of Malawi
Tk: Toka
To: Plateau Tonga/ Valley Tonga
Tu: Tumbuka cluster: Fungwe/ Henga/ Kamanga/ Lambya/ Nyika/ Poka/ Senga/ Tambo/ Wandya/ Wenya/ Yombe

SPELLING & PRONUNCIATION

Since the vernacular names were recorded over more than half a century, during which there was no uniform system of orthography, the sources often show great variety in spelling the same word. One consonant might be represented as "d", "l", or "r" by different writers, another as "b" or "w", another as "b" or "p". Earlier authors wrote "ch", "sh", and "zh", whereas "c", "sy", and "zy" are preferred today.

The falling tone at the end of a word might be written as "a" by one authority, while others might hear it rather as "e" or "i", while "o" and "u" are equally hard to distinguish.

Many were the attempts to reproduce the peculiar sound found in the English word "singing". Early sources write 'ng, 'g, g', and ng. These days the symbol "ŋ" is used for the sound, and placed after "n" in the alphabet.

In addition, local variations in dialect were marked, at a period when there was little intercourse between villages and regions. One informant's "gu" was another's "ku". A "mo" heard at Nanzyila would be "mu" at Maala, twenty miles away. The Ila people, lacking front teeth, pronounced the consonant s as "sh", where others pronounced it as "s" or "sy". Since "sh" is not used today, (except for personal and place-names), the lists below show some unavoidable confusion, and the reader should try both "s" and "sy" when using them.

The same name is often applied to different species. In some cases this may be due to mistaken identification or mispronunciation, in others, to regional differences. In cases where the solution is not obvious, the editor has included all the variations, and left the reader to draw his own conclusions from the sources.

Much work remains to be done, and users of this book should regard the vernacular lists as an approximate guide only.

The best way to show how the list works is by giving an example chosen at random. The entry for *Peltophorum africanum* on p. 113 lists 28 names. English is included as being the most widely spoken "second language" in Zambia, and Afrikaans is found in many of the reference books. In the case of other vernacular names, the language of each is indicated by letters referring to the key on p.1 above.

Thus, a botanist searching for specimens of *P. africanum* in the Ila area and seeking local guidance might ask for I: **mukasala** (10), I/Le/Lo/Tk/To: **munyeele** (3, 3a, 4, 8, 10, 17, 22, 23), or I/To: **mungele** (19).

These letters are explained in the language key above. They indicate that **mukasala** is found only in Ila, **mungele** in Ila and Tonga, while **munyeele** is common to Ila, Lenje, Lozi, Toka,and Tonga. The sources fromwhich these vernacular names were obtained are indicated by the numbers in brackets after each name, which are linked with the references in the "VERNACULAR NAMES sources" on p. 287. Thus,

mukasala is given by 10 (Palgrave 1957)
munyeele is given by 3 (Fanshawe 1995), 3a (Fanshawe 1962, 1968), 4 (Fowler 2000), 8 (Mitchell 1963), 10 (Palgrave 1957), 17 (Torrend 1931), 22 (Watt and Breyer-Brandwijk 1962) and 23 (White 1962).
mungele is given by 19 (Trapnell 2000)
Each of the other names for *P. africanum* is shown with its sources in the same way.

d) The PLANT PARTS used in the treatments are listed.

e) The COUNTRIES in which the use of the plant is reported are listed alphabetically.

f) <u>TRADITIONAL USES</u>: In each country, documented methods of treatment used by traditional healers are given, with the names of sources, as listed in "Citation References" on p. 162

g) <u>TESTS</u>: Brief summaries of any laboratory tests are given.

h) <u>TOXICOLOGY</u>: Details are given where applicable.

SECTION 1: MALARIA REMEDIES

Abrus fruticulosus Wall. ex Wight & Arn. [*A. pulchellus* Wall. ex Thwaites; *A. schimperi* Hochst. ex Baker] (Fabaceae; Papilionoidae)
DESCRIPTION:
Woody shrub, occurring on rocky hillsides and open savannah woodland
VERNACULAR NAMES:
Nm: mgagati, ugegati (22)
PLANT PARTS USED:
leaves
TRADITIONAL USES:
W. Africa: the leaf is decocted and used as a wash to treat fever: *Berhaut (1976) pp. 22, 23, Vergiat (1970) pp. 171-199,* quoted in *Burkill (1995) p. 271*

Abrus precatorius L. (Fabaceae; Papilionoidae)
DESCRIPTION:
Woody twining climber. Widespread by rivers or lakes, or in mixed deciduous woodland, often in thickets.
VERNACULAR NAMES:
I: mutipilili (17) **I/ Le**: mutipi-tipi (17) **Ku**: kalunguti (1) **Ku/ Ny**: nyumbu (1) **Ln**: muzaviji (2, 3, 9a) **Lo**: munkongi (8), mupiti (2, 3, 3d, 9a, 23) **Lo/ N**: syimonyi-monyi (3d, 16) **Lo/ Tk**: mupiti-piti (2, 3, 3d, 9a, 23) **Lo/ To**: mutiti (17, 22) **Lv**: mukenjenge (22), mukenyenge (6, 23), mukube (6) **Nm**: kacence (1a, 22), lufyambo (1a) **Ny**: masowa-pusi (2, 3, 3e, 9a), mkanda-nyulugwe, nsiyapita, ntimba, ulanga-wiyu (9) **Tk/ To**: musolo-solo (2, 3, 9a, 23) **To**: musansi (17) **Tu**: munyunyu (8)
PLANT PARTS USED:
Leaves, roots
TRADITIONAL USES:
Benin: the roots are used to prepare a fever remedy: *Adjanohoun et al. (1989)*
Madagascar: the leaves are decocted to prepare a remedy: *Pernet et al. (1957)*; *Rasoanaivo et al. (1992)*; *Riviére et al. (1995)*
Malawi: a leaf infusion is drunk like tea as a remedy for fevers: *Morris (1996) p. 360*
Tanzania: the leaves are crushed and the juice drunk: *Chhabra et al. (1990)*; a roots decoction is drunk as a febrifuge: *Gessler et al. (1995)*

Abutilon mauritianum (Jacq.) Medic. (Malvaceae)
DESCRIPTION
Semi-woody perennial weed, widespread in tropical Africa
VERNACULAR NAMES:
E: bush mallow, country mallow
PLANT PARTS USED:
leaf sap
TRADITIONAL USES:
Tanzania: the leaf sap is drunk as a remedy for malaria: *Haerdi (1964) p. 87*

Acacia hockii De Wild. [*A. stenocarpa* auct. non Hochst.] (Fabaceae; Mimosoideae)
DESCRIPTION
Thorny shrub or small tree up to 4m. Widespread, in thickets or semi-deciduous woodland.

VERNACULAR NAMES:
B: cibombo (23) **Ku**: kalasa, usongwe, uzimwe (1) **Ln**: katete (3) **Nm**: mnye-nyele (22) **Ny**: kafifi (3e)
PLANT PARTS USED:
leaf sap, roots
TRADITIONAL USES:
<u>W. Africa</u>: a decoction of the roots mixed with leaf sap is used to treat malaria: *Burkill (1995) p. 183; Haerdi (1964) p. 47*

Acacia mellifera (Vahl) Benth. subsp. detinens (Burch.) Brenan (Fabaceae; Mimosoideae)
DESCRIPTION
A rounded shrub or small tree, usually leafless, with hooked thorns. Found in *mopane* country.
VERNACULAR NAMES:
Lo: kakumbwe (3, 3d), muhani, mukoko (22) **Nm**: murugara (22) **Tk**: munganga (23)
PLANT PARTS USED:
Bark, roots
TRADITIONAL USES:
<u>E. Africa</u>: A bark decoction is drunk as a remedy: *Kokwaro (1976) p. 125*
<u>Somalia</u>: a handful of fresh stembark, crushed, boiled for 20 m. in 1 litre of water, 3 small cups/day/3 days: *Samuelsson et al. (1992) cited in Prelude med. plants database (2011)*
<u>Tanzania</u>: the bark or roots are infused to make a healing drink: *Johns et al. (1994)*

Acacia nilotica (L.) Willd. ex Delile subsp. kraussiana (Benth.) Brenan [*A. benthamii* Rochebr.; *A. arabica* Willd.] (Fabaceae; Mimosoideae)
DESCRIPTION
Small spiny tree up to 10m, with spreading flat or rounded top. Found in *mopane* and *Julbernardia* woodland.
VERNACULAR NAMES:
B/ Ku: kafifi (1) **I/ Le/ Lo/ To**: mugoga (19) **I/ To**: muwombe (1b, 19) **K**: mwiba (3) **Ku**: ocenca, uzimwe (1) **K/ Tu**: ocenca (1, 8) **Lo**: mukoto-koto (3d), muwonzo (22), muswe-benga (7a) **Lo/ To**: mukoka (23) **Nm**: mdubilu (22) **Ny**: ciseyo, licongwe, mcezime, mtete, namalonga, ngagaga (9), cisiyo (3e, 7a), denganya (3e), khawa, sehoho (22), uzimwe (3, 3e) **Tk**: m'onzo (23) **To**: mukona (3), muwo-mbege (19, 20), nsyangwa (22) **Tu**: kabilili (20)
PLANT PARTS USED:
bark, gum
TRADITIONAL USES:
<u>E. Africa</u>: a bark decoction is drunk by feverish children : *Kokwaro (1976) p. 126*
<u>Egypt</u>: an infusion of powdered fruit is used to treat fevers: *Boulos (1983)* cited in *Prelude med. plants database (2011)*
<u>W. Africa</u>: the bark is infused and the pod decocted as a febrifuge: *Burkill (1995) p. 189*
<u>Zambia</u>: fresh gum is used in a remedy for malaria: *Haapala et al.(1994) p. 8*
TESTS:
In vitro tests of a bioassay guided fractionation on *P. falciparum* showed that the ethyl acetate extract presented an ID_{50} activity of 1.50 mc/ ml: *El-Tahir et al. (1999a)*
Methanol and aqueous extracts of the tree were tested against *P.falciparum* NF54 and ENT30 strains): *Kirira et al.,(2006)*

TOXICITY:
The above extracts, when tested on lymphocyte proliferation, also revealed low toxicity to human cells. For BST tests on stem bark, see *Adoum (2009) p. 137*. Showed as cytotoxic when tested on snails at 50-100 ppm: *Elsheikh et al.(1990)*

Acacia polyacantha Willd. subsp. campylacantha (Hochst. ex A. Rich.) Brenan (Fabaceae; Mimosoideae)
DESCRIPTION:
Slender spiny flat-topped tree up to 18m high, widespread and gregarious.
VERNACULAR NAMES:
B: cibombo (1, 3c, 14, 23), cunga-nunsyi (3c, 14), kalume-kamukanunsyi, muungu (10), mangobe, myunga-cikoba (20), munga-nunsyi (3, 3c, 14), mutanda-fiwa (3c) **B/ I/ K/ Ku/ Lb/ Le/ Ln/ Ny**: combwe (1, 3, 3e, 4, 8, 10, 14, 17, 19, 23) **B/ K**: cambwe (3a, 9a, 10) **B/ Ku/ Ny/ Tu**: ngobe (1, 3, 9a, 10, 20) **I**: mugoya (10) **I/ K/ Le/ Lo/ Tk/ To**: mukoka (3a, 9a, 10, 14, 19, 22) **I/ Le/ Lo/ To**: mugoga (19) **I/ Lo/ Tk**: mufwefwe (4, 10, 17, 19) **I/ To**: mufwele (19) **K**: kyombe (14), kyombwe (3, 19) **Ku/ Ny**: ngowe (1, 3, 10, 14, 20) **Ln**: musense (6) **Lo**: lilongo (10) mukoto-koto (3, 3d, 14) **Lo/ Tk/ To**: mumbu (3), muunga (23) **Lv**: muzeze (6) **M**: ntalula (10) **Nm**: livindwe, mgupulu, mhungu, miguplu, mugu, ngu (22) **Ny**: citongololo (6), ngowe (3e), maninga, mlonga (9), mgongolo, mtete, nyungwe (9, 10) **Tk**: mukoko, munfene (8) **To**: mufweli (10) muunga (17)
PLANT PARTS USED:
root
TRADITIONAL USES:
Tanzania: A root decoction is drunk as a remedy for malaria: *Haerdi (1964 p. 47*

Acacia robusta Burch. subsp. clavigera (E. Mey.) Brenan [*A. clavigera* E. Mey.] (Fabaceae; Mimosoideae)
DESCRIPTION:
Large tree up to 20m, with dense irregular crown; mopane woodland fringing Zambezi, riverine thickets in Gwembe valley.
VERNACULAR NAMES:
B/ Ku: kafifi (1) **Ku**: mbwasa (1) **Ku/ Ny**: mbawa (1) **Lo**: mumbovwa (3, 3d) **Nm**: livindwe, mhusyi, miriva-nhinga, mliwa-nhwiga, muliwa-nhwiga (22) **Ny**: mzunga-nyewe (3e) **To**: mambomvu (3), munyenye, munyenyene (17), munyenye-ngwe (3, 19, 20, 23)
PLANT PARTS USED:
root
TRADITIONAL USES:
E. Africa: the plant is said to have been used by Arab slave traders to cure their slaves of malaria: *Kokwaro (1976) p. 124*
Tanzania: the roots are used against malaria: *Bally (1937) p. 16; Watt et al. (1962) p. 539*

Acacia seyal Del. var. fistula (Schweinf.) Oliv. (Fabaceae; Mimosoideae)
DESCRIPTION:
Small, slender, hook-thorned, 2 -10m. Found in heavy clay alluvium woodlands in Southern Province.
VERNACULAR NAMES:

Le: munkwa (19) **Ln**: katete (3), muzeze (6) **Lv**: mungonga (6) **Nm**: mlela, ulula-mweyupe (22) **To**: mungatuba (3) **Tu**: mosiya-bolozi, mzungu-newe (8)
PLANT PARTS USED:
Bark, seeds
TRADITIONAL USES:
<u>N.Africa</u>:a decoction of seeds is drunk: *Boulos, L. (1983)*
<u>E. Africa</u>: the bark is used as a febrifuge & emetic: *Holford-Walker (1951)*; *Boury (1962) 14: 6,* quoted in *Burkill (1995) p. 198*

Acacia sieberiana DC. [*A. woodii* Burtt Davy; *A. vermoesenii* De Wild.; *A. sieberiana* DC. var. *vermoesenii* Keay & Brenan] (Fabaceae; Mimosoideae)
DESCRIPTION:
Small-medium tree with irregular crowns, paired spines. Widespread, locally common, in open or closed woodland on clay or sandy alluvium.
VERNACULAR NAMES:
B: munga-nunsyi (3, 3b, 3c, 7a, 10, 14), mutanda-fiwa (3, 3b, 3c, 10, 14) munga-nwinci (20), mungwena (10,20) **B/ Ku**: munga-ntubetube, munga-nunsyi (1) **I**: itubetube (4) **I/ Lo/ To**: muukila (4, 17, 19), muzunda (3, 4) **I/ Le**: combwe, mungotuba (4, 17, 19) **I/ Ku/ Ny/ To/ Tu**: mutube-tube (4, 10, 17, 19, 20), muvatuba (19) **K**: kasele (19, 20), muwiba-utoka (10), mwiba (3, 3b, 14) **Ku**: mpumbu, mtuwe-tuwe (1) **Ku/ Ny**: tui-tui (?); **Lb**: munga-lunsi (10) **Le**: muumvwa-utuba (17) **Ln**: muzenze (3, 3b, 14) **Lo**: mukakane (10), mukate (3d, 14, 17), mungao-musweu (19) **Lo/ Tk/ To**: muunga (23) **Lo/ To**: mutuba-tuba (3, 3b, 3d, 7a, 14, 17) **Ny**: cisi, cisisi, mkansolo, mpangala, mtube-tube, mtwe-twe (3e), mkaiya (3e, 10, 20), tuba-tuba (3, 3b, 7a, 14), mzizi (3, 3b, 3e, 7a, 9, 10, 14) **Ny/ Tu**: mtwe-twe (3e), mungatuba (20) **Tk**: mumbovwa (23) **Tk/ To**: munkila (10) **To**: mumpangala (3, 3b, 14) muunga-utuba (17) **Tu**: icesea (7c), mutuwi-tuwi, mtwe-twe (8)
PLANT PARTS USED:
bark
TRADITIONAL USES:
<u>DRC/ Zaire</u>: to bathe a child, the rootbark is macerated in water: *Kokwaro (1976);* the bark of stem, branches & trunk is decocted & drunk:*Dhetchuvi et al., (1990)*
<u>E. Africa</u>: an infusion of pounded bark is drunk by a child with fever. The child is bathed in the liquid twice daily: *Kokwaro (1976) p. 126*
<u>Uganda</u>: the roots are used to treat malaria: *Tabuti (2008)*

Acacia tortilis (Forsk.) Hayne subsp. spirocarpa (Hochst. ex A. Rich) Brenan [*A. heteracantha* Burch.; *A. maras* Engl.; *A. litakunensis* Burch.; *A. spirocarpoides* Engl.] (Fabaceae; Mimosoideae)
DESCRIPTION:
Evergreen tree, 5m-20m high, widespread in low-altitude areas on deep alluvium, rarer on rocky outcrops, in a variety of types of woodland.
VERNACULAR NAMES:
E: Umbrella thorn; **Afr**: Haakdoring, Haak-en-steek, Withaak; **B/ Ny/ Tu**: fifi (20), kafifi (1, 3e, 20) **I**: buungasiya, mungasia, muunga-usiya (4) **I/ K/ Le/ Lo/ To**: mukoka (2, 4, 14, 17, 19, 20, 23), muunga (19) **Nm**: mgunga, migunga, ngurugunga (22) **Ny**: mzunga (3e,14, 23), nsangu, nsangu-sangu, nyoswa (3e, 14), umsasane (22) **To**: ngoka (22)
PLANT PARTS USED:
leaves, bark, roots.

TRADITIONAL USES:

N. Senegal: a bark infusion is drunk for fever.The leaves are also used: *Boury (1962)*, quoted in *Burkill (1995)*

Kenya: an infusion of crushed roots is used against malaria: *Koch et al. (2005) p. 96*

TESTS

In vitro tests on the twigs (?) revealed a high degree of antiplasmodial activity at IC_{50} 4.8 μg/ml: *Clarkson et al. (2004); Pillay et al.(2008)*

Tests in Kenya in 2005 showed antiplasmodial efficacy at $IC_{50}=10$ μg/ml: *Koch et al. (2008) Kirira et al.(2006)*. The bark contains saponins, which may affect the *in vitro* assay: *Johns et al. (1999)*, cited by *Koch et al. (2008)*

TOXICITY: the fresh leaf and immature pod yield much hydrocyanic acid (166mg/100gm.). However, the mature seeds are non-toxic and enjoyed by stock: *Fanshaw et al., (1967) , Steyn (1935, 1938), Greshoff (1913), Hayman (1941), Watt et al. (1962)*. No reports of toxic effects on humans.

Acacia xanthophloea Benth. (Fabaceae; Mimosoideae)

DESCRIPTION:

Medium to tall, sometimes up to 25m, well-shaped tree. Always in groups, in low-lying swampy areas, along and in rivers, and in the alluvium.

VERNACULAR NAMES:

E: Fever Tree; **Ln:** muzeze (22) **Lv:** muŋonga (22)

PLANT PARTS USED:

root, bark, stem

TRADITIONAL USES:

S. Africa: the Zulu people take the powdered bark of the stem and root as an emetic to treat malaria, and as a prophylactic on entering a malarial area: *Codd 26:14*, quoted in *Watt et al. (1962) p. 552*

Zambia: powdered bark of the root & stem is taken as a prophylactic: *Amico (1977) p. 115*

Acalypha indica L. (Euphorbiaceae)

DESCRIPTION:

Weed up to 2m tall. Widespread, often riverine, usually in shade of thickets, on rocky hillsides or in fields.

VERNACULAR NAMES:

Lo: lombwe-lombwe (9a)

PLANT PARTS USED:

plant

TRADITIONAL USES:

India: the plant has been used as a diaphoretic: *Greshoff (1913)*, quoted in *Watt et al. (1962) pp. 394, 1140*

Acampe pachyglossa Rchb. f. [*A. praemorsa* (Roxb.) Blatter & McCann; *Saccolabium pachyglossum* (Rchb. f.) Rolfe] (Orchidaceae)

DESCRIPTION:

Robust upstanding woody-stemmed orchid, up to 50cm high. Found in woodlands or riverine forests.

VERNACULAR NAMES:

Ny: mwana-wamfepo (9)

PLANT PARTS USED:
juice
TRADITIONAL USES:
<u>Tanzania</u>: the plant juice is drunk as a remedy for malaria: *Haerdi (1964) p. 204*

Acanthospermum australe (Loefl.) Kuntze (Asteraceae)
DESCRIPTION:
An invasive bristly weed. Procumbent stems 10-60cm, found on disturbed, sandy sites below 300m.
VERNACULAR NAMES:
E: Paraguayan bur, Sheep Bur, Spiny Bur
PLANT PARTS USED:
leaves, whole plant
TRADITIONAL USES:
<u>Brazil</u>: A decoction of the whole plant or the leaves is used through the Amazon region to treat malaria: *Brandão et al. (1992) p. 179*

Acanthospermum hispidum DC. (Asteraceae)
DESCRIPTION:
An invasive hairy weed. Erect and widebranched, yellow flowers, up to 50 cm.
VERNACULAR NAMES:
E: Bristly Starbur, Goatshead; **Ny**: nsakambwe, seseresya (9)
PLANT PARTS USED:
whole plant
TRADITIONAL USES:
<u>Ghana</u>: the plant is macerated with a hot pepper, infused and sieved, to drink as required: *Asase et al. (2005)*
<u>Mali</u>: the plant is macerated & decocted to furnish a drink and a steam bath: *Adjanohoun et al. (1981)*
<u>Sudan</u>: an infusion from the herb is taken twice daily to cure malaria: *El-Kamali (1997) p. 527*
TESTS:
Alkaloid extracts were tested *in vitro* against chloroquine resistant and sensitive strains of *P.falciparum*, and produced significant inhibitory activity of IC_{50} 1.23 mg/ml: *Sanon et al.(2003)*
Tests in Benin using dichloromethane extracts of the plants against *P.falciparum* showed efficacy of IC_{50} 7.5mg/ml: *Bero et al.(2009)*
TOXICITY:
The Sanon test also demonstrated weak cytotoxicity against three human cell lines.

Adansonia digitata L. (Malvaceae)
DESCRIPTION:
A large tree, widespread in dry lowland areas of Africa
VERNACULAR NAMES:
E: Baobab, Cream of Tartar Tree, Monkey Bread Tree; **B/ I/ K/ Ku/ Le/ Lo/ Ny/ Tk/ To/ Tu**: mubuyu (1, 3, 3a, 3d, 10, 14, 16, 17, 19, 20, 22, 23) **B/ Ku**: mulambe (1) **I**: ibuzu, mubuzi (4, 10, 17) **I/ Le/ Lo/ N/ Tu**: muuyu (3d, 10, 14, 22, 23) **I/ Lo/ To**: ivuyu (19) **Ku/ Ny/ Tu**: buyu (14, 20), mbuyu (3e, 14, 20) **Lo**: muyu (3d) **Nm**: m'paela, mpela,

mupela, mwanda, ŋwandu (22), mwandu (7c, 22) **Ny**: mkulu-kumba (3, 3a, 3e, 14), mlambe (3e, 9, 14), mlonje (9) **To**: ibbuyu (17, 19)

PLANT PARTS USED:

bark, fruit, leaves, seeds

TRADITIONAL USES:

Angola: dried fruit pulp is made into a drink to remedy fever: *Bonnefoux (ND)*

Benin: seeds are combined with root bark of *Xeroderris stuhlmann* to make a remedy for malaria: *Adjanohoun et al. (1989)*

Congo: bark & branches are macerated for a day to make a drink: *Wellens (1938)*

Central and W. Africa: the white powder surrounding the seeds is made into a drink used as a remedy for fever: *Dragendorff (1898); Karel et al. (1951); Morris (1996); Palgrave (1957); Watt et al. (1962)* ; leaves are used as a diaphoretic & prophylactic against fevers: *Anon. (1906) 4,* quoted in *Watt et al. (1962); Palgrave (1957);* bark is used as a febrifuge and quinine-substitute*: Ainslie sp. no. 12, Irvine (1961),* quoted in *Burkill (1985)*

DRC/Kinshasa): leaves & bark are decocted with leaves of *Carica papaya* & *Ocimum gratissimum* to make a steam bath: *Kasuku et al. (1999)*

E. Africa: the bark is boiled and the steam used as a vapour bath to treat high fever: *Kokwaro (1976)*

Ethiopia: a leaf decoction is drunk against malaria: *Lemordant (1971)*

Europe: the bark, known as "Cortex cael cedra", is marketed as a substitute for quinine: *Braun (1930),* quoted in *Watt et al. (1962)*

Guinea Conakry: leaves are infused to make a remedy: *Pobeguin (1912)*

Madagascar: a leaf decoction is taken as a febrifuge: *Terrac (1947); Pernet et al.,(1957); Rasoanivao et al. (1992)*

Mali: trunk bark is scraped & decocted to make a drink: *Malgras (1992)*

Mozambique: reported to be efficient in treating malarial symptoms: *Bandeira et al. (2001)*

Nigeria: bark scrapings are decocted in 1pt water to make a fever remedy; leaves are powdered and infused to make a prophylactic drink: *Ainslie (1937);* leaves are decocted and drunk 3x daily until recovery: *Ajibesin et al.(2008)*

Puerto Rico: bark is used as a febrifuge: *Loustalot (1949, 1950),* quoted in *Watt et al. (1962)*

Senegal: a leaf decoction is drunk morning and evening: *Aubreville (1950); Potel (2002)*

Sierra Leone: the leaves are used as a prophylactic against malaria: *Anon. (1906),* quoted in *Watt et al., (1962)*

Sudan: fruit pulp and fresh leaves are infused with seame seeds: *El-Kamali (2009)*

S. Africa: the seed pulp is mixed with water and drunk as a remedy for malaria: *Watt et al., (1962).* A seed infusion is drunk to treat fevers: *Palgrave (2002)*

Tanzania: the powdered bark is made into a porridge as a remedy for malaria: *Haerdi (1964)*

Tropical Africa: the bark and fruit are used as a remedy for malaria: *Githens (1948).* For the efficacy of the fruit-pulp, see *Ramadan et al., (1994)*

West Indies: the tree is used as remedy for malaria:*Watt et al. (1962)*

TESTS:

Tests in 1947 and 1951 failed to detect any antimalarial activity: *Spencer et al. (1947); Karel et al. (1951)*

In vitro tests of stem bark against *P. falciparum* showed IC $_{50}$ value of 8.2μg/ml (*Gessler et al. (1994)* For chemical analysis of bark, see *Ramesh et al. (1992).*

TOXICITY:

Doctors working in a Ghanaian hospital in 1967 observed toxic effects in children who had been given a combination of root decoctions, of which *A.digitata* was one: *Haaf, E., (1967)*, cited by *Neuwinger (2000)*. See *Watt et al. (1962)* for various tests. For cytotoxicity score see *Brendler (2010) p. 286* Tests of leaves against Brine shrimp showed LC$_{50}$ value <2 and against Vero cell line 0.1155 *(Brendler et al., 2010)*.

Adenia gummifera (Harv.) Harms [*A. cissampeloides* Harms] (Passifloraceae)
DESCRIPTION:
Large, woody tendril-climber with stems up to 20m thick. Often on anthills, and in evergreen fringing forest.
VERNACULAR NAMES:
E: Snake Climber; **B**: cimboyi (3, 3c), mwilikano (2, 3, 3c), ŋombe-yanina (3c) **Nm**: ngole (1a, 22) **Ny**: mkuta, msamba-mfumu, mwana-wamfepo (9), mlozi (3e, 9), namulozi, ndozi (3e) **Tk**: musyiko (8) **Ny/ Tu**: muwole (3e, 8)
PLANT PARTS USED:
juice
TRADITIONAL USES:
Nigeria: a root decoction is used occasionally to treat malaria (*Vergiat, p. 309*).
Mozambique: a decoction of the roots is drunk, and the patient bathed in steam from boiling leaves as a remedy for malaria: *Watt et al. (1962) p. 828*
S. Africa: the leaves and roots are used to treat fever and biliousness *Githens (1948) p. 98*. The root is chopped and infused with boiling water, to treat the general seediness and depression brought on by malaria: *Bryant (1909) p. 45*

Adenia lobata (Jacq.) Engl. (Passifloraceae)
DESCRIPTION:
Woody tendril-climber up to 6m. high. Found on edge of evergreen fringing forest and on rocky outcrops.
VERNACULAR NAMES:
B: mobole (3c) **Ln**: mucinje (3)
PLANT PARTS USED:
leafy twigs
TRADITIONAL USES:
Côte d'Ivoire: the leafy twigs are crushed and mixed with palm wine to make a drink to reduce fever pains: *Kerharo et al. (1950) p. 38*

Aerva lanata (L.) Juss. Ex J.A.Schult. (Amaranthaceae)
DESCRIPTION:
A perennial herb, up to 2m., widespread in open forest on mountain slopes, and in waste land.
VERNACULAR NAMES:
Ku/Ny: ciwombela, combela (1) **Ku/Tu**: cambela (1, 8) **Lo**: saka-saka (3d) **Ny**: mlomo-wambuya (7c)
PLANT PARTS USED:
leaves, whole plant,
TRADITIONAL USES:
E. Africa: a decoction of leaves is used to wash babies rendered unconscious from malaria. The plant is burned, and the smoke inhaled: *Kokwaro (1976) p. 18*

Nicobar Islands: The leaves are pounded, then boiled in coconut oil and pigs' blood. The liniment is rubbed on the body during fever: *Dagar et al. (1991), p. 116*
TESTS:
In vitro tests in 1994 revealed anti-plasmodial activity (IC $_{50}$ 8.6μg/ml); *Gessler et al. (1994)*

Aframomum alboviolaceum (Ridl.) K. Schum. [*A. biauriculatum* K. Schum.]
(Zingiberaceae)
DESCRIPTION:
A herb with leafy shoots, 5' - 10' high, with white or pinkish flowers. Widespread, usually in savannah
VERNACULAR NAMES:
B: ntungulu, vitungulu (20) **B/ Lb**: citungulu (20) **B/ Lb**: mutungulu (3c, 19, 20) **K**: mutungule (19) **Ln**: mutungunu-musindwa (6), ntungulu (19) **Lo/ Lv**: mutundu (3d, 6) **To**: mutundulu (19) (?)
PLANT PARTS USED:
fruit
TRADITIONAL USES:
Tropical Africa: the fruit pulp is used against fever: *Burkill (2000) p. 560*

Afzelia quanzensis Welw. (Fabaceae; Caesalpinioideaceae)
DESCRIPTION:
Large tree up to 20m high, often flattish crown. Scattered widely but not common.
VERNACULAR NAMES:
E: Pod mahogany; **B/ I/ Lb/ K/ M/ Ny/ To/ Tu**: mupaapa (3a, 3c, 4, 6,7a, 8, 10, 14, 17, 19, 20, 23) **B/ Ku**: mpeta (seeds) (1) **I/ Lo**: mwande (3, 4, 10, 14, 16, 19, 22, 23) **I/ Lo/ Tk/ To**: mukamba (1c, 3a, 4, 8, 10, 17, 19, 23) **I/ To**: mbambara (19) **K**: kulakazi (19), musamba-mfwa (3, 10, 14), nkulakazyi (3, 14) **Lo/ Ln**: mwande (3a, 3d, 7a, 10, 22) **Ln**: kibalebale (10), mwala (3, 6, 10, 14, 19, 21, 22) **Lv**: mubalatobe (10), muvulatowo (6, 22) **N**: mubala (10, 16, 23) **Nm**: kola, mhora, mkola, mkora (22) **Ny**: mkolando, mpapa (3, 3e, 7a, 14), mkongomwa (9), msamba-mfumu (3e, 9), ngolyondo, ngongomwa (10) **To**: mukanda (7a), mukomboolo (17), mwasambala (23)
PLANT PARTS USED:
bark
TRADITIONAL USES:
Malawi: an infusion of the bark is drunk as a remedy for malaria: *Morris (1996) p. 334*
Tanzania: Children weakened by malaria are bathed in a bark infusion: *Haerdi (1964)*

Agelaea pentagyna (Lam.) Baill. [*A. heterophylla* Gilg.] (Connaraceae)
DESCRIPTION:
A scandent shrub or twining liana, with hairy leaves and fruit and fragrant white flowers, found in forest margins from 1000-1500m.
PLANT PARTS USED:
roots
TRADITIONAL USES:
E. Africa: the roots are used to treat fever. Care must be taken with the dosage, since they are believed to be toxic: *Kokwaro (1976) p. 76*

Agelaea sp. (Connaraceae)

PLANT PARTS USED:
juice
TRADITIONAL USES:
Côte d'Ivoire: the clear liquid exuded from the cut liana is used in nasal inhalation as a febrifuge: *Burkill (1985) p. 517; Kerharo p. 170*

Ageratum conyzoides L.(Asteraceae)
DESCRIPTION:
Rampant invasive weed, .5-1m high, with white or mauve flowers.
VERNACULAR NAMES:
E: Goatweed, Chickweed, Whiteweed; **Ku**: kafwaya (1) **Ny**: ciwanda, kafwaya, kapanti (3e)
PLANT PARTS USED:
plant, leaf, sap, stems
TRADITIONAL USES:
Benin:the plant is used to make a drink: *Adjanohoun et al.(1989)* cited in *Prelude med. plants database (2011)*
Central Africa: the leaf is used as a fever remedy: *Githens (1948) p. 76*
Comoros:the whole plant is decocted and drunk: *Adjanohoun et al.(1982)* in *Prelude med. plants database (2011)*
Congo:the plant is used as a febrifuge: *Goossens (1924)* ; *Staner et al. (1937)*
Equatorial Guinea:leaves are macerated to make a drink: *Akendengue (1992)*
Gabon:the fruit and leaves are used to make malaria remedies: *Raponda-Walker et al. (1961)*; *Lamidi et al.(2007)*
India & Mexico: the plant is used as a fever remedy: *Dragendorff,* quoted in *Watt et al. (1962) p. 198*
Madagascar: a decoction of the roots is used against malaria: *Pernet (1957)*; a decoction of stems and leaves is used: *Rasoanaivo et al.(1992)*
Nigeria: a leaf decoction is taken as a febrifuge: *Burkill (1985) p. 444;* a bath is made from the pounded leaves of *A.conyzoides, Vernonia amygdalina,* & black soap: children drink juice of the crushed leaves: *Adjanohoun et al.(1991)* in *Prelude med. plants database (2011)*
N.Africa: the plant is used as antipyretic, sudorific & stimulant: *Boulos (1983)*
Togo: pulped leaves are mixed with lemon juice to drink & rub in: *Adjanohoun et al. (1986)*
Zaire: the leaf sap in decoction is used for a wash for children with fever: *Bouquet (1969) p. 91,* quoted in *Burkill (1985) p. 444* and in *Prelude med. plants database (2011)*
Zambia: the whole plant (is infused ?) as a remedy for fever; the leaves are applied externally to "ague" patients: *Nair (1967) p. 94*
TOXICITY:
Danger of liver damage: *Acamovic et al. (2004) p.2*

Albizia amara (Roxb.) Boiv. subsp. **sericocephala** (Benth.) Brenan (Fabaceae; Mimosoides)
DESCRIPTION:
Open, shapely tree up to 12m. high, flat-topped. Widespread, forming thickets on overgrazed grassland; common on anthills in wetter areas.
VERNACULAR NAMES:
B: cibombwe-sala, mulala-ntanga (3c, 14), mulala-ntete (3, 3c, 14) **B/ K/ Le**: mwikala-nkanga (3, 3b, 3c, 4, 14, 17) **I/ To**: mukala-nkanga (4, 17) **I/ Le**: mukasala (19) **I/ Le/ Lo/ To**: mukangala (3, 3b, 3d, 4, 14, 17, 19, 20, 23) **Ln**: kasongu (3, 3b) **Nm**: mpogolo,

mporogo-bongore (22) **Ny**: mjele-njete (3, 3b), mkalanga, nyele (3e, 14), mkala-nkanga, nsengwa (3e) **To**: kankumbwila (3, 3b, 14), mukalala (19), simwakasala (17)

PLANT PARTS USED:
fruit

TRADITIONAL USES:
Zambia & Zimbabwe: the fruits are used as a remedy for malaria: *Drummond p. 51; Githens (1948) p. 76; Palgrave (2002) p. 257; Storrs (1979) p. 206; Watt et al. (1962) p. 553*

Albizia anthelmintica (A. Rich.) Brongn. [*A. umbalusiana* T. R. Sim] (Fabaceae; Mimosoides)

DESCRIPTION:
Small, much-branched deciduous tree up to 10m high, often several stems from common base, forming thickets in hotter and drier areas.

VERNACULAR NAMES:
B: cibombwe-sala (7a) **Ku**: kansalu-nsalu, muzanga (1) **Lv**: musaka-yaze (6, 22) **Lo**: mukoto-koto, mukwena (23), mwikala-kanga (7a), mwikala-nkanga (3, 3d) **Nm**: mpingu (1a, 22), ngada, ngata (22) **Ny**: cigolo-golo (3e, 7a), cikalolo (8, 9), citale, mpefu, mtanga-tanga (9), kaweleka, masango (6), msase, ngaza (3e), mzanga (3, 3e) **Tk**: mukazuta-bike (23) **To**: condwe, mucongwe (3, 23), mwihangwe (3) **Tu**: munzanga (8)

PLANT PARTS USED:
bark, root, root bark

TRADITIONAL USES:
Kenya: a decoction of bark or root is given as an emetic to treat malaria. Correct dosage is essential as an overdose can kill: *Heine et al.(1988); Beentje (1994) p. 270; Kokwaro (1976) p. 127; Kareru et al.(2007)*; the bark of stem, branch & trunk is decocted to drink and to wash the patient: *Lindsay et al.(1978)*; a cold decoction of pounded roots and bark is drunk: *Nanyingi et al.(2008)*

Nigeria: the bark and stems are decocted: *Iwu (1986b)* cited in *Prelude med. plants database (2011)*

Tanzania: a broth of root bark is used as a febrifuge: *Bally (1938) Bd. 102, Brenan (1949),* quoted in *Watt et al. (1962) p. 554; Johns et al.(1994)*

TOXICITY:
See entry for Kenya above.

Albizia glaberrima (Schum. & Thonn.) Benth. var. glabrescens (Oliv.) Brenan (Fabaceae; Mimosoides)

DESCRIPTION:
Large evergreen tree up to 25m, with flattish crown. Low-altitude mopane woodlands, especially on banks of Kafue and Luapula rivers.

VERNACULAR NAMES:
I: mbaselwenje, mukomba (4, 23), mukonono (17) **I/ Le**: mukololo (17) **Ny**: cikwani, msanku-sanku, mtanganga, njoka (9), msenjele (3, 3e, 9) **To**: mugololo (17)

PLANT PARTS USED:
bark

TRADITIONAL USES:
Nigeria: the bark is rubbed on the skin of fever patients: *Thomas (1945),* quoted in *Burkill (1995) p. 212*

Albizia gummifera (J. F. Gmel.) C. A. Sm. (Fabaceae; Mimosoides)
DESCRIPTION:
Large spreading flat-topped tree, most found in high-altitude forests.
VERNACULAR NAMES:
Not known
PLANT PARTS USED:
bark
TRADITIONAL USES:
E. Africa: a decoction of the bark is used to treat malaria: *Kokwaro (1976) p. 127; Muthaura et al.(2007)*
Kenya: an infusion of the bark is drunk as a remedy for malaria: *Glover et al.(1966), Beentje (1994) p. 270; Burkill (1995) p. 212*

Albizia zygia Macbride (Fabaceae; Mimosoides)
DESCRIPTION:
Large tree up to +20m. High spreading crown and buttressed bole. Found in gallery forest fringing Zambezi.
VERNACULAR NAMES:
K: mweyeye (3)
PLANT PARTS USED:
leaf, bark
TRADITIONAL USES:
Benin: leaves stem are decocted to make a remedy: *Adjanohoun et al. (1989)*
Burkina Faso: leaf decoction is used as drink and to bathe stiff back caused by fever: *Kerharo et al. (1950)*
C.A.Rep:bark from stems, branches & trunk is scraped, pounded with salt, applied locally: *Sillans (1953)*
Côte d'Ivoire: a leaf decoction is used as a lotion & drink: *Kerharo et al.(1950) p. 94*
E. Africa: the bark is decocted or infused or cooked with food as a remedy for malaria: *Kokwaro (1976) p. 128*
Senegal:the patient is bathed with a decoction of bark from stems, branches & trunk: *Kerharo et al.(1974)*
Uganda:the bark is decocted to make a remedy: *Tabuti (2008)*
W. Africa: the powdered bark is rubbed on the skin of patients suffering from 'eruptive fevers': *Dalziel,* quoted in *Watt et al. (1962) p. 558.* The bark is decocted and used as a vapour bath to ease fever pains and stiffness: *Berhaut (1975) pp. 506-9, Bouquet (1969) p. 163, Irvine (1961) pp. 334-5,* quoted in *Burkill (1995) p.217*
Uganda: the bark is macerated and decocted to make a drink: *Kokwaro (1976)*

Allium cepa L. (Alliaceae)
DESCRIPTION:
Planted as vegetable
VERNACULAR NAMES:
E: Onion; **B**: kanyense (3c) **B/ Ku/ Ny**: anyenze (1, 3e) **I**: wanyisi (17) **K/ Ln/ Lv**: sapole (19) **Le**: hanyansi (17) **Lo**: nyenyese (19) **Lo/ To**: nyanise (17, 19) **To**: hanyisi, honyanisi (17)
PLANT PARTS USED:
Juice, bulb, root

TRADITIONAL USES:

Angola: the body is rubbed by an onion pounded to release the juices: *Bossard (1996)*

East Africa: the onion is used as a remedy for fever: *Watt et al. (1962) p. 674; Tabuti (2008)*

Morocco: slices of onion are applied to forehead as remedy: *Bellakhdar (1997)* cited in *Prelude med. plants database (2011)*

Allium sativum L. (Alliaceae)

DESCRIPTION:

Cultivated as "relish"

VERNACULAR NAMES:

E: Garlic; **Nm**: Kitungulu, situngulu (22)

PLANT PARTS USED:

bulbs, leaves

TRADITIONAL USES:

Ethiopia: the bulb is used: *Teklehaymanot et al.(2007)*; fresh leaves are macerated in water: *Gedif et al. 2003)*; dry fresh dry leaves of *Croton macrostachyus* are crushed & pounded, decocted with *A.sativum* bulb roasted with butter left overnight outside,drunk in the morning: *Mesfin et al.(2009)*

Nigeria: principally taken against fevers & chills: *Ndukwu et al., (n.d.)*; cited in *Prelude med. plants database (2011)*

W. Africa: garlic is added to gruel and drunk, as a remedy for feverish chills: *Ainslie sp. no. 19, Akinniyi et al.(1983), Oliver-Bever (1960) pp. 4, 45, (1983) pp. 4, 44, 45,* quoted in *Burkill (1995) p. 490; Dalziel (1937) pp. 485-6*

S. Africa: the Xhosa drink a decoction of garlic leaves and bulbs, mixed with *Artemisia afra* and *Fagara capensis,* as a febrifuge: *Watt et al. (1962) p. 674*; garlic bulbs are used: *Dykman (1908);*

Allophylus africanus Beauv. [*A. transvaalensis* Burtt Davy; *A. melanocarpus* (Sonder) Radlk.; *A. rhodesicus* Exell] (Sapindaceae)

DESCRIPTION:

Shrub 1-2m. high, or tree up to 10m. Found in riverine thickets, forest or open woodland, often on anthills.

VERNACULAR NAMES:

E: African False-currant; **Afr**: Afrika-valstaaibos; **Ku/ Ny**: cisitu (1, 3e) **Lb**: lumombwe (23) **Lo**: mala-oya (3d) **Ny**: kafupa-kacimbwi, mzaule, nsitula (3e), kandula, mpira-nkututu, mpindi-mbi, msangasi, msawa-sawa, mtala-wanda, mtsatsule, mtatu, mwana-wamfepo (9) (?): mulalusya (23)

PLANT PARTS USED:

leaves

TRADITIONAL USS:

Côte d'Ivoire: the leaves are decocted to make a fever remedy for children, to be drunk and used as a lotion: *Dalziel (1937) p. 330; Deighton (1954) 1999,* quoted in *Burkill (2000) p. 7*

Aloe buettneri A. Berger [*A. bulbicaulis* Christian] (Asphodelaceae)

DESCRIPTION:

Perennial herb with rosette of fleshy leaves with sharply-toothed edges, found in grassy places in moist savannah.

VERNACULAR NAMES:

Not known
PLANT PARTS USED:
stock
TRADITIONAL USES:
Burkina Faso: the fleshy stock is chopped small, dried, and roasted to powder. One small spoonful to be sucked twice a day between meals: *de la Pradilla (1988) p. 11*
Togo: A root decoction mixed with root bark of *Entada abyssica* and powdered fruit of *Xylopia aethiopia* is diluted in alcohol as remedy: *Adjanohoun et al., (1986)* ; cited in *Prelude med. plants database (2011)*

Alternanthera nodiflora R. Brit. (Amaranthaceae)
DESCRIPTION:
Decumbent annual weed up to 1m. high with white flowers. Found on moist sandy/clay alluvium.
VERNACULAR NAMES:
E: Common Joyweed; **Ku/ Ny/ Tu**: cambulu- luzi (1, 8)
PLANT PARTS USED:
roots
TRADITIONAL USES:
Côte d'Ivoire: the crushed roots are used against fever pains: *Kerharo et al. (1950) p. 34*

Alternathera pungens Kunth (Amaranthaceae)
DESCRIPTION:
Invasive noxious weed with white flowers and opposed leaves on prostrate hairy stems. Found on sandy soils in pastures or on roadsides.
VERNACULAR NAMES:
E: Khaki weed, Creeping Chaffweed
PLANT PARTS USED:
plant
TRADITIONAL USES:
Côte d'Ivoire: the whole plant is pulped and used in a bath or vapour bath as a fever remedy: *Bouquet (1974) pp. 13-4,* quoted in *Burkill (1985) p. 47;*

Alternanthera sessilis (L.) DC. (Amaranthaceae)
DESCRIPTION:
Perennial noxious weed with white/pinkish flowers, up to 1m. tall, erect or procumbent. Found in ditches, swamps, roadsides, moist fields.
VERNACULAR NAMES:
E: Sessile Joyweed, Alligator weed, Dwarf Copperleaf; **B/ Ku**: kasila-butwilo (1) **Ny**: mkungira (7c)
PLANT PARTS USED:
plant
TRADITIONAL USES:
Sri Lanka: the herb is used as a febrifuge:*Quisumbing (1951),* cited in *Watt et al. (1962) p. 14*

<u>Zambia</u>: the herb is used as a febrifuge: *Amico (1977) p. 105*

Amaranthus spinosus L. (Amaranthaceae)
DESCRIPTION:
An invasive weed, erect, monoecious, multibranched, with tiny prickly greenish flowers, from 100-130cm tall. Found in wastelands, overgrazed pastures, roadsides, ditches.
VERNACULAR NAMES:
E: Pigweed, Thorny/ Prickly/ Spiny Amaranth, Edlebur, Needlebur; **To:** mbowa-wacilungu (22)
PLANT PARTS USED:
plant
TRADITIONAL USES:
<u>Benin</u>:the plant is used to treat fever: *Adjanohoun et al. (1989)*
<u>C.A.R.</u>: the whole plant is decocted to bathe the patient: *Boulesteix et al. (1979)*
<u>Côte d'Ivoire</u>: branches and leaves are used to make a malaria remedy:*Tra-Bi-Fézan (1979)*
<u>Philippines</u>: the plant is used as a sudorific & febrifuge: *Quisumbing (1951)*, quoted in *Watt et al.(1962) p. 16*
<u>Zambia</u>: the whole plant is used to make a sudorific and febrifuge: *Amico (1977) p. 106*
TESTS :
The aqueous extract of *A. spinosus* bark was screened for antimalarial properties in mice inoculated with *erythrocytes* parasitized with *Plasmodium berhei berghei*...and showed a dose-dependent antimalarial activity in a 4-day assay using chloroquine as the reference antimalarial drug. ED_{50} values for the extract and chloroquine were 789.4 and 14.6 mg/kg respectively: *Cai et al., (2003)*
An identical result (ED_{50} 789) was produced three years later by independent tests: *Hilou et al, (2006)*
TOXICITY:
The aqueous extract of the bark has a relatively low toxicity: LD_{50} =1450 mg/kg: *Cai et al., (2003)*

Ampelocissus africanus (Lour.) Merr. [*A. grantii* Bak.] (Vitaceae)
DESCRIPTION:
A widespread woody climber with yellow flowers, up to 6m. high but sometimes prostrate. Found on savannah, especially among rocks.
VERNACULAR NAMES:
Ku: muzinda-ngulube (1) **Lo:** kalimbwe, munsansa, nacisansa (3d) **Ny:** mwana-wamfepo (9) **Tk:** sansa-pati (8) **Tu:** sambilinga (8)
PLANT PARTS USED:
leaves
TRADITIONAL USES:
<u>Sudan</u>: a leaf infusion is mixed with onions and drunk and used as a wash to relieve fever pains in head and neck: *Burkill (2000) p. 287; Dalziel (1937) pp. 300-1*
<u>Tanzania</u>: the leaf sap is drunk as a remedy for malaria: *Burkill (2000) p. 287; Haerdi (1964) p. 111*

TESTS:
In vitro tests in 1994 revealed antimalarial activity of IC_{50} value 9.0mg/ml: *Gessler (1994)*

Anacardium occidentale L. (Anacardiae)
DESCRIPTION:
Much-branched tree up to 12m. high with spreading crown and rough grey bark. Introduced in Copperbelt, Lusaka, Barotseland
VERNACULAR NAMES:
E: Cashew Nut
PLANT PARTS USED:
bark, leaves
TRADITIONAL USES:
Ecuador: a decoction of bark and leaves is used to treat fever: *Coe et al. (1996) p. 96*
Nigeria: a bark decoction is drunk 2x daily: *Ajebesin et al. (2008);* bark & leaves are decocted:*Gill et al.(1986)*; bark is soaked in a bottle of water, & the filtered extract drunk 1x or 2x daily: *Kayode et al. (2009)*; the leaves are used to prepare a malaria cure: *MacDonald et al.(2000)*; a decoction of roots, bark & leaves is drunk 2x daily for 3 days: *Ogie-Odia et al.(2009)*
Zambia: the bark is used as a febrifuge; *Amico (1977) p. 121*

Anisophyllea boehmii Engl. (Anisophylleaceae)
DESCRIPTION:
Small evergree tree up to 15m high with edible reddish fruit, widespread in open woodland in north.
VERNACULAR NAMES:
B/ K/ Lb/ Ln/ Lo/ N: mufungo, mufungu (3d, 4, 7a, 14, 19, 21, 23) **Ln**: luvung'u (21)
Ny: mangondo, mgondo (3e), muhoto (14), mfungo (3e, 7a, 14)
PLANT PARTS USED:
bark
TRADITIONAL USES:
Zambia: an infusion of bark is given for malaria: *Haapala et al.(1994) p. 10*

Annona senegalensis Pers. [*A. chrysophylla* Bojer] (Annonaceae)
DESCRIPTION:
Shrub or small tree up to 8m., found in sandy soils, along rivers, mixed scrub or woodland, rocky outcrops, swamp forest. Edible yellowish fruit.
VERNACULAR NAMES:
E: Wild Custard-apple; **Afr**: Wildesuikerappel; **B/ I/ K/ Lb/ Lo/ Lv/ N/ Tk/ To**: muloolo (3, 3c, 3d, 4, 6, 7a, 9a, 16, 22, 23) **B/ Ku/ Ny**: mpovya (1, 3, 9a, 20) **K**: munkanda (3, 9a)
Lo: lilolo (16), mukono-mgwa (7a), mukono-nga (19) **Lo/ To**: mukono-ngwa (3, 3d, 9a)
Nm: mfira, mkonora, mtope-tope (1a, 22), mtela (1c) **Ny**: ciyuta, mponjela, mposa (9), mpo-uya (7a), mpovya, mtopa (3e), mtanta-nyerere (3, 9a) **To**: mukonyo-mbwa (17), njofa (22)
PLANT PARTS USED:
plant, roots
TRADITIONAL USES:
Benin:leaves of *A.senegalensis* & stems with leaves of *Ximenia Americana* are used to treat malaria: *Adjanohoun et al. (1989)*

Burkina Faso: a large bunch of roots is boiled in 8 lit. of water for 20 min, and half a glass to be drunk while bathing in warm water on 2 consecutive evenings: *de la Pradilla (1988) p. 11; Kerharo et al.(1950)*

Congo-Brazzaville: decoction of leaves of *A.senegalensis, Senna occidentalis* & *Carica papaya* is drunk 2x daily, adults 100ml, children 50ml.: *Diafouka (1997)*

D.R.C.-Kinshasa: a root decoction is drunk by the glass: *Delaude et al. (1971)*

Mali: Leaves are decocted with various other plants to bathe patients: *Haxaire (1979)* cited in *Prelude med. plants database (2011)*

Senegal: the plant is used as a remedy for malaria: *Kerharo et al., (1974)*

Tanzania: the roots are decocted to prepare a malaria remedy: *Chhabra et al. (1987)*

TESTS:

In 2005 the antimalarial efficacy of a methanol extract was tested *in vivo* against *Plasmodium berghei berghei* in mice. At doses of 100mg/kg it produced over 57% suppression of parasitemia, and at 800mg/kg it produced 91.1%: *Ajaiyeoba et al. (2005)*

Anthocleista schweinfurthii Gilg. [*A. zambesiaca* R.E.Fr. non Baker] (Gentianaceae)
DESCRIPTION:
Small spineless tree up to 20m. with smooth grey/brown bark Found in fringing forest, thickets, savannah or rainforest.
VERNACULAR NAMES:
B: mulu-lwalwa (20), mulwa-lwalwa (3c) **Ny**: bekaveka (3e)
PLANT PARTS USED:
bark, roots
TRADITIONAL USES:
N. Transvaal: a decoction of the bark is given to malaria patients in the Drakensberg mountains: *Codd (1951), quoted in Watt et al. (1962) p. 727*
Tanzania: a root decoction is taken against malaria: *Burkill (1995) p.527; Haerdi (1964) p.127*

Aporrhiza nitida Gilg. ex Milne-Redh. (Sapindaceae)
DESCRIPTION:
Medium-large shrub/ tree 10-16m high. Found at low altitudes in evergreen, riverine fringe and swamp forests.
VERNACULAR NAMES:
B/ Ln: wunkomukomu (9a, 21)
PLANT PARTS USED:
bark
TRADITIONAL USES:
Tanzania: the body is bathed with a bark decoction to relieve malaria symptoms: *Burkill (2000) p.8; Haerdi (1964) p. 123*

Arachis hypogaea L. (Fabaceae, Papilionoideae)
DESCRIPTION:
Cultivated locally
VERNACULAR NAMES:
E: groundnut, monkeynut, peanut; **B**: lubalala (3c, 9a) **B/ Ku**: mbalala (1) **B/ Ku/ Ny/ Tu**: nsyaba (1, 3e, 20) **I**: inkonze (4) **I/ K/ Le/ Ln/ Lo**: inyemu (3d, 4, 19) **I/ Le**: inyemu-mafuta (17) **I/ Lo/ To**: indongo (17, 19) **Lb/ Ln**: muninga (19) **Lv**: byelu-byamunyinga,

byelu-byansoke (19) **Ny**: kapansya, matewele (20), ntedza (3e) **To**: indongwe (17) **Tu**: mayowa, siyaba (20)
PLANT PARTS USED:
roots
TRADITIONAL USES:
Zambia: groundnut roots are used as a febrifuge: *Nair (1967) p. 47*

Argemone mexicana L. (Papaveraceae)
DESCRIPTION:
Invasive weed, common in rocky open land or waste land. Green prickly leaves and stems, yellow flowers
VERNACULAR NAMES:
E: Mexican Prickly Poppy; **B/ Ku**: tambala (1) **Ny**: cinyaku (3e), dosa, licongwe, lilime-lyaŋombe, mkuma-jalaga (9)
PLANT PARTS USED:
flower, whole plant
TRADITIONAL USES:
Ethiopia (?): the herb and flower are used as a diaphoretic: *Watt et al. (1962) p. 817*
Mali: a decoction is drunk twice daily for a week. The liquid is also used to bathe child patients: *Graz et al., (2009)*
TESTS:
Clinical trials with traditional healers have proved the efficacy of the treatment. Of 301 patients treated over 28 days, 89% required no second-line treatment, compared with 95% given ACT treatment: *Willcox et al. (2007), Graz et al.(2009).*
TOXICITY:
Sixty-five human deaths linked with *A.mexicana* poisoning were reported from Delhi in 1999: *Verma et al.(2000).* For trials on rats and mice causing toxic symptoms and deaths, see: *Pahwa et al.(1989), El Gamal et al. (2008), Ibrahim et al.(2009)*

Aristolochia albida Duch. [*A. petersiana* Klotsch] (Aristolochiaceae)
DESCRIPTION:
A woody liana, up to 11m. high
VERNACULAR NAMES:
E: Dutchman's Pipe; **Ny**: dululu, kanjoka, matholisa (9) **To**: malundu (5b)
PLANT PARTS USED:
roots
TRADITIONAL USES:
Malawi & Zimbabwe: the roots are infused as a remedy for malaria: *Gelfand et al.(1985) p. 120; Morris (1996) p. 225; Williamson (1975) p. 36*
TOXICITY:
Used as fish-poison. Tested for molluscicidal activity and shown to be active: *Kela et al.(1989)*

Artabotrys monteiroae Oliv. [*A. nitidus* Engl.] (Annonaceae)
DESCRIPTION:
Woody climber with hooked woody tendrils forming self-supporting thickets up to 5 m. high. Found in clumps in *Brachystegia* woodland, in riverine/ fringe/ swamp forests, among rocks and on anthills.

VERNACULAR NAMES:
E: Red hook-berry, Tail-grape; **Afr**: Rooihaakbessie; **B**: lubulu-kutu, mulya-nduba (3c) **To**: incincayo, ntiti (22) (?)
PLANT PARTS USED:
leaf juice, roots, bark
TRADITIONAL USES:
<u>Tanzania</u>: The body is bathed with leaf juice, and a decoction of roots and bark is drunk, against malaria: *Haerdi (1964) p. 37*

Artemisia afra Jacq. ex Willd. (Asteraceae)
DESCRIPTION:
Much-branched shrub up to 2m high, smelling strongly of wormwood; found on Nyika Plateau, at edge of montane forest nr.source of Shire River
VERNACULAR NAMES:
E: African Wormwood; **Nm**: fifi (1a), usyemeli (7c)
PLANT PARTS USED:
whole plant, leaves
TRADITIONAL USES:
<u>Angola</u>: to hasten recovery from malaria, a leaf infusion is drunk: *Bossard (1996)*
<u>E. Africa</u>: In the mountainous areas, it is used as an emetic and febrifuge: *Kokwaro (1976) p. 58; Watt et al.(1962) p. 201*. The leaves are used as a fever remedy in E. and S. Africa: *Githens (1948) p. 78* . The Chagga apply the warmed herb in the treatment of small children suffering from fever: *Watt et al. (1962) p. 201* A. afra was found to be the plant most widely used by traditional practitioners in Tanzania to treat malaria: *Schlage et al. (1999)*
<u>Ethiopia</u>: a drink is made from fresh and dry powdered leaves mixed with butter & coffee, drunk before breakfast for 3 days: *Mesfin et al.(2009)*
<u>S. Africa</u>: the plant is widely used as a remedy for malaria. A double handful of leaves is infused with a quart of hot water, and administered either as enema or emetic or drink for febrile complaints: *Bryant (1909) p. 45.; Dykman(1908); Pappe (1850); Smith (1895)*. It is often taken as an infusion or decoction, often made syrupy with sugar, or mixed with *Lippia javanica*: *Watt et al. (1962) pp. 199, 201, 202, 1051* ; *Kling (1923)*; a poultice of leaves with brandy or vinegar is applied to the stomach
<u>Zimbabwe</u>: plant decoction is drunk as a remedy for fever: *Gelfand et al.(1985) p. 231*
TESTS:
Ethiopian *in vitro* tests against *P.falciparum* gave an IC_{50} of 7µg/ml: *Kassa et al. (1998)*
Tests on lipophilic extracts of aerial parts showed *A.afra* to be the most active of nine plant species traditionally used for malaria in Zimbabwe. The IC_{50} values were 8.9µg/ml against the chloroquine-sensitive strain PoW, and 15.3 µg/ml against the chloroquine-resistant clone Dd2 of *P. falciparum*: *Kraft et al. (2003)*.
Tests on the leaves revealed high antiplasmodial activity (IC_{50}5 µg/ml), yet the active constituents do not appear to be artemisinin-type compounds: *Clarkson et al.(2004)*.
TOXICITY:
Tests on rats in 2006 indicated that *A.afra* is non-toxic when given acutely, has low chronic toxicity potential, and, in high doses, may have a hepaprotective effect: *Mukinda et al. (2007)* Tests of leaves on Brine shrimp showed LC_{50} 0.0697 and on Vero cell line 0.2346: *Brendler et al., (2010)*

Artemisia spp.
PLANT PARTS USED:
plant
TRADITIONAL USES:
Colombia: an infusion or decoction is administered 3x daily to malaria patients until the fever is cured: *Milliken (1997) pp. 22, 25*

Asparagus africanus (Lam.) A. A. Obermeyer (Asparagaceae)
DESCRIPTION:
An invasive climbing thorny weed reaching up to 8m. high, smothering small trees and shrubs.
VERNACULAR NAMES:
E: Orange-fruited asparagus; **B**: akako-bamakanga (3c, 9a) **Lo**: kasonge-kayisi, mulya-lamana (3d) **Ny**: kankande, mgala-mbuti, mtsitsila-manda (9)
PLANT PARTS USED:
roots
TRADITIONAL USES:
Malawi: an infusion of the roots is taken as a remedy for fevers: *Morris (1996) p. 393*
TESTS:
Bioassay-guided fractionations of the root extract led to the isolation of two active compounds, the sapogenin *muzanzagenin* and the lignan *nyasol*: *Schwikkard et al., (2002)*

Asparagus plumosus Baker (Asparagaceae)
DESCRIPTION:
Wiry plant, erect, twining or trailing, up to 2.6m. high, fern-like appearance.
VERNACULAR NAMES:
B: akemya-nsyinge (3c, 9a)
PLANT PARTS USED:
leaves, stem
TRADITIONAL USES:
Tanzania: the Lobedu drink a cold infusion of the leaves and stem for malaria: *Brenan,* quoted in *Watt et al. (1962) p. 689; Burkill (1995) p. 50*

Azanza garckeana (F. Hoffm.) Exell & Hillcoat [*Thespesia garckeana* F. Hoffm.; *T. trilobata* Baker f.; *T. rogersii* S. Moore] (Malvaceae)
DESCRIPTION:
A bushy shrub or small/ medium tree up to 10m. high. Found in almost all types of woodland and bush and roadsides, from sea level to 1700m., widespread but not dominant.
VERNACULAR NAMES:
E: Snot-apple, Tree Hibiscus; **B**: cinga (3c, 10, 14, 23), mukukuma (10) **B/ K/ Ku/ Le/ Ln/ Ny/Tu**: mukole (1, 3, 3c, 10, 14, 17, 19, 20) **I/ To**: mutoba (10, 23), mutobo (1c, 4, 17, 19, 23) **Ku**: matowo, ukole (1) **Lb**: mufufuma, mukome (10) **Le**: mutabo (19) **Lo**: lilongo, luwanika (10) **Lo/ Tk/ To**: munego (10, 14, 17, 19, 22), muneko (3, 3d, 10, 14, 17, 19, 22) **M/ Ny**: ntowo (10) **N**: mundendha (10) **Nm**: mtobo (22) **Ny**: mkole (3, 3e, 14), mtowu (9, 22) **To**: matokwe (19)
PLANT PARTS USED:
root
TRADITIONAL USES:

<u>Malawi</u>: a root decoction is taken as a remedy for fevers: *Morris (1996) p. 400*
TESTS:
Aqueous and organic fractions of roots and leaves were tested *in vitro* against the multi-drug resistant Vietnam-Smith strain of *P. falciparum* (VI/S). Some activity was seen: *Connelly et al.(1996)*

Baccharoides lasiopus (O.Hoffm.) H. Rob. [*Vernonia lasiopus* O. Hoffm.] (Asteraceae)
DESCRIPTION:
Coarse shrub up to 1.5m high
VERNACULAR NAMES:
Not known
PLANT PARTS USED:
plant
TRADITIONAL USES:
<u>Kenya</u>: the Kikuyu people use the plant as a malaria remedy: *Beentje (1994) p. 568*

Balanites aegyptiaca (L.) De Lile (Balanitaceae)
DESCRIPTION:
Small evergreen tree, us. 3-5m. high. Found in dry, wooded grassland and near rivers on flats
VERNACULAR NAMES:
E: Single Greenthorn, Simple-thorned Torchwood; **Afr**: Enkelgroendoring; **B**: katika-yengele (3c), mubambwa-ngoma (3b, 3c, 7a, 9a, 23) **B/ Ku**: mubanga-ngoma (1, 3) **Ku/ Ny**: mbaya-mpondolo (1) **Lo/ To**: mwalabwe (3d, 7a, 14) **Nm**: mbaba-ngoma, mguyu-guyu, mruguyi, myuguyu (22), mwamba-ngoma (1a, 22) **Ny**: nkuyu (7a), syava-nkunzi (8), umnulu (22) **Tk**: muvaba-ngoma (8) **To**: mulya-nzovu (2, 3, 3b, 7a, 14, 17), muwongo (22) **Tu**: nkuyu (8)
PLANT PARTS USED:
seeds, roots
TRADITIONAL USES:
<u>N. Africa</u>: a root extract is infused to treat malaria: *Boulos (1983); cited in Prelude med. plants database (2011)*
<u>Chad</u>: the seed is used as a febrifuge: *Watt et al. (1962) p. 1064*
<u>Kenya</u>: a root infusion is used as an emetic against malaria: *Beentje (1994) p. 378*
TESTS:
Spencer et al. (1947) p. 170: score 0 / +; Root extracts have proved "slightly effective" against experimental malaria: *Watt et al. (1962) p. 1065*
TOXICITY:
Bark, roots and fruit used extensively as a fish poison but no reports of harm to man. Tests on rats in 2009 showed "no serious safety concern" : *Obidah et al., (2009)*. Tests of leaves on Brine shrimp showed LC_{50} -2 and on Vero cell line 0.6446: *Brendler et al., (2010)*. See also *Kela et al.(1989)*

Bauhinia reticulata DC. [*Piliostigma reticulatum*: Hochst.] (Acanthaceae)
DESCRIPTION:
A flowering shrub or small tree up to 10m. high, with spreading intricate canopy. Often found in valleys and humid zones, forming thickets.
VERNACULAR NAMES:

E: Camel's Foot; **B**: umufumbe (22) **Lb**: mufumbe (22)
PLANT PARTS USED:
leaves, bark, root
TRADITIONAL USES:
W. Africa: the leaves, bark, and roots are used as a remedy for malaria: *Githens (1948) p. 79; Watt et al. (1962) p. 560*
Zimbabwe: the plant is used as a remedy for malaria: *Watt et al. (1962) p. 560*

Bersama abyssinica Fresen. subsp.not recorded (Melianthaceae)
DESCRIPTION:
Well-foliaged shapely tree, 7-15m. high. Found along watercourses, wooded ravines, margins of evergreen forest & open woodland, often associated with anthills.
VERNACULAR NAMES:
E: Winged Bersama; **B**: mucenya-mbulo (3c), nakancete (3, 3c, 9a, 23) **K/ Ln**: kamwenge (3) **Ku**: mpange (1) **Ln**: kamwengo (23),kapacike-longolo (3) **Ln/ Lv**: kapaci (6, 22) **Ny**: cenja-kulo (3e), ciwindu, mblaka, mcinji, mkanga, mtutu-muko (9), nkanga (3, 3e)
PLANT PARTS USED:
roots
TRADITIONAL USES:
Zimbabwe: a root infusion is drunk as a remedy for fever: *Gelfand et al.(1985) p. 180*
TOXICITY:
Some reports suggest that ingestion of leaves by livestock can cause salivation, diarrhoea, death: *(Blood, 2007)*. However, brine shrimp tests of a dichloromethane extract of the plant showed a very low toxicity figure: LC_{50} 127.7mgl/kg: *Moshi et al.(2010)*

Bidens pilosa L. (Asteraceae)
DESCRIPTION:
A small herbaceous tropical plant.
VERNACULAR NAMES:
E: Beggarsticks, Beggar Tick, Black Fellows, Black Jack, Bur Marigold, Cobblers' Pegs, Pitchfork, Spanish Needles, Sweethearts; **Afr**: Dulwelskerwel, Knapsekerel, Ouwewenaar, Wewenaar, Wewenaartjies; **B**: Kasokopyo; **Ku**: Kanunka, Kasokopya; **Ku/ Ny**: Kanzota; **Ny**: Cinomba, Cisosoci, Kaliputi, Kalivuti
PLANT PARTS USED:
leaves/aerial parts, roots,
TRADITIONAL USES:
Brazil: the roots and aerial parts are used to treat malaria and the liver complications believed to result from that disease: *Krettli et al.(2001)*
Cameroon: powdered leaves are rubbed in, decocted as enema, and filtered as eye-drops, to treat malaria: *Wome (1985)*
E. Africa: the roots are chewed or decocted as a remedy for malaria: *Kokwaro (1976)*
Ethiopia: Leaves powdered with *Ocimum gratissimum* leaves in 5 l.water to bathe child; powdered with *Chenopodium ambrosioides* leaves and palm oil to rub child: *Desta (1993)* ; cited in *Prelude med. plants database (2011)*
Rwanda: the leaves are used to treat malaria: *Van Puyvelde et al. (1975); Chagnon, (1984)*
TESTS:
Early tests produced negative results (*Spencer et al.1947)*, but recent tests on the leaves indicated a high degree of antiplasmodial activity (IC_{50} = 5µg/ ml): *Clarkson et al., (2004).*

In vivo tests showed showed a reduction of parasitaemia by 36-60%: *Krettli et al. (2000).* See also *Brandão et al. (1997).* The active ingredients are polyacetylene and methoxylated flavonoids (*Oliveira et al., 2004*) which are not very stable.

In vitro tests of the roots show IC_{50} ranges from 10-17 µg/ml, and *in vivo* (in mice) 60% inhibition was caused at ED_{50} 485 mg/ kg. The activity is greater for wild collected plants than for cultivated plants: *Andrade-Neto et al.(2004).* Research in Cameroon has also shown that it can reverse resistance to chloroquine: *Mbacham et al. (2005).*

TOXICITY:
None found by *in vivo* tests: *Chagnon (1984)*

Biophytum umbraculum Welw.[*B. petersianum* Klotsc.] (Oxalidaceae)
DESCRIPTION:
Widespread invasive weed with hairy stem, spreading leaves, 10-20cm. high.
VERNACULAR NAMES:
Ny: khunyata, mpembezu, namwalinyala, ngosa, nyalisi (9)
PLANT PARTS USED:
roots
TRADITIONAL USES:
E. Africa: a young child with fever is washed thrice daily with a cold infusion of the pounded roots: *Kokwaro (1976) p. 170*

Bixa orellana L. (Dipterocarpaceae)
DESCRIPTION:
Dense, spreading shrub or small tree, widely planted as ornamental, profusely fruiting with prickly red/orange seed-pods.
VERNACULAR NAMES:
E: Annatto; **Ku/ Ny**: kale (1, 3e)
PLANT PARTS USED:
seeds, roots, shoots
TRADITIONAL USES:
Brazil: The seeds and roots are used in the Amazon region as a remedy against malaria: *Brandao et al. (1992),* or the plant is infused or decocted, and drunk 3x daily until the cure is complete: *Milliken (1997) pp. 22, 25; (1997a) p. 219*
Peru: A decoction of shoots or seeds is drunk as a general febrifuge here as well as in the Brazilian state of Amazonas: *Duke (1994), cited in Milliken (1997a) p. 219*
Zambia: the leaves and fruit are used to treat fever: *Githens (1948) p. 79*

Blighia unijugata Baker (Sapindaceae)
DESCRIPTION:
Small/medium tree, up to 20m. high. Found at medium to low altitudes, in evergreen and riverine fringe forest and open woodland, often associated with anthills. Fiery red/ pink leaves and fruit.
VERNACULAR NAMES:
E: Triangle-tops; **Afr**: Driehoektolletjies
PLANT PARTS USED:
roots, bark, leaves
TRADITIONAL USES:

<u>E. Africa</u>: an infusion of the pounded roots is drunk twice daily as a remedy for fever: *Kokwaro (1976) p. 199.* The bark, leaves and roots are used to make a fever remedy: *Burkill (2000) p. 12*
<u>Côte d'Ivoire</u>: bark pulped in enema, or macerated as draught, is administered as a febrifuge: *Burkill (2000) p. 12; Kerharo et al. (1950) p. 165 [as Phialodiscus unijugatus]*
<u>Nigeria</u>: the root is used as a febrifuge: *Oliver-Bever (1960) p. 20,* quoted in *Burkill (2000) p. 12*
<u>Tanzania</u>: the root is used in fever medicine: *Wigg,* quoted in *Burkill (2000) p. 12*
<u>Zaire</u>: the leaves are used in a vapour bath for feverish children: *Bouquet (1969) p. 221,* quoted in *Burkill (2000) p. 12*
TOXICITY:
Clinical tests on rats in 2008 showed that "no death or clinical sign of toxicity was observed in any of the groups. No severe damage to the liver but the possibility of cellular lipotoxicity" *Oderinde et al. (2009)*

Boophane disticha Herb. [*Ammocharis taveliana* Schinz] (Amaryllidaceae)
DESCRIPTION:
A bulb with fanshaped grey-green fan of leaves, scented pink/purple flowers. Widespread on open woodland and grassland, where withered inflorescences form windblown "tumbleweeds"
VERNACULAR NAMES:
E: Century plant, Candelabra flower, Cape Poison bulb, Sore-eye flower, Tumbleweed; **Afr**: Gifbol, Malgif, Perdeskop, Seerooglelie; **Ny**: katupe, matakala (9) **Tk**: puli-zangala (8) **Ny/ Tu**: ndiyu (3e, 8)
PLANT PARTS USED:
bulbs
TRADITIONAL USES:
Extracts of the bulb have proved "slightly effective" against experimental malaria: *Watt et al. (1962) p. 23*
TOXICITY:
Poisonous to sheep and cattle, and deaths to humans reported: *Watt et al. (1962), Botha et al., (2005)*

Boscia angustifolia A. Rich. var. corymbosa (Gilg.) De Wolf [*B. corymbosa* Gilg.] (Capparaceae)
DESCRIPTION:
Small evergreen tree up to 8m. Found in all types of woodland in semi-arid areas.
VERNACULAR NAMES:
E: Rough-leaved Shepherds-tree; **Afr**: Skurweblaarwitgat; **B**: mwembembe (3, 3c, 3b, 14) **B/ I/ To**: musasa (3b, 4, 14) **I**: musoyo (4) **K/ Ln**: mukandacine (3, 14) **Ku**: lufwiti (1) Ku/ **Ny**: muntungwa (1) **Ln**: mukanda-cine (3b), musosu (21) **Lo**: kabumbu-mutemwa, musyosya (14), kabumbu, mutemwa (3b) **Lo/ To**: mubita (3b, 14) **Ny**: mhaka, mpetu, muka (9) **Tk**: mubite (23) **To**: umbombwe, musansa (14)
PLANT PARTS USED:
leafy twigs, bark
TRADITIONAL USES:
<u>Burkina Faso</u>: a decoction of leafy twigs and bark is drunk and used to wash twice daily for 3 days or until the fever is healed: *de la Pradilla (1988) p. 12*

<u>E. Africa</u>: the bark is boiled in water and the liquid drunk 3 times daily as a remedy for malaria: *Gelfand et al.(1985) p. 290; Kokwaro (1976) p. 45*
TESTS:
Plant extracts were tested *in vitro* against *P.falciparum*: *Bah et al. (2007)*

Brachystegia boehmii Taub. [*B. woodiana* Harms.] (Fabaceae, Caesalpinioideae)
DESCRIPTION:
Sturdy, small spreading tree up to 16m., common in open deciduous woodland, frequently dominant, particularly on poorly-drained soils. Leaves bright red when new.
VERNACULAR NAMES:
E: Prince of Wales's feathers; **B**: ingyansa (10), kapapa (10, 20) **B/ I/ K/ Lb/ Le/ Ln/ Ny**: musamba (3, 3a, 3c, 4, 9a, 10, 14, 17, 19, 23) **B/ Lb/ Ny/ Tu**: ngansa (3, 3a, 3c, 3e, 9a, 10, 14, 23) **B/ Ku**: ŋanza (1) **B/ Lb/ Ln/ Lo/ Ny/ Tk/ To**: muwombo (3, 3a, 3c, 3e, 9a, 10, 14, 23) **B/ I/ Lb/ Lo/ To**: mubombo (1c, 3, 3d, 4, 10, 12, 14, 17, 19, 23) **I**: inganza, umbomba (4) **I/ K/ Lb/ Lo/ Ln/ Tk/ Lo/ To**: mubomba (3a, 9a) **Ku**: ziye, ziye-nganza (1) **Ku/ Ny**: mufunda-nzinzi (1) **Ku/ Ny/ Tu**: ciyombo (10, 20) **M**: muzombo (10) **Nm**: luhombo, miyombo, muyenzi, mwiyenzi, mwiyombo, myenze (22) **Ny**: mcenga, mkoma-nyanda (3e, 14), mfunda-zizi (14), mfunda-nzizi (3e), mombo (10), msamba (3, 3e, 14), ngansa (3e) **Tu**: msenda-luzi (10)
PLANT PARTS USED:
bark
TRADITIONAL USES:
<u>Tanzania</u>: a bark decoction is drunk as a malaria remedy: *Haerdi (1964) p. 43*

Brachystegia spiciformis Benth. [*B. randii* Baker f.] (Fabaceae, Caesalpinioideae)
DESCRIPTION:
Medium-large tree, 8-15m. high, dominant over large areas. Handsome crown with heavy spiralling branches.
VERNACULAR NAMES:
B: kampela (3c, 14) **B/ Lb/ Le/ Ln/ N**: muputu (3, 3a, 3c, 7a, 9a, 10, 14, 23) **I**: mupansi, mutu (4) mumpanse, musyewe (10) **I/ Le/ Tk**: mpansi (23) **I/ Lo/ Tk/ To**: musewe (3, 3a, 4, 7a, 9a, 10, 14, 23) **K**: mpazyi (3, 7a, 14) **K/ Le**: mpanzi (10) **K/ Ln**: manga (3, 7a, 10, 14, 19, 20, 23) K/ Ln/ Lv: mumanga (3a, 9a, 10, 14, 19, 20, 23) Le: mupanse (20) **Ln/ Lv/ N**: mupuci (3, 6, 10, 14, 19, 20, 21, 22) **Lo**: mutuyu (19, 20, 22) **Ln/ Lo/ M/ N/ Tk**: muputi (3, 3d, 7a, 10, 14), mutuya (3, 3a, 3d, 7a, 9a, 10, 14, 23) **Lv**: mutundwa (6, 22) **M**: muputu (20) **M/ Ny**: mukuti (10) **N**: muhamba (10) **Nm**: gwangaka, miyombo, miyombo-botha, molo, mtundu, muzombo (22) **Ny**: cumbe (9, 10), kamponi, mcenga, msumbuti, mvukwe, tsamba (9), mkuti (3), mputi (3e, 7a), msale (10), puti (3a, 9a, 10, 14) **Ny/ To**: mupapa (10, 23) **Ny/ Tu**: mpapa (9, 10) **Tk**: musiwe (8, 10) **To**: muumba, mutinza (10) **Tu**: cumba (10), kaponi (8), msewe, mzimdiwi (22)
PLANT PARTS USED:
roots, leaves
TRADITIONAL USES:
<u>Tanzania</u>: a root decoction mixed with leaf sap is drunk for malaria: *Haerdi (1964) p. 43*
<u>Zambia</u>: an infusion of chopped roots is used to wash the body in cases of light fever: *Haapala et al.(1994) p. 11*

Bridelia duvigneaudi J. Leon. (Euphorbiaceae)
DESCRIPTION:
Shrub/ small tree up to 4m. high, with lax spreading branches, velvety leaves and blackcurrant fruit. Common in understorey of wetter areas of miombo woodland in northern half of Zambia.
VERNACULAR NAMES:
B: cituta (3c), kalamba-bwato (3, 3c, 7a) **Ln**: Mufunji (13a) **Lo**: muni-minga(7a), munwa-mema (3d), nubwa-mena (3) **Ny**: mpasa, mtela (3, 3e), mwanjane (3e, 7a)
PLANT PARTS USED:
leaves
TRADITIONAL USES:
D.R.Congo, Zambia: the leaves are used by the Bemba to cure fever: *Smith (2004) p. 51*

Bridelia ferruginea Benth. (Euphorbiaceae)
DESCRIPTION:
Found only in Mwinilunga District. Similar to *B. duvigneaudii* in appearance.
VERNACULAR NAMES:
Not known
PLANT PARTS USED:
leaves, twigs, bark
TRADITIONAL USES:
W. Africa: decoctions of leaves, leafy twigs, and bark are commonly used as febrifuge: *Bouquet (1974) p. 83,* quoted in *Burkill (1994) p.36; Kerharo et al. (1950) p. 69*

Brucea antidysenterica j.f.mill (Simaroubaceae)
DESCRIPTION:
Shrub or tree up to 5/6m, often as secondary growth in deforested areas and at forest edges.
VERNACULAR NAMES
Not known
PLANT PARTS USED:
bark, roots, fruit
TRADITIONAL USES:
Ethiopia: the bark, roots and fruit are used as a remedy for fever: *Burkill (2000); De Wildeman (1949), Subramanian (1955),* quoted in *Watt et al. (1962); Grace et al. (2008)*
TESTS, ANTI-MALARIAL:
The bark, which has a bitter taste, contains quassinoid compounds known as bruceolides, notably bruceantin and bruceanic acids. [*Aubreville (1950); Oliver-Bever (1960, 1983)*]], and several other bitters and other substances: *Watt et al.(1962).* Root extracts gave negative results in 1947, [*Spencer et al. (1947)*]; however, subsequent clinical trials have shown bruceantin to be highly active against *P.falciparum,* with IC_{50} 0.0008 µg/ml: *O'Neill et al. (1986)*
TOXICITY:
The fruit ex Ethiopia yields 22% of fixed oil and a small amount of volatile acids, as well as 1% of a resin, a bitter principle, a yellow colouring matter, and dextrose: *Power (1914); Watt et al. (1962).* The fruit is said to be fatal when eaten by livestock, especially sheep: *Burkill (2000); Getahun (1975).* Although bruceantin proved toxic in *in vitro* tests againsy

9KB mammalian cells, this was at a higher concentration than in the antimalarial tests,which suggests that it may be possible to find a quassinoid with good antimalarial activity and low mammalian cytotoxicity: *O'Neill et al. (1986)*

Burkea africana Hook. (Fabaceae, Caesalpinioideae)
DESCRIPTION:
Small/medium tree up to 15m. Crown light, Fissured bole, hairy twigs and buds. Widespread in several woodland types, esp. sandy soils. Often dominant at edge of dambos.
VERNACULAR NAMES:
E: Wild seringa; **Afr**: Wildesering; **B**: mukoso (3, 3c, 7a, 14) **B/ K/ Ku/ M/ Ny/ Tu**:kapanga (1, 2, 3, 3a, 3c, 3e, 9a, 10, 14, 20, 23) **B/ K**: kayimbi (10) **I**: oombe (4) **I/ Lo/ Ln/ Lv**: musese (3d, 4, 10, 14, 17, 19, 22) **I/ Le/ Lo/ Tk/ To**: musesye, musyesye (3a, 4, 7a, 8, 10, 16, 19) **K**: kabulwe-bulwe, nakayumbe (10) **Ku**: kawigi (1) **Ku/ Ny**: kawizi (1, 3e) **Lb**: muwanga, umune-nene, umuwanga (22) **Le**: museese (17) **Ln**: musese-wezenzele (10) **Ln/ Lo**: wezenzele (6, 22) **Ln/ Lo/ To**: musesye (2, 3, 9a, 14, 23) **Ln/ Lv**: msese (6, 22) **Lo**: mwandakasi (3d, 14) **Lo/N**: mubulwe-bulwe (3d, 7a, 10, 14), mukazihebe (3d, 10, 14), syacibula (19) **N**: mwandakati (10) **Nm**: mgando, mugando (22) **Nm/ Ny**: mkalati (9, 10, 22) **Ny**: kawizi 3, 14), mkoso (3e, 7a, 14), msase (2, 3), ngalati (3a, 3e, 9a, 14), nyalati (7a) **Tk**: sicibula, mwandakazi (10) **To**: musenya (3a, 9a), mwansa-buso (19), mwansasa-buso (23), mwansusa-buso (10) **Tu**: musase (8), kawidzu, musase-mamuna (10)
PLANT PARTS USED:
plant
TRADITIONAL USES:
Upper Volta: the leaves make a draught to treat fevers: *Burkill (1995) p. 72; Kerharo et al. (1950) p. 102*
Zimbabwe: the leaves are mixed with those of *Lippia javanica* and boiled, and the steam inhaled as a remedy for fever: *Gelfand et al.(1985) p. 133*

Caesalpinia pulcherrima (L.) Schwartz (Fabaceae, Caesalpinioidae)
DESCRIPTION:
Ornamental tropical shrub about 2m. high, with few scattered prickles, orange/ yellow flowers
VERNACULAR NAMES:
E: Barbados Pride
PLANT PARTS USED:
flowers, roots
TRADITIONAL USES:
Brazil: the flowers and roots are decocted to treat fever: *Di Stasis et al. (1994) p. 533*
Zambia: the roots and flowers are used as a cure for fever: *Nair (1967) p. 43*

Calendula officinalis L. (Asteraceae)
DESCRIPTION:
Widely cultivated as ornamental flower
VERNACULAR NAMES:
E: Pot Marigold
Not known
PLANT PARTS USED:
leaves and flowers

TRADITIONAL USES:
<u>Zambia</u>: the leaves and flowers are used as a diaphoretic: *Nair (1967) p. 94*

Calotropis procera (Ait.) R. Br. [*Asclepias procera* Aiton] (Apocynaceae)
DESCRIPTION:
Shrub about 3m. high, with stout woody stems. The greenish globes look inviting, but break open to reveal toxic latex. Common in villages along the Lower Luapula, but always planted or naturalised.
VERNACULAR NAMES:
E: Giant/ Swamp Milkweed, Roostertree, Sodom Apple; **Nm**: mpumbula (22) **Ny**: citonje, letaunde, mpira (9)
PLANT PARTS USED:
bark, root bark, flower buds
TRADITIONAL USES:
<u>Bénin</u>: fresh leaves and stems are infused to treat fever: *Adjanohoun et al. (1989)*
<u>Central Sahara</u>: fresh leaves are infused to treat fever: *Hammiche et al. (2006)*
<u>India</u>: the bark is used as a tonic diaphoretic: *Watt et al. (1962) p. 126*. The flower buds and root bark are crushed to make tablets to treat malaria: *Singh et al. (1993) p. 71*. Another recipe uses the buds mixed with salt and pepper to make pills, taken twice daily for three days: *Anis & al.(1986) p. 53*
TOXICITY:
Leaves can cause severe skin eruptions: *Behl et al. (2002)*

Cannabis sativa L. (Cannabaceae)
DESCRIPTION:
Widely planted in villages; annual erect herb with feathery leaves, up to 5m
VERNACULAR NAMES:
E: Ganja, Hashish, Hemp, Marijuana; **B**: ibange (3c, 7a) **B/ Ku/ Ny**: camba (1, 3e) I: lubange (4) **Ku/ Ny**: mbanje (1, 3e, 7a) **Lo**: libangwe (3d), matokwani
PLANT PARTS USED:
plant, pipe dottle
TRADITIONAL USES:
<u>Zambia</u>: the oil residue left in pipes after smoking cannabis is mixed with water and drunk thrice daily as a remedy for malaria: *Haapala et al.(1994) p. 12*
<u>Zimbabwe</u>: the plant is used as a remedy for malaria: *Watt et al. (1962) p. 762*
TOXICITY:
See Watt et al. (1992) pp 759-772

Capparis sepiaria L. [*C. laurifolia* Gilg & Gilg- Ben.; *C. citrifolia* Lam.; *C. subglabra* (Oliv.) Gilg & Gilg- Ben.] (Cappariaceae)
DESCRIPTION:
A scrambling shrub with sharp hooked thorns and white flowers, 5' – 6' high, found in dry deciduous woodland or thickets and riverine fringes.
VERNACULAR NAMES:
E: Hedge Caper-bush; **Afr**: Heiningkapperbos; **B/ Ku**: kalongo (1) **Ny**: mgezi-ngono (3e)
PLANT PARTS USED:
plant
TRADITIONAL USES:

<u>Zambia</u>: the shrub is used as a febrifuge: *Nair (1967) p. 49*
TOXICITY:
Non-toxic: *Rajesh et al., (2010)*

Capparis tomentosa Lam. (Cappariaceae)
DESCRIPTION:
A woody scrambler up to 10m. high, or a dense thorny impenetrable bush with orange spherical berries. Found in a variety of habitats, often on termite mounds
VERNACULAR NAMES:
E: Woolly Caper-bush; **Afr**: Wollerige Kapperbos; **B/ Ku**: kalongo (1) **B/ Tu**: cunga-nkobwe (20) **B/ Ku/ Ku**: galango, kalango (1) **I**: ikalangwe (17) **I/ Lo**: ciweze-weze (3d), ciwezeze (2, 3, 4, 17, 23) **I/ To**: conswe, conzwe, mulya-balisina (3, 4, 17, 23) **K**: cikalangwe (2, 3) **Ku**: kalu-kulu (1) **Le**: likalango (17), mukalango (23) **Ln**: muzezi (2, 3, 23) **Lo**: ciweze-weze, mupapi (3d), mukonangwe (19, 23) **Lv**: musango (23) **Ny**: kalanga, kali-lesya, kapala-pala, mfwete (3e) **To**: ciwezyezyi (17), mukorongwe (22), ngizi (23) **Tu**: galanga (8)
PLANT PARTS USED:
root bark
TRADITIONAL USES:
<u>S. Africa</u>: The root bark is a Zulu remedy for malaria: *Watt et al. (1962) p. 162*
TOXICITY:
Root decoctions were used to poison lion-bait: *White (1962)* Tests of leaf decoctions on sheep, goats and Zebu calves proved fatal: *Ahmed et al. (1980, 1981)*

Capsicum annuum var. frutescens Kuntze. (Solanaceae)
VERNACULAR NAMES:
E: Red Pepper; **B/ I/ Lb/ Le/ M/ To/ Tu**: mpili-pili (9a, 19, 20) **I**: impili-pili, iŋomba (4) **I/ Le/ To**: maŋomba (3, 4, 17) **Lo**: mbili-bili (19)
PLANT PARTS USED:
fruit
TRADITIONAL USES:
<u>Burkina Faso</u>: the fruit is eaten as a part of the regular diet to prevent malaria: *de la Pradilla (1988) p. 13*
<u>D.R.Congo</u>: the fruits are prepared as a fever-remedy: *Ndukwu et al. (n.d.)*
<u>Morocco</u>: A decoction of pounded fruit is drunk 3x daily (ad. 100ml., chn. 50ml.); leaves pounded & mixed with water to make an enema: *Diafouka (1997)*

Cardiospermum grandiflorum Sw. (Sapindaceae)
DESCRIPTION:
An invasive climbing vine, with hairy stems and leaves. spreading 6m. White flowers have pleasant smell. Seeds in papery capsules. Found on disturbed land, moist gullies.
VERNACULAR NAMES:
Balloon vine, Heart seed.
PLANT PARTS USED:
leaf
TRADITIONAL USES:
<u>Côte d'Ivoire</u>: a leaf decoction is taken by draught for fever: *Visser p. 50,* quoted in *Burkill (2000) p. 14*
<u>E. Africa</u>: fresh leaves are infused with water as a remedy for fever: *Kokwaro (1976) p. 199*

TOXICITY:

The leaves contain *Cardiospermin* as the *p*-hydroxybenzoate [60444–99–7] and *p*-hydroxycinnamate [79197–19–6] and *Cardiospermin sulfate* [78856–87–8] : *Harborne et al. (1999)*

Cardiospermum halicacabum L.var. halicababum (Sapindaceae)
DESCRIPTION:
Invasive climbing vine invading forest margins, river banks and urban open spaces. Can smother a small tree.
VERNACULAR NAMES:
E: Balloon Vine, Heart Seed, Love-in-a-puff; Ku: kankululu, katangala-tuze, kunze (1) Ny: cikumba, cipati-kila (3e) To: kankolo (3, 9a) Tu: cifatukila (8)
PLANT PARTS USED:
leaf, root
TRADITIONAL USES:
Antilles and East Indies: the leaf and root are used as a diaphoretic: *Dragendorff,* quoted in *Watt et al. (1962) p. 930*
India:The root is decocted as a diaphoretic: *Burkill (2000) p. 15; Quisumbing (1951) pp. 547-9,* quoted in *Watt et al. (1962) p. 930*
TOXICITY:
Plants contain cyanogenic lipids, but no damage to humans reported. See report of Committee for veterinary medical products, in *EMEA (2000)*

Carica papaya L. (Caricaceae)
DESCRIPTION:
The pawpaw tree is widely grown in the tropics, and much used for medical purposes.
VERNACULAR NAMES:
E: Melon Tree, Pawpaw; **B**: Ipapawo (3c), Lipapau (20); **B/ Ku/ Ny**: Cipapaya (1, 3e); **B/ M**: Mapapaya, Mapapayu (20); **I/ Le/ To**: Ipopo (17); **Ln/ Lv/ Lv**: Papau (18, 19); **Lo**: Mampubila (18, 19; **Nm**: Mubabaju (22); **Ny**: Papaya (9), Papayi (3e); **Ny/ Tu**: Mapapayi, Vipapayi (20); **To**: Papopo (18, 19)
PLANT PARTS USED:
leaves, roots
TRADITIONAL USES:
Brazil: a leaf(?) decoction is administered 3x daily to malaria patients: *Milliken (1997a)*
Burkina Faso: 3 bunches of leaves are boiled for 10 min. in 10 lit. of water. To be used to wash each evening for 3 to 5 days: *de la Pradilla (1988)*
Cameroon: the mature leaves and seeds are used to treat malaria: *Titanji (2008)*
Congo-Brazzaville: the leaves of *C.papaya* are decocted with leaves, roots or fruits of a variety of plants, inc. *Senna occidentalis, Lantana camara , Cymbopogon citratus, Musa sapientium,* and *Annona senegalensis* to make a remedy; or the leaves of *Musa sapientium* are boiled to make a steam bath; *Diafouka (1997)*
E. Indies: the leaf has been used as a febrifuge and laxative: *Sorgdrager (1942)*
Ethiopia: a leaf decoction is used to treat malaria: *Gedif et al. (2002)*
Malawi: a leaf decoction is used as a remedy for fevers: *Williamson (1975); Morris (1996)*
Nigeria: the leaves are used as a febrifuge: *Ainslie sp. no. 76,* quoted in *Burkill (1985);*

Chew seeds 2x daily, drink decoction of unripe fruit withunripe pineapple, lime, 10cm sugar cane, 6 teabags in 4 l. water: *Obute (nd)*

Zambia: a roots infusion is drunk or mixed with porridge: *Vongo (1999)*

TESTS:

Spenser's tests for antimalarial activity proved negative: *Spencer et al. (1947).*

In vitro tests on seeds and rind extract in 2001 showed IC$_{50}$ 15.19 mg/ml, but Ngemenya recorded IC$_{50}$ at 60mg/ml *(Bhat et al. 2001; Ngemenya et al. (2004); Titanji et al. (2008)*

Phytochemical screening in Nigeria in 2008 revealed potent antioxidant activity, IC$_{50}$ =0.58 mg/ml: *Ayoola et al.,(2008).*

TOXICITY:

Both vegetative and reproductive organs of the plant yield hydrocyanic acid: *Quisumbing (1951), Webb (1948)*, cited by *Watt et al. (1962)*. The alkaloid *carpaine* is poisonous in large doses, and the enzyme *papain* is powerfully toxic when injected intravenously: *Sorgdrager (1942)*, cited by *Watt et al. (1962); Osol and Farrar (1955)*. For chemical analyses, see *Watt et al. (1962)*. See also www.ansci.cornell.edu./plants/medicinal/ papaya.html

For tests on rats, see *Udoh et al. (1999)*

Carissa edulis Vahl [*Arduina edulis* (Forssk.) Spreng.] (Apocynaceae)

DESCRIPTION:

A dense or semi-scandent evergreen spinous shrub up to 5 m. high, with purple edible fruit.

VERNACULAR NAMES:

B: mukomfwe (3, 3c, 23) **B/ K/ Ln/ Lo**: mufombwa (3, 3c, 23) **I/ K**: mubambwa-ngoma (4, 23) **K**: mubanga-ngoma (3) **Lo**: mufuwe (3d), munyo-ngolwa (3d, 7a) **Nm**: mfuje-anje (1a, 22) **Ny**: mbango-nge (3e, 7a), mkanga-mwazi, mkhala, mkololo, mpambula (9), msyara-kunzi (3) **Ny/Tu**: nya-koko (3e, 8) Tk: katenje (8)

PLANT PARTS USED:

plant

TRADITIONAL USES:

E. Africa: a decoction of the roots is used to treat malaria: *Kokwaro (1976) p. 26*

Madagascar: a root decoction is taken as a febrifuge: *Rasoanaivo et al. (1992) p. 119*

Sudan: a root decoction is taken twice dailty in 5 – 10 mil doses: *El-Kamali (1997) p. 527*

TESTS:

Methanol and aqueous extracts were tested against resistant strains of *P.falciparum*: *Kirira et al., (2006)*

TOXICITY:

Tests of leaves on Vero cell line showed LC$_{50}$ -1: *Brendler et al., (2010)*

Casearia battiscombei R.Fr. (Salicaceae)

DESCRIPTION:

Large tree up to 40m, found between 1000 and 2000m, in evergreen forest and at forest margins.

VERNACULAR NAMES:

E: Forest sword-leaf

PLANT PARTS USED:

roots

TRADITIONAL USES:

Mozambique: the root is used to treat malaria: *Mulhovo (1999)*

Cassia abbreviata Oliver subsp. abbreviata [*C. beareana* Holmes; *C. granitica* Baker f.; *C. abbreviata* Oliver var. *granitica* (Baker f.) Baker f.] (Fabaceae, Caesalpinioideae)
DESCRIPTION:
A shrub or smallish tree 3 – 10m. high. Found at low to medium altitudes in open bushveld, woodland, or wooded grassland, along rivers, on hillsides, and on anthills.
VERNACULAR NAMES:
E: Long-tail Cassia, Long-pod cassia; **Afr**: Sjambokpeul; **B**: munsoka-nsoka (3, 3b, 3c, 7a, 10, 14, 20, 23), mununka-nsyimba (3c), musamba-mfwa (3c, 14) **B/ Ku**: musoka-nsoka (1, 20) **I/ K/ Le/ Lo/ Tk/ To**: mululwe (3, 3b, 3d, 4, 5, 7a, 10, 14, 17, 23) **K**: mululu (14), mululwe (10) **Ku**: khoswe, koswe (1) **Ku/ Ny**: mleza (1, 3e) **Lb**: akafungu (22), mufungu (10) **Lo**: kapatati (3d, 7a, 14) **Nm**: mrunda-runda, munzoka (22) **Ny**: mcala-mira, mkwapu-kwapu, mluma-nyame (9, 10), mkoswe (3, 3b, 3e, 7a, 14), mlembe-lembe, mzabaza, nyoka (3e), mnyoka (3, 3b, 14), mtanta-nyelele (3e, 14) **Tk**: malulwe (8) **To**: inkonkoni (?), lumanyama (22) **Tu**: leza (10), mulembe-lembe (8)
PLANT PARTS USED:
roots, bark
TRADITIONAL USES:
Tanzania: a root and bark decoction mixed with root shavings is eaten as a malaria remedy. It is said to contain tannin: *Anon. (1911), Bostock (1907) p. 273,* quoted in *Watt et al. (1962) p. 566; Bally (1937) no. 10; Githens (1948) p. 81; Haerdi (1964) p. 44*
TESTS:
Aqueous and organic fractions of roots and leaves were tested *in vitro* against the multi-drug resistant Vietnam-Smith strain of *P. falciparum* (VI/S). High activity at $IC_{50} < 3mg/ml$ was seen: *Connelly et al.(1996)*
TOXICITY:
Not dangerous: *Parry (1992)*

Cassytha filiformis L. (Lauraceae)
DESCRIPTION:
A twining parasitic herb with scales in place of leaves.
VERNACULAR NAMES:
E: Love Vine, Devil's Gut, Dodder; **B**: lunsyi-nsamba (3, 3c) **K**: kula-kasi (8) **Lo**: luze (3d) **Nm**: mlangamia (1a, 22) **Ny**: lwando (9), msaka-dzinje, saka-sinji (3e, 9), mzewa, sanga-sinje (3e) **Tk**: zazambi (8) **To**: sibone-nwisiku (3) **Tu**: zyewa (8)
PLANT PARTS USED:
sap
TRADITIONAL USES:
Benin:the plant is used to treat malaria: *Adjanohoun et al.(1989)*
Congo:the plant is used to treat malaria: *Staner et al.(1937)*
Madagascar: stems & leaves are decocted as a febrifuge: *Rasoanaivo et al.(1992)*
Tanzania: the sap is drunk as a malaria remedy: *Burkill (1995) pp. 40; Haerdi (1964) p. 39*
TOXICITY:
Not dangerous. $LD_{50} > 500mg/ml$: *Babayi et al. (2007)*

Catha edulis (Vahl) Forssk. ex Endl. (Celastraceae)
DESCRIPTION:

A shrub or tree 2-15m in height, occurring at the margins of, and in, high-altitude evergreen forest, in open woodland, and on rocky wooded hillsides. Used widely as an inebriant stimulant and narcotic.

VERNACULAR NAMES:
E: Abyssinian Tea, Bushman's Tea, Chirinda Redwood, Somali Kat, Khat, Wild Tea; **Afr**: Spelonkentee; **Ny**: Mdima-madzi, Mdima-manzi, Mutsawari; **To**: Siboneu-wisiku; **Tu**: Ikwa

PLANT PARTS USED:
bark.

TRADITIONAL USES:
E. Africa: the green bark from the youngest branches is chewed as a remedy for malaria: *Kokwaro (1976)*

TESTS, ANTI-MALARIAL:
In vitro tests of the roots in 2004 revealed a high degree of antiplasmodial activity (IC $_{50}$ 0.63 µg/ml): *Clarkson et al. (2004)*. For analysis, see *Watt et al. (1962)*

TOXICITY:
Student's *t*-test with *p<0.05* considered significant gave toxicity score of 0: *Van der Venter et al.(2008)*. No reports of toxic effects on humans.

Catharanthus roseus (L.) G. Don [*Vinca rosea* L.] (Apocynaceae)
DESCRIPTION:
A fleshy perennial growing to 80cm high, with mauve flowers. Widely established by escaping from gardens.

VERNACULAR NAMES:
E: Madagascar periwinkle, Rosy periwinkle

PLANT PARTS USED:
root

TRADITIONAL USES:
Queensland: a decoction from the root is said to be febrifuge, and showed positive results when tested: *Spencer et al. p. 154; Webb p. 232,* quoted in *Watt et al. (1962) pp. 85, 88*

TOXICITY:
Highly toxic.All parts of the plant, and the root bark in particular, contain active alkaloids of the indole type. Inhalation or ingestion can cause hallucinations and acute poisoning , with seizures and inflammation of the eyes, nose and mouth: *Russell et al. (1997)* For LD$_{50}$ score on leaves, see *Brendler (2010) P. 286*

In vitro bacterial & mammalian cell assays showed that the plant caused both DNA damage & chromosomal aberrations: *Fennell et al.(2004)*

Showed as cytotoxic when tested on snails at 50-100 ppm: *Elsheikh et al.(1990)*

Catunaregam obovata (Hochst.) A. E. Gonc. *s. l.* [*C. spinosa* (Thunb.) Tirveng subsp. *spinosa* (misapplied name); *Xeromphis obovata* (Hochst.) Keay] (Rubiaceae)
DESCRIPTION:
A spinous shrub or small tree up to 5.5m high, widespread in drier areas.

VERNACULAR NAMES:
E: Thicket-thorn; **I**: mutunga-bambala (4) **I/ K/ Le/ Lo/ To**: matunga-mbambala (3, 17) **Ku**: cipembele, mulasa-kubili (1) **Ku/Ny**: mucinga-zuba (1) **Lo**: mwanda-bala (3d) **Nm**: minuasa-ungu, moca-ngoko (22) **Ny**: cipembele (3, 3e, 9), cipisya-wago, cigwenembe, msondoka (9), mlasaku-wili (3e) **Tk**: matunga-mbawala (8) **To**: mutunga-babala (22)

PLANT PARTS USED:

roots
TRADITIONAL USES:
Malawi: the roots are used as a remedy for fevers: *Morris (1996) p. 455*
Zimbabwe: a root infusion is drunk, & used to bathe the body : *Gelfand et al.(1985) p. 229*

Cayratia gracilis (Guill. & Perr.) Suesseng. [*Cissus gracilis* Guill. & Perr.] (Vitaceae)
DESCRIPTION:
Slender climber with angular grooved stems, tiny greenish flowers, black currant-like fruits. Widespread in forest and savannah areas.
VERNACULAR NAMES:
Nm: mwungera (7c) **Ny**: cikumba (3e), mtelevele, mwana-mpefo, ncofu (9)
PLANT PARTS USED:
leafy twigs
TRADITIONAL USES:
Burkina Faso: 3 bunches of leafy twigs are boiled for 10 min. in 10 lit. of water. Patients wash once daily for 3 days: *de la Pradilla (1988) p. 16*

Ceiba pentandra (L.) Gaertn. (Malvaceae)
DESCRIPTION:
Tall erect tree 15m high. Found in Luangwa valley, edge of Central Barotse plain, Livingstone.
VERNACULAR NAMES:
E: Kapok, Silk-cotton; **B**: busufu (1) **Ny**: mpilila (3e)
PLANT PARTS USED:
bark, leaves
TRADITIONAL USES:
Benin:the leaves are used to make a drink: *Adjanohoun et al.(1989)*
Congo-Brazzaville: the leaves are pulped to make a massage medium: *Bouquet (1969)*
Nigeria: a bark decoction is taken as a febrifuge: *Ainslie sp. no. 82,* quoted in *Burkill (1985) p. 281*; the leaves are infused: *Odugmeni et al.(2007)*
TOXICITY:
"Very low toxicity profile in all the tested animals & is relatively safe for herbal oral medication" with LD_{50} >5000mg/ml: *Sarkiyayi et al.(2009)*

Celtis africana Burm. f. (Celtidaceae)
DESCRIPTION:
Deciduous shrub or tree up to 13m high in Zambia, widespread distribution.
VERNACULAR NAMES:
E: White Stinkwood; **Afr**: Witstinkhout
PLANT PARTS USED:
bark
TRADITIONAL USES:
Nigeria: the bark is pounded as a fever remedy: *Irvine (1961) p. 418,* quoted in *Burkill (2000) p.219;*

Centella asiatica (L.) Urb. (Araliaceae)
DESCRIPTION:
Small herbaceous annual plant, found in ditches and in low wet areas.

VERNACULAR NAMES:
E: Asiatic Pennywort, Indian Pennywort; **Ny**: namsanganya, tengwa (9)
PLANT PARTS USED:
roots, leaves
TRADITIONAL USES:
Guinea:the leaves are used to treat malaria: *Sugiyama et al.(1992)*
Tanzania: the roots are decocted as a remedy for small children with malaria: *Burkill (2000) p.229; Haerdi (1964) p. 163*; the aerial parts are mixed with local beer, or decocted with the aerial parts of *Aspilia mossambicensis*: *Mainen et al.(2009)*
Zambia: the leaves are used (in a decoction?) in fever: *Nair (1967) p. 90*
TOXICITY:
For cases of hepatoxicity following ingestion of *C. asiatica* or of contact dermatitis following topical use, see *Brendler (2001) p. 81*. For LD_{50} score, see *ibid. p. 286*

Chamaecrista mimosoides (L.) Greene [*Cassia mimosoides* L.] (Fabaceae, Caesalpinioideae)
DESCRIPTION:
Herb or low shrub, 1.7m high, very polymorphic, found in wasteland, roadsides, forest clearings of the wooded savannah, commonly on sandy soil, widely dispersed.
VERNACULAR NAMES:
E: Tea senna; **B/ Ku**: soyo(1) **Ny**: ngwala-ngalate (9) **Tk**: lungu-nuwa (8)
PLANT PARTS USED:
twigs
TRADITIONAL USES:
Burkina Faso: 5 bunches of leafy/flowering twigs are boiled for 15 min. in 10 lit. of water. Adults drink and wash in the liquid, twice daily for 4 days; infants are washed in it: *de la Pradilla (1988) p. 14*

Chenopodium ambrosoides L. (Chenopodiaceae)
DESCRIPTION:
Annual invasive herb with small yellow flowers up to 1m high
VERNACULAR NAMES:
E: Wormseed, Mexican tea, Jesuit's tea, Jerusalem tea, Spanish tea; **Ku**: kamcenga (1) **Ny**: kanyuku, njoka, nunkamani (9)
PLANT PARTS USED:
plant, leaves
TRADITIONAL USES:
Angola:the plant is infused to brew a drink: *Bossard (1996)*
Burundi: the leaves are decocted to drink, & the juice of the leaves is sniffed and dropped into the ears: *Baerts et al.(1989)* juice expressed from the leaves is used for drinks & enemas: *Poligenis-Bikendako (1990)*
Cameroon:leaf-tips are gound and infused to treat malaria:*Jiofack et al.(2009)*
Congo/ Brazzaville: the leaves are decocted for baths for child patients: *Bouquet (1969)*; the leaves are pounded & filtered to be drunk 2x daily, or ground in palm oil with *Bidens pilosa* leaves to make a body-rub: *Diafouka (1997)* ;
Congo/ Kinshasa: the plant is used: *Ngalamulume et al.(1995)*; the plant is decocted or macerated to make a drink, pounded & macerated with *Capsicum frutescens* fruit to make

an enema, and the leaves are decocted with those of *Hymenocardia acida, Vitex madiensis, Annona senegalensis,* & *Aframomum sp.* as a prophylactic: *Kasuku et al.(1999)*
Congo/Zaire:leaves are ground, macerated in water & filtered to make an enema: *Wome (1985); Adjanohoun et al., (1988), Nyakabwa et al.(1990)*
Madagascar:oil from the leaves is used: *Pernet et al.(1957)* ; the stems & leaves are decocted for drinks & inhalation: *Rasoanaivo et al.(1992)*
N. Africa:leaves & shoots from Rabat market are added to soup as febrifuge: *Boulos (1983)*
Nigeria:the plant is pounded or infused to make a febrifuge: *Ainslie(1937)*
S. Africa: an infusion of the plant is reputed to be diaphoretic: *Watt et al.(1962) pp. 188*
Uganda: the leaves are decocted to make a steam bath: *Tabuti et al.(2003)*
the aerial parts are decocted to make an inhalant & to bathe: *Adjanohoun et al.(1993)*
W. Africa: a hot infusion of the whole plant is used for fever in Gabon and Nigeria: *Ainslie sp. no. 87, Cavaco (1963) a20,* quoted in *Burkill (1985) p. 366.* In Sierra Leone the young leaves are used to treat fever: *Deighton 2702,* quoted in *Burkill (1985) p. 366*
Zimbabwe: the leaves are pounded to make ointment, which is rubbed on the body of an infant running a high temperature: *Gelfand et al.(1985) p. 120; Watt et al. (1962) p. 187*
TESTS:
Ascaridole, a terpene isolated from the plant, is one of the few naturally occurring endoperoxides, and exhibits some antiplasmodial activity: *Schwikkard et al., (2002) p. 679*
TOXICITY:
Oil derived from seeds can be toxic toxic to livestock. See *Cornell University website* and *Monzote et al.(2007)*

Chlorophytum sphacelatum (Baker) Kativu var. milanjianum (Rendle) Nordal
[*Anthericum whytei* Baker] (Anthericaceae)
DESCRIPTION:
Perennial herb up to 90cm
VERNACULAR NAMES:
Not known
PLANT PARTS USED:
leaves
TRADITIONAL USES:
Tanzania: a leaf decoction is used to bathe the body in malaria: *Haerdi (1964) p. 197*

Chrysanthellum indicum DC. subsp. afro- americanum B. L. Turner [*C. americanum* (L.) Vatke] (Asteraceae)
DESCRIPTION:
Annual herb 3-45cm, prostrate or erect.
VERNACULAR NAMES:
E: Golden chamomile; **Ny**: nyasuna (1)
PLANT PARTS USED:
plant
TRADITIONAL USES:
Burkina Faso: a handful of the dried plant is boiled for 10 min. in a litre of water, and used as a wash once daily for 3 days. The plant contains flavonoids and sapoinins: *de la Pradilla (1988) p. 16*
TOXICITY:
None

Cilalwe tree
VERNACULAR NAMES:
I: cilalwe (4)
PLANT PARTS USED:
leaves
TRADITIONAL USES:
Zambia: the leaves are chewed, and the bitter juice squirted into the nostrils and ears of malaria patients: *Fowler (2000) pp. 109, 812*

Cissampelos pareira L. var. orbiculata (DC.) Miq. (Menispermaceae)
DESCRIPTION:
A woody vine climbing in secondary growth, resembling a spiral ribbon round the tree.
VERNACULAR NAMES:
E: False pareira brava, False pareira root, Velvet leaf; **Afr**: Dawidjies; **Nm**: Mlaga-laga, Ukulwanti
PLANT PARTS USED:
leaves, roots.
TRADITIONAL USES:
Angola:a root decoction is used to treat malaria: *Bonnefoux (1937)*
Burundi: leaves are decocted to make a steam bath: *Polygenis-Bikendako (1990)*
Ecuador: a decoction of leaves and roots is taken to treat fever: *Coe et al. (1996)*
Kenya: the root bark is decocted to make a drink: *Muthaura et al.(2007)*
Madagascar: the roots are decocted to make a malaria remedy: *Boiteau (1986). Pernet et al.(1957), Rasoanivao et al. (1992);* The alkaloids were found to increase the effects of chloroquine: *Rasoanivao et al. (1991)*
Tanzania:the roots are decocted or powdered as a febrifuge: *Gessler et al.(1995)*
S. Africa: the leaves and roots are used as a febrifuge: *Githens (1948)*
Sudan:the roots are decocted or infused to make a drink: *El-Kamali (2009)*
TESTS, ANTI-MALARIAL:
Tests in 1941 gave conflicting results for anti-malarial activities (*Sapeika 1941*), but the Spencer tests in 1947 produced one of his rare positive results: *Spencer et al. (1947): score ++ / ++.* For analysis, see *Watt et al.(1962)*
In vitro tests in 1994 gave a high level of antiplasmodial activity (IC_{50} 0.38μg/ml) [Confusion with *C. mucronata*?): *Gessler et al. (1994)*
Aqueous and methanol extracts were tested *in vitro* against chloroquine sensitive and resistant strains of *P. falciparum*, and showed activity at IC_{50} 5.85 mg/ml: *Rukunga et al. (2008)*
TOXICITY:
Used as a fish-poison in the West Indies and the Philippines, in Zambia it was found to be non-toxic in tests on cattle and sheep: *Northern R. (1931)*, cited in *Watt et al. (1962)*
In tests on rats, the plant did not show any sign of toxicity and mortality up to a dose level of 1000 mg/kg: *Amresh et al., (2007)*

Cissus aralioides L. var. orbiculata (DC.) Miq. (Vitaceae)
DESCRIPTION:
A lofty climber, woody at base, greenish white flowers, with caudex up to 20cm, widespread in tropical Africa.

VERNACULAR NAMES:
Not known
PLANT PARTS USED:
stem
TRADITIONAL USES:
Congo/Zaire: the stem is used to make an embrocation for fever pains: *Bouquet (1969) p. 244,* quoted in *Burkill (2000) p. 291*
TOXICITY:
The root is used as an arrow poison in DRC: *Metafro infosys: www.metafro.be*

Cissus cornifolia (Baker) Planch [*C. lonicerifolia* C. A. Sm.] (Vitaceae)
DESCRIPTION:
Scandent shrub or small tree growing from large tuberous rootstock, with erect or sub-erect shoots appearing each year after the fires in savannahs. Green/yellow flowers, black/purple fruits.
VERNACULAR NAMES:
E: Ivy grape; Ku: cituzi, mutunila (1) Lo: mutuka-mbamba (3d) Lo/Tk: mumina-nzoka (3, 23) Ny: cinto-mbolozi, kana-walisisi, mgongo, mpeska (3e), mbulu-mbunji, mpelesya, mwana-wamfepo (9)
PLANT PARTS USED:
roots
TRADITIONAL USES:
Tanzania: the roots are decocted to make a remedy for malaria: *Burkill (2000) p. 293; Haerdi (1964)   p. 112*
TOXICITY:
Tests in 2010 showed that a methanol leaf extract used on mice and rats was relatively safe at the analgesic and anti-inflammatory doses used: *Yaro et al.(2010)*

Cissus integrifolia (Baker) Planch. (Vitaceae)
DESCRIPTION:
A vigorous woody climber reaching the top of trees up to 9m. high.
VERNACULAR NAMES:
B/ Ku: mwanya (1) **Ny**: mpelesya, mthambe, mwana-wamfepo, nthambi-tambi (9)
PLANT PARTS USED:
roots
TRADITIONAL USES:
Malawi: the roots are powdered or infused as a remedy for fevers: *Morris (1996) p. 499*

Clausena anisata (Willd.) Hook. f. ex Benth. [*C. inequalis* (DC.) Benth.] (Rutaceae)
DESCRIPTION:
Shrub or small tree up to 13m. high, widespread in areas of higher rainfall, colonising forest edges.
VERNACULAR NAMES:
E: Horsewood, Maggot-killer; **Ny**: kalenga-ntunzi (3e, 9), mkangano, mpungabwi, mpunga-ziwanda (9), msamba-senya (3, 3e, 9)
PLANT PARTS USED:
leaves, roots
TRADITIONAL USES:

E. Africa: the roots are pounded and put into soup, which is drunk as a remedy for malaria: *Kokwaro (1976) p. 195*

Ethiopia:the aerial parts are used to treat malaria: *Geyid et al.(2005)*

Kenya: a root decoction is used to treat malaria: *Beentje (1994)*

Nigeria: the roots & leaves are decocted: *Ainslie(1937); Muthaura et al.(2007)*

S. Africa: the root is used as a fever remedy: *Githens (1948) p. 82*. Babies are held in the steam from boiling leaves as a remedy: *Morris (1996) p. 499; Palgrave (2002) p. 421.*

Tanzania: leaf juice is drunk and rubbed in as a remedy for malarial convulsions in children: *Haerdi (1964) p. 117.* In the Kwai area, the steam from a leaf decoction is used on a febrile patient: *Braun (1927) p. 45,* quoted in *Watt et al. (1962) p. 918.*

TOXICITY:

See *Okunade et al. (1987)* for tests on rats. Brine shrimp tests gave a mildly toxic figure of LC_{50} 60.5 mg/ml: *Moshi et al.(2010)*

Clematis brachiata Thunb. [*C. hirsuta* Thunb.; *C. sinensis* sensu R. E. Fr. non Fresen.] (Ranunculaceae)

DESCRIPTION:

A widespread deciduous climber/ scrambler up to 5m., with slender twisting woody stems and masses of creamy white flowers. Found at roadsides, rocky or grassy sites.

VERNACULAR NAMES:

E: Traveller's Joy; **Ku**: cikondi (1) **Le**: mukoka (22) **Ny**: cisa-scamavu, kongwe, lyundumula, mbulambula (9), ntukwinama (3e) **To**: kalatongo (3), maamba, miyenu (22)

PLANT PARTS USED:

plant, roots

TRADITIONAL USES:

Burundi: the leafy stems are decocted to treat malaria: *Durand (1959), Baerts et al.(1989)*

Ethiopia: The plant is used against malaria: *Lemordant (1971)*; the juice of fresh leaves is rubbed on the body: *Teklehaymanot et al.(2007)*

E. Africa: the roots are made into broth as a remedy for malaria: *Durand (1959), Kokwaro (1976) p. 181; Muthaura et al.(2007)*; a root decoction is taken "to provoke a diarrhoea": *Glover et al.(1966)*

S. Africa: a hot decoction of the plant is used to steam the malaria patient, and afterwards drunk: *Watt et al. (1962) p. 878*

TOXICITY :

Tests suggest that leaf extract « has selective toxic effect on the haematological parameters and the functional indices of the liver and kidney of male rats ....is not completely safe » : *Afolayan et al. (2009)*

Cleome gynandra L. [*Gynandropsis gynandra* (L.) Briq.] (Capparaceae)

DESCRIPTION:

A widespread annual flower

VERNACULAR NAMES:

E: African cabbage, Bastard mustard, Cat's whiskers, African Spider flower/ wisp; **B/ K/ Le/ Tu**: lubanga (1, 3, 3c, 19, 20) **Ku**: suntha, sunta-konde (1) **Lo**: lutembwe, namanga (3d) **M**: musunta (20) **M/ Tu**: muzimwa, wuzimwa (20) **Nm**: gagani (1a, 22), muangi (22) **Ny**: nsila (9), msunta (3) **Ny/ To**: lyuni (3) **Ny/ Tu**: luni (9, 20), sunta (20) **To**: cisyungwa, isyungwa (19), syungwa (3)

PLANT PARTS USED:

leaves, stems, roots
TRADITIONAL USES:
Bénin: the stems & leaves are used to make a remedy for malaria, or decocted with *Portulaca oleracea* & *Ocimum gratissimum*: *Adjanohoun et al.(1989)*
India: The leaf juice mixed with crystalline sugar is used to treat malaria in forest regions: *Singh, V. K. & al. (1993) p. 71*
Zambia: a decoction of the roots is given in fevers: *Nair (1967) p. 49*
TESTS:
Spencer (1947): score 0: *Spencer et al. p. 156*
TOXICITY :
Ethanolic extract of aerial parts found to diminish significantly (P<0.05) the immune system of albino rats : *Kori et al. (2009)*

Clerodendrum capitatum (Willd.) Schum. & Thonn. (Lamiaceae)
DESCRIPTION:
A sparsely-branched, weak-stemmed shrub, about 2m high. Widespread in *mutesyi* forest, thickets, and on anthills.
VERNACULAR NAMES:
B: mulepula-mpapa (3, 3c, 9a), naka-ncete (22), musayi(3c) **Ku**: kayimbu (1) **Ku/ Ny**: ngwevula (1, 3e) **To**: syamanya (3)
PLANT PARTS USED:
twigs
TRADITIONAL USES:
Côte d'Ivoire: the leafy twigs are decocted and drunk: *Kerharo et al. (1950) p. 231,* or put into baths as a febrifuge: *Bouquet (1974) p. 172,* quoted in *Burkill (2000) p. 249*

Clerodendrum eriophyllum Gurke [*C. glabrum* pro parte sensu Coates Palgrave (1983) in Trees of Southern Africa] (Lamiaceae)
DESCRIPTION:
A shrub / small tree up to 9.5m high. Found in woodland, thickets, anthills, thin rocky soil
VERNACULAR NAMES:
E: Hairy tinderwood
PLANT PARTS USED:
leaves
TRADITIONAL USES:
Kenya: the Kamba people use the leaves to treat malaria: *Beentje (1994) p. 613*
S. Africa: a decoction of leaves is used as a remedy for fevers: *Kokwaro (1976) p. 285; Palgrave (2002) p. 989; Watt et al. (1962) p. 1047*
TESTS:
In vitro tests on plant extract showed score of LC$_{50}$ 9.51-10.56 mg/ml: *Muthaura et al. (2007)*

Clerodendrum glabrum E. Mey. [*C. rehmannii* Gurke; *C. ovale* Klotzsch] (Lamiaceae)
DESCRIPTION:
A widespread, much-branched deciduous shrub or small tree up to 9m. high, with smelly white/mauve flowers.
VERNACULAR NAMES:
E: Smooth tinderwood; **Afr**: Gladdetontelhout

PLANT PARTS USED:
leaves, roots
TRADITIONAL USES:
Congo/Zaire: the leaves are decocted to remedy malaria: *Staner et al.(1937)*
Sierra Leone: a leaf decoction is drunk for fever: *Deighton 3349,* quoted in *Burkill (2000) p. 251*
S. Africa: the Zulu people use the leaf as a fever remedy: *Watt et al. (1962) p. 1047*
Zaire: the leaves and roots are used as a fever remedy: *Githens (1948) p. 82*

Clerodendrum sp.
PLANT PARTS USED:
roots, leaves
TRADITIONAL USES:
Tanzania: a root decoction mixed with leaf juice is drunk as a malaria remedy: *Haerdi (1964) p. 152*

Clerodendrum ternatum Schinz [*C. lanceolatum* Gurke] (Lamiaceae)
DESCRIPTION:
A rhizomatous suffrutex from an extensive woody rootstock, stems up to 60cm. high.
VERNACULAR NAMES:
E: Glorybower; **Ku:** civumbe (1, 3e) **Tk:** kalizya-lapami (8) **Tu:** katanda-imbo (8)
PLANT PARTS USED:
roots
TRADITIONAL USES:
Zimbabwe: an infusion of the roots is drunk and also used as an enema against fever: *Gelfand et al.(1985) p. 212*

Clutia abyssinica Jaub. & Spach [*C. glabrescens* Knauf; *C. pedicellaris* (Pax) Hutch.] (Euphorbiaceae)
DESCRIPTION:
Shrub/ small tree up to 2m, found in clearings and margins of evergreen forest and high - rainfall woodland
VERNACULAR NAMES:
E: Large/ Smooth-fruited Lightning Bush; **Lo:** mukuli-kuli (3d), mulokwambula (16) **Lo/ N:** mutuku (3d, 16) **Ny:** ciyuta (9)
PLANT PARTS USED:
roots, leaves
TRADITIONAL USES:
Burundi: the roots are decocted to make a drink: *Baerts et al. (1989)*
E. Africa: the roots are boiled in soup which is drunk. The patient is given a vapour bath with steam from boiled leaves circulating under a blanket: *Kokwaro (1976) p. 88; Muthaura et al. 2007.* In Tanzania a cupful of leaf decoction is taken 3x or more daily. On a journey, the leaves may be chewed. Children to take smaller doses: *Harjula(1980) .* In Kenya, the plant is used by the Kikuyu people to treat malaria: *Njoroge et al.(2006)*

Cocculus hirsutus (L.) Diels (Menispermaceae)
DESCRIPTION:

Herbaceous or woody climber up to 5m high. Found in woodland and on anthills in drier areas.
VERNACULAR NAMES:
E: Broom creeper, Inkberry, Monkeyrope; **B/ Ku**: mansisye (1) **I**: ngulikila (4, 23) **Lo**: sikoka (3) **Ny**: cipapata, msisisi (3e) **To**: itende (3) **Tu**: musisyi, musyisyi (8)
PLANT PARTS USED:
plant
TRADITIONAL USES:
India: the plant has been used as a remedy for fevers: *Watt et al. (1962) p. 757*
TOXICITY:
Acute toxicity, evaluated in rats, found to be higher than 3000 emg/kg, so not dangerous to humans: *Ganapaty et al.(2002)*

Coffea arabica L. (Rubiaceae)
DESCRIPTION:
Escaped from plantations, and growing wild in vicinity
VERNACULAR NAMES:
Coffee
PLANT PARTS USED:
beans
TRADITIONAL USES:
Ecuador: the beans are used against fever: *Coe et al. (1996) p. 104*

Combretum adenogonium Steud. ex . Rich. [*C. ghasalense* Engl. & Diels; *C. ternifolium* Engl. & Diels; *C. tetraphyllum* Diels; *C. fragrans* F. Hoffm.] (Combretaceae)
DESCRIPTION:
Small/medium tree, 10-15m high, found at medium to low altitudes in dry woodland, fringing vleis or pans, often on anthills.
VERNACULAR NAMES:
E : Four-leaved Bushwillow ; **B**: kakolo (20), mufuka (7c, 23) **B/ I/ K/ Ku/ Lb/ Le/ Lo/ M/ Ny/ To/ Tu**: mulama (1, 4, 17, 19, 20, 23) **B/ Ku**: kalama (1) **I/ To**: mukuta-bulongo (4, 14, 17) **Ln**: kakolo (3, 3b, 14) **Ln/ Lv**: mombomba (6, 22) **Lo**: musyembe, mulala-bapalwe (3d, 14), mulamana (3, 3b, 3d) **Lo/ Tk**: mulamane (22, 23) **Nm**: mluja-minzi, muja-minzi, muluja (22) **Ny**: cinama, fiti, kadale, mpakasa, muramo, murima (9), kakolo (1, 3e), kalama (2, 6, 14), kambaku-mbaku (3e, 14) **Ny/ Tu**: kalama (3e, 20) **To**: mukuta-bulonga (3, 3b), mumvuzya-syoti (17) **Tu**: tulama (20)
PLANT PARTS USED:
twigs, leaves
TRADITIONAL USES:
Burkina Faso: 50 gr of dried leafy twigs, or a bunch 18cm x 5 cm, (half this for children) boiled in water for 10 min., to produce 2 glasses of liquid. Wash twice daily for 2 days only: *de la Pradilla (1988) p. 17*
Guinea-Bissau: the leaves form part of a fever-medicine: *D'orey,* quoted in *Burkill (1985) p. 396*
TESTS:
80% MeOH extract of the leaves was tested against *P. falciparum*, and showed antiplasmodial activity: *Maregesi et al. (2010)*

Combretum collinum subsp. elgonense Fresen (Exell) Okafor [*C. mechowianum* O. Hoffm.] (Combretaceae)
DESCRIPTION:
Savannah shrub/ tree up to 30ft high, common in wet areas
VERNACULAR NAMES:
B: kaunda (3c), munondwe (3, 3b, 3c, 23) **B/ K**: mufuka (3, 3b, 3c, 7a, 14) **I**: mubunde (20), mukuta-bulongo, muunza (1c, 4), muzatavame (23) **I/ K/ To**: mulama (4, 20, 23) **Ln**: katumbwangu (22), mundumbwa-kav aya (6) **Ln/ Lv**: cilungu-viva, mweya (6, 22) **Lo**: mububu (23), mufufu (19), mulamana (3, 3b, 3d, 7a, 14) **Lv**: katombwangu (6) **Nm**: mfunfwa, mlandala, mulandala, muruzya-minzi (22) **Ny**: kakunguni, kangolo, mlowe, mweti (3e), kalama (3e, 14), mtebe-lebe, mkute (3, 3b, 3e, 14) **Tk**: mutobo (23) **To**: mukunza (3, 3b, 7a, 14, 23)
PLANT PARTS USED:
roots, leaves
TRADITIONAL USES:
Tanzania: a decoction of the roots and leaves is drunk and a steam-bath of the leaves is taken as a malaria cure: *Haerdi (1964) p. 103*

Combretum microphyllum Klotsch [*C. paniculatum* Vent. subsp.*microphyllum* (Klotsch) Wickens] (Combretaceae)
DESCRIPTION:
Shrub 1.3-2m high, or climber up to 8m, found in thickets and riverine bush in low-lying areas.
VERNACULAR NAMES:
E: Flame Climbing bushwillow, Burning-bush combretum; **Afr**: Vlamklimop; **I/ Le/ Ny/ To**: moobe-zuba (1c, 19, 20) **Lo**: kacinga-mwezi (3d) **Ny**: mkotamo (9) **Ny/ To**: citungulu (3e, 6,17, 22), moobazyuba (17), mutombolo (3), sisumbi (3)
PLANT PARTS USED:
roots, twigs
TRADITIONAL USES:
Burkina Faso: 4 bunches of leafy twigs are boiled in 10 lit. of water for 15 min., and used as a drink and a wash twice daily: *de la Pradilla (1988) p. 18*
E. Africa: a root decoction is drunk as a remedy for fever: *Burkill (1985) p. 404; Kokwaro (1976) p. 56*

Combretum molle R. Br. ex G.Don [*C. velutinum* DC.; *C. atelanthum* Diels; *C. gueinzii* Sonder; *C. holosericeum* Sonder; *C. splendens* Engl.] (Combretaceae)
DESCRIPTION:
Small tree up to 13m high, dense rounded crown and small golden fruits. Widespread, often in *Acacia-Combretum* or *Combretum-Pterocarpus* woodland, especially on rejuvenated soils, and on anthills
VERNACULAR NAMES:
E : Velvet Bushwillow; **Afr** : Fluweelbosswilg; **B**: kalume-kamulama (23), kaunda (14), monta-mfumu (7a, 14), kunda, mulondwe (10) **B/ K/ Le/ Nm/ Ny**: mulama (1a, 3, 10, 14, 22, 23) **I**: mukuku-lama (4, 17) **I/ Le**: kakuku-lama (4, 5, 10, 17, 23) **I/ Tk/ To**: nkalanga (10, 17, 23) **Ln**: cihuma (14), mulamata (6) **Lo/ Lv/ N**: mufula (3, 3d, 7a, 14, 23) **Lv**: muhuma (6) **Nm**: mkuruya-mnenfuwa (7c), mlama (7c, 22), mnama (1a, 22) **Ny**: cangongo, cinama, kadale, kakungi, mpakasa, msimbiti, napini (9), cisimbiti, kadale (9, 10),

kakunguni (3e, 10, 14), kalama (3e, 7a, 14), mkute (3e, 14) **Tk**: munkangala (8) **To**: mulanga (7a), mulanga-mupati, nkalangu (17) **Tu**: isumbu (22)

PLANT PARTS USED:

roots, leaves

TRADITIONAL USES:

E. Africa: a boiled root decoction is drunk as a fever remedy: *Kokwaro (1976) p. 55,* & used as a prophylactic: *Bussmann et al.(2006)*

Ivory Coast: a leaf decoction is used 2x daily to drink & to inhale: *Kone et al.(2002)*

Mali: the plant is used by the female herbalists of Bamabo to treat malaria: *Traore (n.d.)* ; leaf decoctions are used to make drinks, to bathe, and as steam bath: *Grønhaug et al.(2008)*

Senegal:leaves are infused for some days, until they change colour, when they are mixed into porridge:*Van Der Steur (1994)*

Tanzania: the root is used to treat fevers: *Mbuya et al.(1994)*

Togo:leaves & stem are used to make an enema: *Adjanohoun et al.(1986)*

Uganda:the leaves are infused to treat malaria: *Tabuti (2008)*

Zambia: the Lenje use a decoction of the leaves to bathe a feverish child: *Palgrave (2002) p. 806; Watt et al. (1962) p. 193*

TESTS:

Extracts of the stem bark showed antimalarial activity, associated with two hydrolysable tannins, of which one was identified as the ellagitannin punicalagin: *Schwikkard et al. (2002) p. 680*

TESTS:

MeOH leaf extracts were evaluated *in vitro* against the multi-resistant strain (W2) of *P. falciparum*, and showed moderate antiplasmodial activity (IC$_{50}$ 5.7mc/ml): *Gansané et al. (2010)*

TOXICITY: The above tests revealed low toxicity against K562S cells.

Commelina africana L. (Commelinaceae)

DESCRIPTION:

Spreading perennial herb up to 0.5m high, with canary-yellow flowers. Flourishes in sandy soil in rocky areas.

VERNACULAR NAMES:

E: Yellow commelina; **Afr**: Geeleendagsblom; **Ny**: khovani, mwana-wamfefo (9)

PLANT PARTS USED:

plant

TRADITIONAL USES:

E. Africa: an infusion of the plant is used to bathe fever patients: *Kokwaro (1976) p. 231; Magogo & Glover 904,* quoted in *Burkill (1985) p 429*

Commelina forskaolii Vahl (Commelinaceae)

DESCRIPTION:

A mat-forming herb, annual or perennial, up to 30cm high.

VERNACULAR NAMES:

E: Rat's Ear

PLANT PARTS USED:

plant

TRADITIONAL USES:

E. Africa: an infusion of the plant is used as a wash to reduce fever: *Kokwaro (1976) p. 232*
TOXICITY:
Roots need careful preparation: *Maundu et al. (1999)*

Commelina imberbis Hassk. (Commelinaceae)
DESCRIPTION:
Weedy annual, with long straggling stems up to 2m long, rooting at the nodes. Dark green leaves and blue flowers
VERNACULAR NAMES:
Not known
PLANT PARTS USED:
plant
TRADITIONAL USES:
E. Africa: an infusion of the whole plant is used as a wash to reduce fever: *Kokwaro (1976) p. 232; Magogo & Glover 904*, quoted in *Burkill (1985) p. 434*

Commiphora africana (A. Rich.) Engl. var. africana [*C. pilosa* (Engl.) Engl.] (Burseraceae)
DESCRIPTION:
Small spinous shrub from 0.3 – 4m high, or tree up to 8m. Found in deciduous thickets in drier areas.
VERNACULAR NAMES:
E: Poison-grub Corkwood; **Afr**: Gifkaneedood; **M**: citonto (10) **Nm**: m'gazu, mponda, msagazi, msyagasyi (22) **Ny**: cokolola, kamponi, khobo (9) **Tk**: mubwabwa-usiya(8) **To**: mubwa-bwa (22)
PLANT PARTS USED:
root bark, resin, bark, roots
TRADITIONAL USES:
Benin:the leafy stems are mixed with the ripe fruit of *Xylopia aethiopica* to treat malaria: *Adjanohoun et al.(1989)*
Burkina Faso: dried and powdered root bark, one spoonful to be taken after meals thrice daily. The resins contain triterpenoides: *de la Pradilla (1988) p. 18*
E. Africa & Malawi: the resin is made into a plaster and applied to the head, or mixed with fat and used as a body lotion, to reduce fever: *Githens (1948) p. 83; Morris (1996) p. 237; Watt et al. (1962) p. 152*
E. Africa: the barke and roots are used to make a steam bath for fever patients: *Kokwaro (1976) p. 43*
Tanzania: the dried powdered bark is eaten as a mash to cure malaria: *Burkill (1985) p. 305; Haerdi (1964) p. 119*
Tropical Africa: the resin is used as a fever cure: *Githens (1948) p. 83*

Commiphora madagascariensis Jacq. (Burseraceae)
DESCRIPTION:
A spinous shrub up to 5m high, with dark grey or plum-coloured branchlets. Found in deciduous thickets in L. Mweru basin.
VERNACULAR NAMES:
E: Myrrh(?); **B**: cinungu (23)
PLANT PARTS USED:
fruit

TRADITIONAL USES:

E. Africa: as a cure for fever, fruits are crushed and left overnight. A glass or two of the infusion is drunk on an empty stomach, to cause vomiting of bile: *Kokwaro (1976) p. 43*

Conyza bonariensis (L.) Cronq. (Asteraceae)
DESCRIPTION:
Widespread annual weed of roadsides and wasteland, stems hidden by the leaves.
VERNACULAR NAMES:
E: Hairy horseweed, Flax-leaved Fleabane, Asthmaweed, South American Horseweed
PLANT PARTS USED:
leaves
TRADITIONAL USES:
E. Africa: the leaves are squashed and inhaled as a remedy for fever: *Kokwaro (1976) p. 62*

Conyza pyrrhopappa Sch. Bip. (Asteraceae)
DESCRIPTION:
Woody herb or shrub, up to 2.7m high
VERNACULAR NAMES:
Not known
PLANT PARTS USED:
leaves
TRADITIONAL USES:
E. Africa: pounded leaves are infused in warm water and drunk as a remedy for malaria: *Glover et al.(1966); Burkill (1985) p. 459; Kokwaro (1976); p. 62*

Conyza sp. aff. Conyza volkensii (Asteraceae)
VERNACULAR NAMES:
Ny: kabwani, msanda, zungu-mbwa (3e)
PLANT PARTS USED:
leaves
TRADITIONAL USES:
Tanzania: small children with malaria drink leaf juice and their bodies are rubbed with the juice: *Haerdi (1964) p. 166*

Corchorus olitorius L. (Malvaceae)
DESCRIPTION:
Annual/ perennial invasive herb, up to 3.5m, with yellow petals 6 –8mm long
VERNACULAR NAMES:
E: Red Jute, Tossa Jute, Jews' Mallow, Nalita Jute, Bush Okra; **B**: lusaka-saka (2, 3) **B/ Ku**: mutelele, tindi-ngoma (1) **B/ I/ Ku/ Ny**: delele (1, 1c)
PLANT PARTS USED:
seeds, leaves
TRADITIONAL USES:
Nigeria: the seeds are used in a fever remedy: *Ainslie sp. no. 91,* quoted in *Burkill (2000) p.196*
Zambia: the leaves are infused as a febrifuge: *Nair (1967) p. 52*

TOXICITY:
Seeds contaminate cereal grain, contain a cardiac glycoside which can cause death: *Negm et al. (1980)*

Cordia goetzei Gurke [*C. ravae* Chiov.] (Boraginaceae)
DESCRIPTION:
A shrub, often tending to scramble, or small tree up to 7m. high, found at low altitudes, most often in riverine fringes, also in thickets and woodland, usually near watercourses.
VERNACULAR NAMES:
E: Grey-bark Cordia; **I**: kabangaluulu (4, 5, 17, 23), lubangaluulu (4, 17), mubangalala (4, 12) **Le**: musato (17) **To**: nazilyango (17)
PLANT PARTS USED:
roots
TRADITIONAL USES:
Tanzania: a root decoction is drunk to treat malaria: *Haerdi (1964) p. 150*

Cordia sinensis Lam. [*C. rothii* Roem & Schult.; *C. gharaf* Ehrenb. ex Aschers.; *C. Ovalis* sensu Palmer & Pittman (1973) Trees of Southern Africa)] (Boraginaceae)
DESCRIPTION:
A shrub up to 4m. high or a small bushy tree up to 8m., often with slender drooping branches; found generally at low altitudes in hot dry woodland, along river banks, on anthills, and in littoral scrub.
VERNACULAR NAMES:
E: Grey-leaved Saucer-berry; **Afr**: Grysblaarpieringbessie.
PLANT PARTS USED:
roots, leaves
TRADITIONAL USES:
E. Africa: the roots are boiled in milk as a remedy for malaria: *Kokwaro (1976) p. 39*
Senegal: the leaves are used, either alone or mixed with those of other drug-plants, as a fever remedy: *Burkill (1985) p. 290; Kerharo (1974) p. 249*

Costus spectabilis (Fenzl.) K. Schum. (Costaceae)
DESCRIPTION:
A succulent perennial herb sometimes completely flattened on the ground, with spectacular yellow flowers. Found in wooded grassland and deciduous woodland, often near anthills or among rocks.
VERNACULAR NAMES:
E: Yellow Trumpet
PLANT PARTS USED:
leaves
TRADITIONAL USES:
E. Africa: the green leaves are chewed as a remedy for fever: *Kokwaro (1976) p. 244*

Craterispermum schweinfurthii Hiern [*C. laurinum* auct. non (Poir.) Benth.; *C. reticulatum* De Wild.] (Rubiaceae)
DESCRIPTION:
Large shrub to medium tree 2 – 15m. high, found in evergreen forest, riverine fringes, and forested ravines.

VERNACULAR NAMES:
E: Porridge-sticks; **B**: mukambila (3c), museswa (3, 3c, 20) **Ln**: mucenke (3)
PLANT PARTS USED:
plant
TRADITIONAL USES:
Mozambique: the rootbark is used to treat malaria: *Mulhovo (1999)*
W. Africa: the plant is used as a remedy for any mild fever: *Dalziel (1937),* quoted in *Watt et al. (1962) p. 898*

Crocosmia aurea (Hook.) Planch. (Iridaceae)
DESCRIPTION:
Evergreen or perennial herb, with spectacular orange flowers, growing from basal underground corms.
VERNACULAR NAMES:
E: Montbrecia, Golden ballerina, Coppertips, Falling stars.
PLANT PARTS USED:
leaves, corms
TRADITIONAL USES:
Tanzania: the leaf sap and a decoction of the corms are drunk as a malaria remedy: *Burkill (1994) p. 423; Haerdi (1964) p. 201*

Crossopteryx febrifuga (Afzel. ex G.Don) Benth. [*C. africana* K.Schum.] (Rubiaceae)
DESCRIPTION:
A small deciduous tree with rounded crown and drooping branches,usually less than 7m. high but occasionally up to 15m. Widespread, especially in hotter and drier areas; in *mopane* woodland, basaltic outcrops, and rocky hills.
VERNACULAR NAMES:
E: Sand Crown-berry, Crystal-bark; **Afr**: Sandkroonbessie; B/ Ku/ Ny: mwavi (1, 3, 3e) **I**: muleya-mbezo (4, 5) **K**: mutoba-kongwe (3) **Lo/ Ny**: kangwe, kapulu-koso (7a) mulenga (22) **K/ Lo/ Tk/ To**: mulya-mbesu (3, 3d, 19, 20, 23) **Ku/ Tu**: kamwavi (1, 8) **Nm**: msanjwambeke, msasa-mbeke (1a, 22), msanza, msanza-mbeki, ngubalu (22) **Ny**: dangwe, goloka, mkako, mkunda-ngulwe (9), mgowo-gowo, mlamafupa (3e), kongwe (3) **To**: mulwa-bensu (7a) mulya-mbyo (3)
PLANT PARTS USED:
bark, leaves
TRADITIONAL USES:
Burkina Faso: a handful of bark scrapings is boiled in 10 lit. of water for 20 min.; drunk & used to wash, twice daily for 3 days. The plant contains alkaloids & saponosides: *de la Pradilla (1988) p.18*
C.A.R.:stem bark is decocted to make a bath: *Boulesteix et al.(1979)*
Central & S. Africa: the bark is used as a remedy for fever: *Githens (1948) p. 82.* Parts of the tree provide a remedy for fever: *Palgrave (2002) p. 1038; Watt et al. (1962) p. 898*
Congo/ Kinshasa: the bark is crushed & decocted & filtered to make a drink (1/2 glass) or enema, and drops in nose: *Delaude et al.(1971)*; the roots are pounded & macerated in palm wine: *Kasuku et al.(1999)*
Congo/ Zaire: the leaves are infused to make a friction rub: *Lubini (1990)*
Côte d'Ivoire: the plant has a great reputation as a febrifuge. The bark is decocted and drunk, or used to bathe the patient: *Kerharo et al. (1950) p. 201; Bouquet et al.(1974)*

E. Africa:ground leaves are soaked for a day in cold water & decocted, or ground roots are soaked for a day & 3 tsp. drunk cold as emetic:*Heine et al.(1988b); Kokwaro(1976)* . Roots & leaves are decocted to provide a bath: *Masinde (1996)*

Ethiopia:pounded leaves are decocted with a roast buttered garlic bulb, left outside all night, & drunk next morning: *Mesfin et al.(2009)*

Ghana, Guinea: the shrub is used as a remedy for fever: *Gelfand et al.(1985) p. 290*

Malawi, Mozambique: a decoction of the bark is used to treat fever: *Morris (1996) p. 456; Aebi p. 1013,* quoted in *Watt et al. (1962) p. 898; Williamson (1975) p. 298*

Mali:a leaf & stem decoction is used to treat malaria: *Malgras (1992)*

Nigeria: a bark decoction is used to make a febrifuge: *Ainslie (1937)*

Tanzania: scrapings of the fresh roots are eaten against malaria: *Haerdi (1964), Hedberg, p. 245* ; a febrifugal drink is made from the stem bark, mixed with root bark of *Senna didymobotrya* & *Abrus precatorius* & decocted: *Gessler et al.(1995)*

Tropical Africa: the bark is used as remedy for malaria and fever: *Githens (1948) p. 84; Watt et al. (1962) p. 898*

Zambia: the bark is infused as remedy for malaria: *Haapala et al.(1994) p.15; Amico (1977) p. 124*

Zimbabwe: the leaves are used as an enema to cure fever: *Gelfand et al.(1985) p. 224*

TESTS:

Early tests gave negative results: *Spencer et al. p. 156; Watt et al. (1962) p. 898.* However, more recent *in vivo* tests against mice revealed supression of parasitaemia similar to that of chloroquine and pyrimathamine *(Elufioye et al., 2004)*

Crotalaria natalitia Meisn. (Fabaceae, Papilionoideae)
DESCRIPTION:
A shrubby herb or shrub up to 2m. high, widespread, often found in *Brachystegia* woodland.
VERNACULAR NAMES:
E: Pioneer Rattle-pod; **Ny:** luwere, mazamba, naka-sewe, ngalawa, ntembenuko, kakunde-zonde (1)
PLANT PARTS USED:
roots, leaves
TRADITIONAL USES:
Tanzania: a root decoction mixed with leaf juice is drunk as a remedy for malarial convulsions in children: *Haerdi (1964) . 53*

Crotalaria ochroleuca G. Don. (Fabaceae, Papilionoideae)
DESCRIPTION:
Erect annual or short-lived perennial herb, up to 2.5m high, bearing yellow-green flowers and cylindric pods, found in damp grass, floodplains, vleis and roadsides.
VERNACULAR NAMES:
E: Slender-leaf Rattlebox
PLANT PARTS USED:
leaves
TRADITIONAL USES:
W. Africa: the leaves are used as a febrifuge: *Burkill (1995) pp. 314, 639*

Crotalaria recta Steud. ex A. Rich (Fabaceae, Papilionoideae)
DESCRIPTION:

Erect, robust shrubby herb up to 3m high, with stout hollow branchlets. Found in open *Acacia campylacantha* and *Brachystegia* woodland.

VERNACULAR NAMES:

E: Rattlebox; **Ny**: ciwere, ngalawa, ndolola, mndyo-ndyolo, sanduka (9)

PLANT PARTS USED:

roots, leaves

TRADITIONAL USES:

Tanzania: leaf-sap and a root decoction are given to children in malarial rigor: *Burkill (1995) p. 319; Haerdi (1964) p. 53*

Croton gratissimus Burch. [*C. zambesicus* Burch.] (Euphorbiaceae)

DESCRIPTION:

A shrub or small tree reaching 10m in height, occurring over a wide range of altitudes and in a variety of woodland types, often associated with rocky outcrops.

VERNACULAR NAMES:

E: Lavender Croton; **Ku**: Katanda-vibanda; **Lo**: Mufyelo, Mukena, Muka-nunkila; **Lo/ Tk/ To**: Kanunkila; **Ny**: Kananga, Mtanda; **Tk**: Moonzu; **To**: Mungayi; **Tu**: Mutanda-viwanda

PLANT PARTS USED:

whole plant, leaves, shoots, roots, bark.

TRADITIONAL USES:

Nigeria: an infusion of bark is used against malaria: *Ainslie sp. no. 119, as C. amabilis,* quoted in *Burkill (1994)*

Nigeria and Sierra Leone: a leaf decoction is used as a wash and taken internally for fever: *Burkill (1994); Dalziel (1937)*

S. Africa: the bark is used as a remedy for fevers: *Githens (1948)*

Sudan: the shoots and roots are used as a febrifuge: *El-Hamidi (1970),* quoted in *Burkill (1994)*

Transvaal: the Sotho use the plant as a remedy for fevers: *Watt et al.(1930)*

Zambia: the whole plant is used as a remedy for fevers: *Nair (1967)*

TESTS:

In vitro tests on the leaves in 2004 reveal a high degree of antiplasmodial activity (IC_{50} 3.5μg/ml): *Clarkson et al. (2004); Pillay et al. (2008)*

In vivo tests on root extracts in 2009 revealed up to 77.27% suppression of parasitaemia in mice: *Okokon et al. (2009)*

TOXICITY:

There is some doubt about its toxicity: although Bryant listed it among the poisonous plants, with qualifications, it is valued as an important stock-feed in Namibia: *Bryant (1909); Heering et al. (1913); Watt et al.(1962).* No reports of any toxic effects on humans.

Croton macrostachys Hochst. ex Del. (Euphorbiaceae)

DESCRIPTION:

Small tree, 5 – 17m. high. Found in ravines at edge of montane forest at Mucinga Escarpment.

VERNACULAR NAMES:

E : Woodland Croton, Forest Fever-tree, Broad-leaved Croton; **Ny**: ciwalika, mbwani, mtutu, nakawalika, tenza (9) **Tu**: mukuru-guru (22)

PLANT PARTS USED:

roots

TRADITIONAL USES:

E. Africa: juice from the boiled roots is drunk as a remedy for malaria by the Kikuyu people: *Beentje (1994) p. 192; Kokwaro (1976) p. 89. Burkill (1994) p.488*

Ethiopia: In Gondar, the plant is used as a fever remedy: *Geyid et al. (2005)*; in Debre Libanos, *C.macrostachys* leaves are boiled with other plants to treat malaria: *Teklehaymanot et al.(2007)*; in Wonago Woreda, fresh and dry leaves are pounded and boiled to make a remedy: *Mesfin et al. (2009)* ; the roots are also used: *Mirutse et al.(2007)*

TOXICITY:

Brine shrimp tests showed high toxicity figure LC_{50} 13.40mg/ml: *Moshi et al.(2004)*

Croton megalobotrys Muell. Arg. [C. gubouga S. Moore] (Euphorbiaceae)

DESCRIPTION :

A shrubby to medium-sized leafy tree reaching 15m, occurring at medium to low altitudes on alluvial flats, and usually a constituent of riverine and swamp fringe forest or thickets

VERNACULAR NAMES:

E: Fever-berry; **Afr**: Grootkorsbessle; **B/ Ny**: cimono-mono (3, 3c, 3e) **I/ Ln**: mupupu (4, 6) **I/ Tk/ To**: mutuwa (2, 3, 4, 8) **Ku/ Ny**: mtutu (1, 9), munanga (1) **Ln**: mufumpu (22) **Lo**: mono-mono, mubuli (3d) **Lo/ Tk**: mutuwa-tuwa (2, 3, 3d, 19, 23) **Lv**: cilama (6, 22), mumfumpu (6), mupupu (22) **Ny**: cinkombwa, mnanga (3e), mfumpu, muko (9), mtonga, mtoowa (22)

PLANT PARTS USED:

bark, seeds

TRADITIONAL USES:

Ethiopia (Gondar)**:** The plant is used against malaria: *Geyid (2005)*

S. Africa: the bark and seeds were developed as a popular remedy by a Dr. Maberley, who described how an old prospector had been cured of "a violent attack of bilious fever" by beans administered by "an old Kafir doctor" in exchange for a pair of greyhounds. He gave some beans to Maberley, who used them with great success at a hospital for miners. Pills were made from the powdered seeds and bark, with the addition of opium: *Maberley p. 874; Palgrave (2002) p. 399*

TOXICITY:

The toxicity and parasiticidal activity of **C. megalobotrys** seed oil was assessed in *Plasmodium berghei* infected mice. **Croton** oil showed acute toxicity with an LD_{50} of 11.778mg/g in 24 hours and 10.937mg/g in 7 days. It caused enlargement of several organs: splenomegaly, cardiomegaly, hepatomegaly, pulmonary enlargement and kidney enlargement. It also caused diarrhoea, which was severe in prestarved mice. From histological findings, **Croton** oil caused leukocyte infiltration into the cardiac tissue, vacuolar hepatopathy, glomerulo-nephritis and reduced cellularity of lymph nodules in the spleen: *Shumba et al. (1996)*

Croton menyhartii Pax (Euphorbiaceae)

DESCRIPTION:

A shrub or small tree up to 4m. in hight, occurring at medium to low altitudes in riverine thicket, often on wooded rocky outcrops in *mopane* woodland, and in thorn scrub and bushveld on sandy soil.

VERNACULAR NAMES:

E: Rough-leaved Croton; **Afr**: Skurweblaarkoorsbessie; **Lo**: Kapuka

PLANT PARTS USED:

roots
TRADITIONAL USES:
E. Africa: a root decoction is drunk as a remedy for malaria: *Kokwaro(1976)*
TESTS, ANTI-MALARIAL:
In vitro tests in 2004 revealed a high degree of antiplasmodial activity in the leaves (IC_{50}, 1.7µg/ ml): *Clarkson et al. (2004).* [Tests in 2001 revealed similar activity in *C. pseudopulchellus: Prozesky et al. (2001),* quoted by *Clarkson et al.(2004)*]
TOXICITY:
None recorded

Croton sylvaticus Hochst. ex Krauss (Euphorbiaceae)
DESCRIPTION:
Tree varying from 7 – 13m high, but can reach 30m. Found in forest or dense woodland.
VERNACULAR NAMES:
E: Forest Croton; **Afr**: Boskoorsbessie
PLANT PARTS USED:
plant
TRADITIONAL USES:
Kenya: parts are used against malaria: *Beentje (1994) p. 193*
TOXICITY:
Highly toxic.*In vitro* bacterial & mammalian cell assays showed that the plant caused both DNA damage & chromosomal aberrations: *Fennell et al.(2004)*

Cucumella engleri (Gilg.) C. Jeffrey (Cucurbitaceae)
DESCRIPTION:
Vine
VERNACULAR NAMES:
Not known
PLANT PARTS USED:
roots
TRADITIONAL USES:
E. Africa: an infusion of the roots is drunk as a remedy for malaria. It has the side-effect of causing acute diarrhoea: *Kokwaro (1976) p. 80*
TOXICITY:
See *Kokwaro (1976)* p. 80

Cussonia arborea Hochst. ex A.Rich. [*C. kirkii* Seem.] (Araliaceae)
DESCRIPTION:
Deciduous tree, 3 – 10m high, occuring in *miombo* woodland, often among rocks.
VERNACULAR NAMES:
E: Octopus Cabbage-tree; **B**: citebi-tebi (3b, 3c, 14), mutebi-tebi (23), mutewe-tewe (10, 20) **I/ To**: mulimpwinini (17) **Lo/ Ln**: cikasa (3, 3b, 14) **Nm**: mdula-msongwa, mlelwa-huwa, mnyonga-mpembe, myonga-mpembe (22), yagi-yanzovu (1a, 22), mkalya (7c) **Ny**: candimbo, nanpapwa (9), cipombo, mpanda-njovu (3e, 14), mbwa-bwa (3, 3b, 3e, 9, 10, 14) **Tk**: mutiti (8) **To**: mubayi, musansi (17, 19, 20), mutimbwinini (3, 3b, 10, 14, 17, 23)
PLANT PARTS USED:
plant
TRADITIONAL USES:

<u>Tanzania</u>: the plant is used as a remedy for fever: *Tanner 5246 Kew,* quoted in *Burkill (1985) p. 212*

Cussonia spicata Thunb. [*C. kraussii* Hochst.] (Araliaceae)
DESCRIPTION:
A thickset tree with much-branched crown, 3 – 10m high, found on Nyika Plateau, near source of Shire river.
VERNACULAR NAMES:
E: Cabbage-tree
PLANT PARTS USED:
roots
TRADITIONAL USES:
<u>S. Africa</u>: the succulent roots are macerated as remedy for malaria: *Bryant (1909) p. 44; Palgrave (2002) p. 851*

Cymbopogon citratus (DC.) Stapf. (Poaceae)
DESCRIPTION:
A common pantropical grass, used to prepare a lemon-flavoured tea, as a beverage as well as a medicine.
VERNACULAR NAMES:
E: Lemongrass; **Ln**: Ipupa; **Ny**: Uteka
PLANT PARTS USED:
Leaves, roots, whole plant
TRADITIONAL USES:
<u>Benin</u>: leaves & roots are decocted to make a drink: *Natabou Degbé (1991)*
<u>Burkina Faso</u>: a small bunch of *Cymbopogon citratus* is steeped for 10 minutes in 1 litre of water in which a small bunch of *Senna occidentalis* twigs has been boiled for ten minutes, the fire having been extinguished. Two litres a day to be drunk between meals, for 4 days: *de la Pradilla (1988)*
<u>Cameroon</u>:the whole plant is decocted to make a drink: *Vanhee (1996)*
<u>Congo/Zaire</u>: the leaves are infused to treat malaria: *Dhetchuvi et al.(1990)* ; leaves are decocted to make a steam bath: *Nyakabwa et al.(1990)*
<u>Ecuador</u>: an infusion of the leaves is used to treat fever: *Coe et al. (1996)*
<u>Ghana</u>:the fresh leaves are decocted to make a drink: *Asase et al.(2009)*
<u>Guatemala</u>: fever patients are bathed in water in which the leaves of the grass have been boiled: *Comerford (1996)*
<u>Mexico</u>: fever patients are bathed in water in which the leaves of the grass have been boiled: *Frei (1998)*
<u>Nigeria</u>: a leaf decoction is drunk 3x daily for 5 days: *Ajibesin et al.(2008)*; sometimes onions and/or honey are added: *Idu et al.(2006)*; *Ogie-Odia et al.(2009)*
<u>Senegal</u>: leaves are infused to treat malaria: *Kerharo et al.* (1974); leaves & roots are decocted: *Potel (2002)*
<u>W. Africa</u>: the leaves are infused as a febrifuge and sudorific:*Ainslie sp. 122, Diarra (1977), Iwu (1986), Walker (1961),* quoted in *Burkill (1994); Dalziel (1937); Kerharo (1973, 1974)*
TESTS:
In vivo tests on *P. berghei* in mice found 93% chemosuppression from an aqueous extract administered subcutaneously, and 90% chemosuppression from a chloroform extract

administered orally. This indicates that *C. citratus* exhibits both schizontocidal and prophylactic activities: *Obih et al.(1986)*

Its main activity seems to be anti-inflammatory and anti-pyretic rather than anti-malarial. The essential oil does have insect-repellent properties (*Pousset, 2004*). The related species *C. nardus* revealed a high degree of anti-malarial activity (IC$_{50}$ 5.8µg/ml) when tested: Clarkson et al., 2004)

TOXICITY:

Although the plant contains hydrocyanic acid, it is non-toxic, and there are no reports of toxic effects on humans.

Cymbopogon nardus (L.) Rendle (*C. validus* (Stapf) Stapf ex Burtt Davy) (Poaceae, Graminiaceae)

DESCRIPTION:

Aromatic grass with red base stems, growing up to 2m high

VERNACULAR NAMES:

E: Citronella grass, Nard grass

PLANT PARTS USED:

oil

TRADITIONAL USES:

S. Africa: Citronella oil is considered to be febrifugal: *Burkill (1994) p. 214; Watt et al. (1962) p. 471*

Cynoglossum lanceolatum subsp. geometricum (Baker & Wright) Brand [*C. geometricum* Baker & Wright] (Boraginaceae)

DESCRIPTION:

Common short-lived perennial herb with annual stems up to 1.8m high, white/blue flowers, found in roadsides and various open habitats.

VERNACULAR NAMES:

E: Hounds' tongue

PLANT PARTS USED:

leaves

TRADITIONAL USES:

E. Africa: vapour from crushed leaves is inhaled to relieve for fever: *Kokwaro (1976) p. 41*

Ethiopia:the plant is used to treat malaria: *Giday (2001)*

Cyperus articulatus L. (Cyperaceae)

DESCRIPTION:

A sedge-grass up to 6 feet high, growing in clumps from rhyzomes, found in damp, marshy and flooded areas.

VERNACULAR NAMES:

E: Jointed flat sedge, Piri-piri; **N**: cihoyoka (8) **Ny**: kauju, mlulu (9) **Tk**: manyanzya (8) **Tu**: tokwe (8)

PLANT PARTS USED:

rhizomes

TRADITIONAL USES:

Angola: the roots are decocted to make a drink: *Bossard (1996)*

Congo/Brazzaville: a new baby is scarified & rubbed with pulped roots: *Bouquet (1969)*; the aerial parts are pounded to make a friction rub: *Diafouka (1997)*

Congo/ Zaire: the rhizome is pounded & filtered to make an enema for a child: *Wome (1985)*
Congo/ Kinshasa:patients are massaged with juice from the roots: *Kasuku et al.(1999)*
Kenya:the tuber is infused in hot or cold water to make a drink: *Muthaura et al.(2007)*
Tanzania: rhizomes are decocted to treat malaria: *Burkill (1985) p.611; Haerdi (1964) p. 206*
TESTS:
Aqueous and methanol extracts were tested *in vitro* against chloroquine sensitive and resistant strains of *P. falciparum*, and showed activity at IC_{50} 4.84 mg/ml: *Rukunga et al. (2008)*

Cyperus rotundus L. (Cyperaceae)
DESCRIPTION:
Widespread troublesome weed, growing from a fibrous root system over a wide range of soil types and temperatures.
VERNACULAR NAMES:
E: Nutgrass, Purple nutsedge, Coco grass
PLANT PARTS USED:
rhizomes
TRADITIONAL USES:
India: a decoction of the rhizome is prepared with stem bits of *Tinospora cordifolia* and dried ginger: *Singh, V.K. & al. (1993), p. 71*
Sudan: fresh roots are given to children as a diaphoretic: *Bebawi et al.(1991)*

Cyphostemma buchananii (Planch.) Desc. ex Wild & R.B.Drumm.[*Cissus buchananii* Planch.](Vitaceae)
DESCRIPTION:
Climbing herb with tendrils and hairy stems; small round red fruit. Found in woodland and at forest edges.
VERNACULAR NAMES:
E: Wild grape; **Ln**: nkululumbi (6, 22); **Ny**: liboha (6, 9a, 22), mwana-wamfepo, namwali-cece, ncofu (9)
PLANT PARTS USED:
whole plant
Mozambique: the plant is used to treat malaria: *Mulhovo (1999)*

Cyphostemma junceum (Webb) Desc. ex Wild & R. B. Drumm. (Vitaceae)
DESCRIPTION:
An erect perennial herb, up to 1.3m high, with greenish/red flowers and purple/black fruit, usually found in *Brachystegia* woodland.
VERNACULAR NAMES:
Ny: mnuwake-munda, mwana-nkale, mwana-wamfepo, mwini-munda, ndemika-ngongo (9)
PLANT PARTS USED:
roots
TRADITIONAL USES:
Zimbabwe: the tuberous roots are powdered and used as an enema to cure fever: *Gelfand et al.(1985) p. 182*

Cyphostemma subciliatum (Bak.) Desc. ex Wild & R. B. Drumm. (Vitaceae)

DESCRIPTION:
A climbing herb with a tuberous root, greenish/yellow flowers, red fruit, leaves often showing purplish colouring. Found at Lusaka, Petauke, Mwinilunga.
VERNACULAR NAMES:
Ny: ncofu, mwana-wamfepo (9)
PLANT PARTS USED:
roots
TRADITIONAL USES:
Malawi: the powdered roots are used as a remedy for malaria: *Morris (1996) p. 503*

Cyrtorchis sp. (Orchidaceae)
VERNACULAR NAMES:
Not known
PLANT PARTS USED:
juice
TRADITIONAL USES:
Tanzania: the juice is drunk as a remedy for malaria: *Haerdi (1964) p. 205*

Dahlia variabilis Desf. (Asteraceae)
PLANT PARTS USED:
plant
TRADITIONAL USES:
Zambia: the tuber is used to produce a diaphoretic: *Nair (1967) p. 94*

Dalbergia boehmii Taub. [*D. elata* Harms] (Fabaceae, Papilionoideae)
DESCRIPTION:
A large shrub to medium tree, up to 20m high, found in various types of woodland and forest.
VERNACULAR NAMES:
E: Large-leaved Dalbergia; **I**: mulyaŋombe? (20), mundalilo, muswati (4, 19, 20) **Ku**: nthombolozi (1) **Nm**: magabu (7c) **Ny**: cinyewe, lapeha, mpereka, mpelele (9), mcindula (3e) **To**: mulamba (19)
PLANT PARTS USED:
plant
TRADITIONAL USES:
Tanzania: a root decoction , or a bark decoction mixed with *Stereospermum kunthianum,* is used against fever: *Haerdi (1964) p. 54*
Guinea/ Senegal: the twigs are steeped in water which is used to wash fever patients: *Ferry (1974) sp. no. 170,* quoted in *Burkill (1995) p.324*

Dalbergia hostilis Benth. (Fabaceae, Papilionoideae)
DESCRIPTION:
A climber with single-seeded small pods, found in savannah woodland vegetation, and riverine areas
VERNACULAR NAMES:
Not known
PLANT PARTS USED:
plant

TRADITIONAL USES:
W. Africa: the leaves are used to make a fever wash: *Burkill (1995) pp. 236, 639*

Dalbergia nitidula Welw. ex Baker (Fabaceae, Papilionoideae)
DESCRIPTION:
A small tree 3 – 9m high, with rough dark grey bark. Widespread: usually found in *miombo* on sandy soils or rocky escarpments.
VERNACULAR NAMES:
E: Purple-wood Flatbean; **Afr**: Pershoutplatboontjie; **B/ Ny**: kabulasese (3e, 7a, 14, 23) **B/ Ny**: kalongwe (1, 3, 3c, 7a, 14, 20, 23) **I/ To**: mukanda-njase (4, 14, 23, 22) **K/Ln/ To**: kafundula (3, 14) **K**: muyimbe-yimbe (14) **Lb**: lubeba (23) **Ln**: mujimbe-jimbe (6), muhonga-ndumba, mutata-imbululu (14) **Lo/ To**: mukonkoto (3, 3d, 7a, 14) **Lv**: musakayaze (6)? **Ny**: cikanga, cipisya-wago, mlungwe, mpinda-nkwangwa, namasimba, ndula-nkwangwa, ntanda-nyerere (9), mvungwe, mzembe (3e, 14), mkolasinga (3, 3e, 9, 14), mulembela (3) **To**: mutata- mbeba (14), ndelele (3, 14), njelele (23)
PLANT PARTS USED:
plant
TRADITIONAL USES:
Tanzania: a root decoction is drunk against malaria: *Haerdi (1964) p. 55*
TOXICITY:
The roots are toxic: *Palgrave (2002).* Brine shrimp tests gave an LC_{50} figure of LC50 0.87mg/ml: *Moshi et al.(2004)*

Datura metel L. (Solanaceae)
DESCRIPTION:
A shrub-like annual herb growing up to 3ft high with oval green/violet leaves and large flowers varying from white to violet
VERNACULAR NAMES:
E: Angels' Trumpet, Devil's Trumpet, Jimpson Weed, Purple Horn-of-plenty, Thorn Apple; **Nm**: iwonge-wonge (7c)
PLANT PARTS USED: plant
TRADITIONAL USES:
India: seeds are burnt, and the ash mixed with cloves and honey to treat malaria: *Singh & al. (1993), p. 71; Anis & al. (1986) p. 53*
Senegal: feverish infants are laid on a bed of leaves and massaged with leaves: *Burkill (2000) p.105; Kerharo (1974) p. 303*
TOXICITY:
All parts of the plant contain dangerous levels of tropane alkyloids, and are highly toxic to humans and animals if ingested: *Watt et al.(1962) pp 948-958*

Delonix regia (Boj. ex Hook.) Raf. (Fabaceae, Caesalpinioideae)
DESCRIPTION:
Tree from 9 – 12m high, with wide-spreading crown giving shade, and spectacular red flowers. Widely grown in tropics.
VERNACULAR NAMES:
E: Flamboyant, Flame Tree, Peacock Flower
PLANT PARTS USED:
bark

TRADITIONAL USES:

<u>Indochina</u>: the febrifugal quality of the bark is recognised: *Burkill (1935) pp. 777-8; Burkill (1995) p. 100*

<u>Senegal</u>: the bark is used to treat intermittent fevers: *Berhaut (1975) pp. 370-2,* quoted in *Burkill (1995) p. 639*

Desmodium barbatum (L.) Benth. (Fabaceae, Papilionoideae)

DESCRIPTION:

Short-lived hairy erect woody perennial growing up to 1m high. Found in grassland, woodland, sandy roadsides, old cultivation.

VERNACULAR NAMES:

E: Hairy Beggar-weed, **Ny**: citiwi, ndepa (9), kasulu (3e) Tk: kankuma (8)

PLANT PARTS USED:

leaves, roots

TRADITIONAL USES:

<u>Ecuador</u>: a decoction of the leaf and root is used to treat malaria and other fevers: *Coe et al. (1996) p. 100*

Desmodium gangeticum (L.) DC. (Fabaceae, Papilionoideae)

DESCRIPTION:

Herb or shrub up to 2m. high, found at Lake Bangweulu, on burnt ground in grassy open woodland.

VERNACULAR NAMES:

B: kalambatila (3, 9a)

PLANT PARTS USED:

roots, whole plant

TRADITIONAL USES:

<u>India/ Senegal</u>: the whole plant, including the root, is regarded as a useful febrifuge: *Avasthi (1955) pp. 625, 628, (155b) p. 272,* quoted in *Watt et al. (1962) p. 594; Berhaut (1976) pp. 198-9, Oliver-Bever (1960) pp. 24, 60; (1983) p. 52; Sastri (1952) p.41,* quoted in *Burkill (1995) pp. 334-5; Burkill (1935) p. 793; Dalziel (1937) p. 239;*

<u>Zambia</u>: the root (*Nair 1967 p. 47)* or the whole plant *(Amico (1977) p. 117)* is used as a febrifuge.

Desmodium triflorum (L.) DC (Fabaceae, Papilionoideae)

DESCRIPTION:

Creeping much-branched herb forming dense mats

VERNACULAR NAMES:

E: Creeping Tick Trefoil, Three-flower Beggarweed, Tick Clover

PLANT PARTS USED:

leaves, roots

TRADITIONAL USES:

<u>Ecuador</u>: a decoction of the leaf and root is used to treat malaria and other fevers: *Coe et al. (1996) p. 100*

Dichrostachys cinerea (L.) Wight & Arn. [*D. glomerata* (Forsk.) Chiov. subsp. *nyassana* (Taub.) Brenan; *D. nyassana* Taub.] (Fabaceae, Mimosoideae)

DESCRIPTION:

Shrub or small acacia-like tree up to 5 – 6m high, found on variety of soils in wooded grassland, and sometimes forming impenetrable thickets and shedding twisted and intricate pods from May – Sep.

VERNACULAR NAMES:

E: Bell Mimosa, Chinese Lantern, Kalahari Christmas-tree, Sickle bush, Sickle Pod; **Afr**: Sekelbos; **B**: kansalu-nsalu (3c, 7a, 14, 20), mupangala, nsalu-nsalu (20) **B/ I/ K/ Le/ Ny/ To**: katenge (3, 3b, 3c, 3e, 4, 7a, 10, 14, 17, 20, 23) **B/ Ku/ Ny/ Tu**: lumpangala (1, 3e, 20) **B/ Ny**: mumpangala (10, 23) **I**: kaluya (1c), muyeeye (4, 17) **I/ Tk/ To**: mweeye (8, 10, 17) **K**: kansalu-salu, katenga (10) **K/ Ln/ Lv**: kaweyi (6, 14, 22) **Ku/ Ny/ Tu**: lusolo (1, 3e, 8, 20) **Lo/ Ny/ To**: musele-sele (3, 3b, 3d, 7a, 10, 14, 19) **Ln/ Lv**: mubanga (6, 14, 22) **Lo**: cisole, musyale (10), omujete (22) **M/ Ny**: mpangala (9) **Nm**: mtundulu (1a, 22), ntihenga (?) (22) **Ny**: cipangala (9, 10), cipisya-wago, ciseyo, mpinda-nkwanga, nafangale, ndula-nkwangwa (9), citongololo, kanenge, nsalo (3e), kampangala (10), lupangala (3, 3b), nthenga (22), nyalu-mpangala (6, 22) **Tk**: mucakali (8), ŋwini-ciyeye (10) **To**: nakalenge (10) **Tu**: impangala (7c), lafangala (8)

PLANT PARTS USED:

Fruits, leaves

TRADITIONAL USES:

<u>Burkina Faso</u>: a handful of fruits is boiled for 15 min. in 5 lit. of water; used to drink & as a wash twice daily for 3 days: *de la Pradilla (1988) p.19*

<u>Congo-Brazzaville</u>: a decoction in water is drunk 2x daily: *Diafouka (1997)*

<u>Mali</u>:to bathe a child with fever, the fruit is decocted in water: *Malgras (1992)*

<u>Nigeria</u>: the plant is used to prepare a remedy for malaria: *Saadou (1979)*

<u>Tanzania</u>: the leaves are infused to make a remedial drink: *Chhabra et al. (1990)*

TOXICITY: tests on goats in 2004 found the tannin content harmless: *Mlambo et al.(2004)*

Dicoma anomala Sond. subsp. anomala [*D. anomala* Sond. subsp. *cirsioides* (Harv.) Wild] (Asteraceae)

DESCRIPTION:

A prostrate, decumbent or erect perenniak herb with stems from an underground tuber. White, mauve or pink flower heads. Found in stony grasslands, hillsides, or flat grassland, and in savannah.

VERNACULAR NAMES:

E: Fever Bush, Stomach Bush; **Afr**: Maagbitterwortel, Kalwerbossies, Koorsbossie, Gryshout, Maagbossie; **B**: lutanda (9a), nyinu (22) **Ny**: palibe-kantu (9)

PLANT PARTS USED:

roots

TRADITIONAL USES:

<u>Zimbabwe</u>: a root infusion is used to wash the body of a patient running a high temperature: *Gelfand et al.(1985) p. 234*

<u>S. Africa</u>: the roots are decocted and blended with gin and *melkbos* as a remedy for fever: *Watt et al. (1962) p. 22*

Dicoma sessiliflora Harv. subsp. sessiflora (Asteraceae)

DESCRIPTION:

An erect perennial herb with pithy stems, 30 – 110cm high, found in dry fields

VERNACULAR NAMES:

Ku/ Ny: lubanda (1)

PLANT PARTS USED:
whole plant
TRADITIONAL USES:
<u>Nigeria</u>: the whole plant is strongly bitter and used as a febrifuge, particularly for children: *Ainslie sp. no. 134,* quoted in *Burkill (1985) p. 465*

Dioscorea bulbifera L. (Dioscoreaceae)
DESCRIPTION:
An invasive perennial vine with broad leaves, growing from underground tubers
VERNACULAR NAMES:
E: Air potato, Bitter Yam; **Ln**: kalilya (20), kalya-lya (19) **Lo**: litu (20) **Lv**: kalangiya, nguvu (19) **Nm**: ndiga (1a, 22) **Ny**: cilazi, lipeta, mpama (9) **To**: sise (19)
PLANT PARTS USED:
tuber
TRADITIONAL USES:
<u>Zimbabwe</u>: the tuber is used as a diaphoretic. No details given: *Gelfand et al.(1985) p. 95*
TOXICITY:
When used as famine food, tubers must be boiled.

Dioscorea quartiniana A. Rich. (Dioscoreaceae)
DESCRIPTION:
Slender perennial climber up to 6m high. Wiry stems with soft hairs and ivory-white pendulous flowers
VERNACULAR NAMES:
Lo: sikubabe (3)
PLANT PARTS USED:
roots
TRADITIONAL USES:
<u>E. Africa</u>: the tuberous roots are chopped and soaked in cold water. One glass of the infusion is drunk every other day over a period, to reduce fever: *Kokwaro (1976) p. 235*

Diospyros mespiliformis Hochst ex A. DC. (Ebenaceae)
DESCRIPTION:
Evergreen tree up to 25m high or more; fairly dense rounded crown, smooth bark. Widespread, in fringe forest, on termite mounds on Plateau
VERNACULAR NAMES:
E: Jackal-berry, African Ebony; **Afr**: Jakkalsbessie ; **B**: mcenja (20) **B/ I/ K/ Ku/ Lb/ Le/ Lo/ Ny/ To/ Tu**: mucenje, mucenja (1, 3, 3a, 3c, 3d, 4, 5, 14, 17, 19, 20, 22, 23) **I**: icenje, incenje (4) **K**: mukyengya (3, 14), munondo (19) **Ku**: nkulo (1) **Ku/ Tu**: mucenga (20) **Ln**: mutowa (6) **Lo**: mukucumu (22), mupako (3d, 14) **Lo/ Ln/ Lv**: mutomwa (3, 3d, 14, 23) **Nm**: mnubulu, mnumbulu, msindi (22) **Ny**: mcenja (6, 14), mcenja-sumu (3, 14), mcenje, msumwa, ntheza (9), mvimbe (14) **To**: mgula (22), mucenje-mulonga (17), mukula (19) **Tu**: narenje (20)
PLANT PARTS USED:
roots, leaves, fruit, twigs, bark
TRADITIONAL USES:
<u>Angola</u>: young leaves or leaves & seeds are infused to treat malaria: *Bonnefoux (nd), Batalha et al.(1979)*

Burkina Faso:a leaf decoction is used to make a drink: *Kerharo et al.(1950)*

Malawi & S. Africa: the leaves, twigs and bark are used to make decoctions and other hot preparations as remedies for fever: *De Wildeman (1949) p. 1,* quoted in *Watt et al. (1962) p. 389; Morris (1996) p. 289; Palgrave (2002) p. 905*

Senegal: A leaf decoction is used as a wash: *Dalziel (1937) pp. 347-8;* the leaves & ripe fruit are dried, powdered and decocted as a malaria remedy: *Kerharo et al., (1974)*

Tanzania: a root decoction, with leaf sap, is drunk as remedy for malaria: *Burkill (1994) p.11; Haerdi (1964) p. 115*

W. Africa: the leaves and roots are used as a remedies for fever: *Burkill (1994) pp. 10, 11; Gelfand et al.(1985) p. 291.* In Nigeria and Ivory Coast, a leaf infusion is used as a febrifuge: *Ainslie sp. no. 137, Bouquet (1974) p. 80,* quoted in *Burkill (1994) pp. 10, 11.* In Bénin, the plant is infused as a remedy: *Natabou Dégbé (1991),* or the bark is pounded & decocted: *Adjanohoun et al. (1989)* . In Guinea Conakry, a leaf infusion is given to drink: *Pobéguin(1911)*

Diospyros senensis Klotzsch (Ebenaceae)

DESCRIPTION:

A rigidly spiky shrub or small tree 3 – 5m high, occurring in hot dry areas, in deciduous woodland, at the edges of pans and in riverine forests

VERNACULAR NAMES:

E: Spiny Jackal-berry; **B/ Ku**: kafulu-kalume (1) **Ku**: kaonga, kawonga (1) **Ny**: cimwande, kawonga (3e), cirima, mdima, mpinji-pinji, mtunga-njila (9) **To**: cimandwe (3)

PLANT PARTS USED:

leaves, roots

TRADITIONAL USES:

Mozambique: The leaves and roots are used to treat malaria :*Mulhovo (1999)*

Diospyros squarrosa Klotzsch (Ebenaceae)

DESCRIPTION:

A deciduous shrub or small tree up to 8m high, found in deciduous woodland and thickets, and rocky hills

VERNACULAR NAMES:

E: Rigid Star-apple; **Ny**: kafotoko (9), msindila (3, 3e, 9)

PLANT PARTS USED:

leaves, bark, roots

TRADITIONAL USES:

Tanzania: an inhalant is made from the decocted leaves and bark, and the roots are decocted to make a drink, as malaria remedies: *Haerdi (1964) p. 115*

Diplorhynchus condylocarpon (Muell. Arg.) Pichon [*Diplorrhynchus mossambicensis* Benth.] (Apocynaceae)

DESCRIPTION:

Small semi-deciduous tree up to 12m high, widespread in open woodland, escarpment hill-country, and at edges of sandy dambos

VERNACULAR NAMES:

E: Horn-pod Tree, Wild Rubber; **Afr**: Horingpeultjieboom; **B**: ndale (22) **B/ K/ Lb/ Lo/ M**: mwenge (1, 3, 3b, 3c, 7a, 10, 14, 19, 23) **B/ Ku/ Ny/ Tu**: mtowa (1, 3, 3b, 6, 7a, 10, 14, 20) **B/ M**: muntalembe (10, 22) **I/ Le**: muto (5, 17) **I/ To**: munto, muntowa (1c, 4, 5, 6,

10, 17, 19) **K**: mubuge (10), mubuga (10, 19) **Ku/ Ny/ To/ Tk/ Tu**: mutowa (1, 3, 3b, 7a, 8, 10, 14, 23) **Ln**: mudyi (21), muzi (3, 3b, 14) **Ln/ Lo**: mulya (3, 3b, 3d, 7a, 10, 14, 16, 19, 22, 23) **Ln/ Lv**: muli (6, 22) **Lo**: musengwa (19) **Lo/ N**: mubuli (10, 22) **Lv**: mtomoni (22), mubuliya, mufumbelele (10), mwidi (6) **Nm**: mbele-neli (1a, 22), msangati, msonga, msongati, msunguti (22) **Ny**: mtombozi (3e, 6, 9, 10), mtomoni (9), mtowa, mwenje (3e) **Tk**: muntowo (10) **Tu**: itelembe (7c), mulimbo, mnya-nyata (10), muthoo (8), ntowa (20)

PLANT PARTS USED:

roots

TRADITIONAL USES:

E. Tanzania: a root decoction or the juice of the leaves is drunk to treat malaria & fevers: *Chhabra et al. (1987) p. 263; Gessler et al. (1995) p. 137.*

Zambia: a root infusion is used to bathe the patient, and as a drink: Vongo (1999)

Dodonaea viscosa Jacq. [*D. angustifolia* L.f.; *D. linearis* E. Mey.] (Sapindaceae)
DESCRIPTION:

Shrub or small tree, often multi-stemmed, 3 – 4m high, found on rocky hillsides and by streams

VERNACULAR NAMES:

E: Sand-olive; **Afr**: Sondolien; **B**: kakonta (3c, 9a), mubaba-nkonde (3, 3c, 9a), muswite (20) **Le**: tsekatseki (22) **Le/ To**: cindamulongo (23) **Ny**: mkwani, msuce, susuti (3e), mlaka, nakamoto, nandolo, nkanga-mani (9), tsutsuti (3)

PLANT PARTS USED:

leaves, wood

TRADITIONAL USES:

Burundi:the leaves are decocted or pressed for malaria remedies; another recipe boils the bark of trunk, branches & stems to make a decoction: *Baerts et al. (1989)* ; *Polygenis-Bigendako (1990)*

Ethiopia:the seeds are decocted to make a drink: *Tadesse (1988)* ; *Giday et al. (2007)*

Gabon: A leaf or wood decoction is used as a fever remedy: *Walker (1953) p. 310, (1961) p. 387,* quoted in *Burkill (2000) p.19* ; leaf juice is used as nose-drops: *Lemordant (1971)* ; 1 cup powdered fruit + 7 cups boiled water, decocted until = 1cup, & milk added to make a drink: *Tadesse (1994)*

Madagascar: the leaf is decocted as a remedy for fever: *Githens (1948) p. 86* ; *Terrac (1947)*; *Pernet et al.(1957)*; *Rasoanaivo et al. (1992)*

Malawi: a leaf infusion is used as a remedy for fevers: *Morris (1996) p. 469*

S. Africa: a (leaf) infusion has been taken internally as a febrifuge: *Pappe (1847)*; *Watt et al. (1962) p. 931*; the leaf tips are infused: *Dykman (1908)* ; *Kling (1923)* ; *Laidler (1928)* ; tea from 3x tspns to be sipped 3x daily: *Thring et al. 2006)*

TESTS:

 Extracts have proved ineffective against experimental malaria: Spencer tests, p. 169: score 0 / 0: *Spencer et al. p. 169*

TOXICITY:

In Brazil an outbreak of acute hepatic insufficiency in which 14 dairy animals died occurred after consumption of *D. viscosa*. The clinical course lasted a few hours and was characterized by apathy and staggering disorders. The main postmortem findings occurred in livers as enhanced lobular patterns and hepatocellular necrosis that ranged from centrilobular to massive. Similar clinical and pathological effects were experimentally induced by dosing a heifer with 30 g/kg of the plant's green leaves: *Colodel et al., (2003)*

Dorstenia psilurus Welw. (Moraceae)
DESCRIPTION:
A perennial herb, stems trailing and rooting, leafy parts erect up to 2ft high, found in forests.
VERNACULAR NAMES:
Not known
PLANT PARTS USED:
rhizomes
TRADITIONAL USES:
Tanzania: rhizomes are crushed and macerated in water as both a drink and a body-wash for children with malaria: *Haerdi (1964) p. 69*

Drymaria cordata (L.) Willd. ex Roem. & Schult. (Moraceae)
DESCRIPTION:
Annual/ perennial herb up to 0.6m high, with white flowers and black globular fruit, having glandular hairs which stick to clothing
VERNACULAR NAMES:
E: Tropical Chickweed, West Indian Chickweed, White Snow
PLANT PARTS USED:
sap
TRADITIONAL USES:
India: the sap is said to be anti-febrile: *Sastri (1952) pp. 113-4*

Duranta erecta L. [*D.repens* L.] (Verbenaceae)
DESCRIPTION:
Sprawling vine-like shrub/ small tree up to 5.5m high, forming multi-stemmed clumps with trailing branches and spectacular blue/ violet/ purple flowers and showy yellow fruits hanging in bunches. Invades woodland, bushveld, and urban spaces.
VERNACULAR NAMES:
E: Forget-me-not Tree, Pigeonberry, Skyflower; **Afr**: Vergeet-my-nie boom
PLANT PARTS USED:
fruit
TRADITIONAL USES:
Bangladesh: fruits are used to treat malaria: *Whistler (2000)*
S. Africa: the fruit has been used as a febrifuge: *Burkill (2000) p. 257; Quisumbing (1951) 16,* quoted in *Watt et al. (1962) p. 1048*
TESTS:
Highly effective in *in vivo* tests against mosquito larvae: *Nikkon et al. (2010)*
TOXICITY:
Poisoning with alkaloid-type reactions and some deaths reported. Chloroform soluble fractions showed toxic effects on rats: *Nikkon et al.(2008)*

Elaeodendron transvaalense (Burtt Davy) R. H. Archer [*Cassine transvaalense* (Burtt Davy) Codd; *Pseudocassine transvaalensis* (Burtt Davy) Bredell; *Crocoxylon transvaalense* (Burtt Davy) N. Robson] (Celastraceae)
DESCRIPTION:

Shrub/ bushy tree up to 5m high, occasionally 10–15m. Widespread but not common, found in bushveld andwoodland, along streams and on anthills.
VERNACULAR NAMES:
E: Bushveld Saffron, Anthill Saffron; **Afr**: Bosveldsaffraan
PLANT PARTS USED:
bark
TRADITIONAL USES:
S. Africa: the bark is infused to make an enema as remedy for fevers: *Palgrave (2002) p. 621*; *Grace et al.(2003)*

Elephantopus scaber L. subsp. plurisetus (O. Hoffm.) Philipson (Asteraceae)
DESCRIPTION:
An erect herb up to 80cm high with forked stems and stiff branches, long-haired or rough to touch, with white to purple and blue flowers
VERNACULAR NAMES:
E: Prickly-leaved Elephants-foot
PLANT PARTS USED:
whole plant
TRADITIONAL USES:
Zambia: the whole plant is used as a febrifuge: *Nair (1967) p. 95*
TOXICITY:
"Acute toxicity studies revealed the non-toxic nature of the crude extract": *Daisy et al.(2009)*

Elephantorrhiza goetzei (Harms) Harms [*E. rubescens* Gibbs; *E. elongata* Burtt Davy] (Fabaceae, Mimosoideae)
DESCRIPTION:
Shrub/small tree about 3m high, found in various types of scrub and woodland, often associated with rocky outcrops
VERNACULAR NAMES:
E: Narrow-pod Elephant-root; **Afr**: Smalpeulbasboontjie; **B/ Ku**: citeta (1) **Lo**: kakola-mvula (3d) **Ny**: candima, cikundulima, mcakima, mceramila, nkhumba (9), citeta (3, 3e, 9)
PLANT PARTS USED:
roots
TRADITIONAL USES:
Zimbabwe: an infusion of the roots is drunk, and used to immerse the naked patient, as a remedy for fever: *Gelfand et al. (1985) p. 143*

Eleusine coracana (L.) Gaertn. (Poaceae)
VERNACULAR NAMES:
E: Finger millet; **B**: amale(1), ubule (1, 20), male (3c) **B/ Tu**: bule (1, 3c, 20) **I/ Ku/ Ny/ To**: mabele (1, 3, 3e, 18, 19) **K/ Lb/ Lo/ N**: luku, ruku (18, 19) **Lb/ Le**: ma-au, mao, mawo (18, 19) **Ln**: kacayi (18, 19) **Lo**: lukesya (18, 19), mahangu (3d) **Lv**: baluku (18, 19) **Ny**: kambale, lupoko(3) **To**: kafwamba, sinene (19) **Tu**: kambala (20)
PLANT PARTS USED:
whole plant
TRADITIONAL USES:
Vietnam: the plant is used as a diaphoretic: *De Wildeman (1949) p. 2,* quoted in *Watt et al. (1962) p. 472*

TOXICITY:
" variety of fungi-harbouring finger millet seeds are potentially toxigenic & ...hazardous directly to man": *Penugonda et al.(2010)*

Eleusine indica (L.) Gaertn. (Poaceae)
DESCRIPTION:
Invasive perennial grass in disturbed places.
VERNACULAR NAMES:
E: Indian Goosegrass, Wiregrass, Crowfoot grass; **B**: bulolo (3c, 9a) **B/ M**: kalolo (20) **Ku**: dulu (1) **Le/To/ Tu**: lukata (4, 8, 17) **Lo**: likwaluku, sikwaluku (19) **Ny**: cigombe, cinsangwi, cipika-mongu, kangodza (9) **Tk**: kakata (8)
PLANT PARTS USED:
roots, whole plant
TRADITIONAL USES:
Brazil: the plant is used as a remedy for malaria: *De Wildeman (1949) p. 2,* quoted in *Watt et al. (1962) p. 472*
Ecuador: a root decoction is used against fever: *Coe et al. (1996) p. 107*
W. Africa: the crushed plant is used to bathe feverish children: *Vergiat pp. 84, 85,* quoted in *Burkill (1994) p. 240*
Zambia: the plant is used as a febrifuge: *Amico (1977) p. 134*
TOXICITY:
Classed as toxic in US FDA poisonous plants datadase. "may contain toxic levels of HCN": *Aplin (1968)*

Emilia coccinea (Sims) G. Don (Asteraceae)
DESCRIPTION:
Annual herb having scarlet tassel-shaped flower heads, growing to 0.6m by 0.3m
VERNACULAR NAMES:
E: Scarlet Magic, Tassel Flower
PLANT PARTS USED:
leaves
TRADITIONAL USES:
Nigeria: a leaf decoction is used as a febrifuge: *Ainslie sp. no. 144,* quoted in *Burkill (1985) p.469*

Entada abyssinica Steud. ex A. Rich. (Fabaceae, Mimosoideae)
DESCRIPTION:
Small tree up to 10m high, dense crown and widespreading branches. Widespread in several woodland types, often on rejuvenated Upper Valley soils.
VERNACULAR NAMES:
B: cibombwesala, sambanteta (3c), mulala-ntanga (3c, 7a, 14), mulala-ntete (3, 3c, 7a, 14), mula-tanga (23) **I**: mkalata (4, 23), muswati, mulya-ŋombe (17) **I/ K/ To**: mufwate (4, 17, 23) **K**: cipunga-ŋombe (19, 23), mukumbwa-ŋombe (3, 7a, 14), mwila (23) **Ku**: cikumba-mbowo, citeta-mala (1) **Le**: lipunga-ŋombe (17) **Ln**: musenze (3, 14) **Lo**: fumbwa-musowa (3d, 7a, 14), sipumba-ŋombe (16) **Nm**: mfutamula (7c) **Ny**: cikumbambu, mtimba-mvulu, nyanyata (3e), cisekele, congololo (3e, 7a, 14) **Tk**: mutimba-vula (8) **To**: cisekele, congololo (7a), muse-nzenze, munyele (14) **Tu**: cikumba-mboo (8)
PLANT PARTS USED:

bark, roots, leaves
TRADITIONAL USES:
<u>Angola</u>: to treat children, the roots are pounded and decocted, to drink cold on an empty stomach: *Bossard, (1996)*
<u>Côte d'Ivoire</u>: a leaf decoction or infusion is administered *per os*: *Kerharo et al. (1950) p. 95*
<u>Tanzania</u>: the powdered leaves are used as a remedy for fever: *Dalziel (1937), De Wildeman (1946) p. 650,* quoted in *Watt et al. (1962) p. 597 .* The leaf sap or a root decoction is drunk as a remedy for malaria: *Haerdi (1964) p. 49*
<u>W. Africa</u>: the bark, roots, and leaves are used as a febrifuge: *Akinniyi, Oliver-Bever (1960) p. 62,* quoted in *Burkill (1995) p. 228*
TOXICITY:
Juice of the bark + Cambum was inserted under the eyelids as ordeal poison: *Agroforestree database (2009)*

Entada africana Guill. & Perr. (Fabaceae, Mimosoideae)
DESCRIPTION:
Small low-branching tree, up to 7 – 12m high, with a narrow open crown, found in high-rainfall savannah areas, and very sensitive to bush-fires
VERNACULAR NAMES:
Not known
PLANT PARTS USED:
roots, leaves, seeds
TRADITIONAL USES:
<u>W. Africa</u>: the roots, leaves, and seeds are all used as a febrifuge: *Adam, Ainslie, Akinniyi,* quoted in *Burkill (1995) p. 230.* In Mali, the roots are powdered & decocted to make a remedy taken orally, and the rootbark and trunkbark decocted make a drink and to bathe the patient: *Malgras, (1992)*
TOXICITY:
A leaf infusion at 1:1000 killed *Carassius auratus* within 12 hrs: *Agroforestree database (2009)* .Tests on mice of stem bark & leaf extracts showed average toxicity, LD_{50} 146.7 and 249.9mg/kg $^{-1}$: *Tibiri et al. (2007)*

Entada gigas (L.) Fawcett & Rendle [*E. phaseoloides* Merr.; *E. scandens* Benth.] (Fabaceae, Mimosoideae)
DESCRIPTION:
A flowering liana having large seed-pods
VERNACULAR NAMES:
E: Sea Heart; **Ln**: ikua-kua (2, 3) **Nm**: godogo (1a)
PLANT PARTS USED:
whole plant
TRADITIONAL USES:
<u>Philippines</u>: the plant has been used occasionally as a febrifuge: *Quisumbing (1951),* quoted in *Watt et al.(1962) p. 598*
TOXICITY:
Used widely as fish-poison. Toxic saponins may be removed by roasting & boiling: *Watt et al.(1962)* pp. 598-600

Erianthemum dregei (Eckl. & Zeyh.) Tiegh [*Loranthus dregei* Eckl. & Zeyh.]
(Loranthaceae)
DESCRIPTION:
Branched parasitic shrub with spreading or pendant stems, up to 1.5m., and orange/red
berries. Found in riverine fringes, miombo woodland at higher altitudes, and forest edges
VERNACULAR NAMES:
Ny: kalemelera, kalisace (9)
PLANT PARTS USED:
leaves
TRADITIONAL USES:
Tanzania: the leaf sap is drunk as a remedy for malaria: *Haerdi (1964) p. 109*

Eriosema psoraleoides (Lam.) Don (Fabaceae, Papilionoideae)
DESCRIPTION:
Much-branched bushy perennial, 1 - 2.5m, with golden-yellow flowers and hairy pods. Found
in high-rainfall areas and in disturbed ground
VERNACULAR NAMES:
E: Canary Pea; **B:** mutunga-nkomo (3c) **Lb:** kapalupalu (23) **Ny:** kapiyo-piyo (6) **Tk:** kabeti
(8)
PLANT PARTS USED:
roots
TRADITIONAL USES:
Nigeria:the leaves are decocted to make a remedy to drink: *Adjanohoun et al. (1991)*
Tanzania: a root decoction is taken for malaria: *Burkill (1995) pp. 348, 639; Haerdi (1964).
57*

Erythrina abyssinica Lam. ex DC [*E. tomentosa* R. Br. ex A. Rich.] (Fabaceae,
Papilionoideae)
DESCRIPTION:
Small/medium tree up to 15m high, with rounded crown and corky bark. Widespread, fire-
resistant, hence often found on site of destroyed forests, on rejuvenated soils, and on
coarse riverine alluvium
VERNACULAR NAMES:
E: Red-hot poker,Coral-tree; **B/ N/ Ny/ To:** mulunguti (2, 3, 3a, 3c, 7a, 9a, 10, 14, 23) **B/
Lb:** cisangwa (3c, 10, 23) **B/ Le:** mutiti (2, 3c, 23) **I:** masole (4) **Ln:** ciputamazala, cisunga
(2, 3, 14) **Lo:** munjindu (3d, 7a) **N/ Ny:** mwale (2, 3, 10) **Nm:** mhalalwa-huba, mkala-
lwanghuba, mkala-wahuba, mlelwa-huw pili-pili (22), mungu (1a, 22) **Ny:** cisungwa,
mtambe, mtengo-wamgoma, muwale, nenepa (9), mlindimila (9, 10), mlunguti, msungwa
(3e), mwale (7a, 14) **Tk:** muthithi (8) **To:** musungula (3a, 9a)
PLANT PARTS USED:
bark, roots
TRADITIONAL USES:
E. Africa: the bark and root are used as a remedy for malaria: *Bally (1937) p. 17, Githens
(1948) p. 88*
TOXICITY:
Seeds are poisonous:*Watt et al.(1962).* LD > 1gm/kg when tested on mice: *J. Ethn. 1984
v.12 p.239*

Erythrocephalum zambesianum Oliv. & Hiern (Asteraceae)
DESCRIPTION:
Prostrate herb 10 – 12cm high with white-cottony stem and oval leaves.
VERNACULAR NAMES:
Ku: mpala-matongwe (1) **Ny**: cimwembe-waukulu, naka- nkhwali, ncaca (9), mkalakate (3e)
PLANT PARTS USED:
roots, leaves
TRADITIONAL USES:
Tanzania: a root decoction mixed with leaf juice is drunk as a remedy for malaria: *Haerdi (1964) p.168*

Erythrophleum suaveolens (Guell. & Perr.) Brenan [*E. guineense* G. Don] (Fabaceae, Caesalpinioideae)
DESCRIPTION:
Tall tree up to 20m high, sometimes slightly buttressed; in seepage and low-altitude rain-forest.
VERNACULAR NAMES:
E: Ordeal Tree; **B**: musepa (20), musyipa (23) **B/ Le**: mwafi (2, 3, 3b, 3c, 7a, 19, 20, 23) **B/ Ny**: mwavi (pl. myavi) (9, 20) **Ln**: musipa (23) **Lv**: mukoso (22) **Nm**: mkola (1a) **Ny**: mwayi (9) **Tk**: mumpola-lundu (8)
PLANT PARTS USED:
whole plant, bark
TRADITIONAL USES:
E. Africa: small portions of a weak macerate are used to treat malaria: *Burkill (1995) p. 119.* A bark decoction is taken for malaria – small doses only, since it is highly poisonous: *Haerdi (1964) p. 45*
TOXICITY:
Highly poisonous: *Haerdi (1964) p. 45*

Eucalyptus grandis Hill ex Maiden (Myrtaceae)
DESCRIPTION:
Tall straight tree up to 46m high; smooth whitish bark. Invades forest gaps, plantations and watercourses
VERNACULAR NAMES:
E: Saligna Gum; **Ny**: bulugamu (9)
PLANT PARTS USED:
leaves
TRADITIONAL USES:
Malawi: an leaf infusion is used as a vapour bath to treat fevers: *Morris (1996) p. 418*

Eucalyptus spp.
VERNACULAR NAMES:
Ku: mbulugamu (1) **Ny**: bulugamu (3e) [= Blue-gum?]
PLANT PARTS USED:
leaves
TRADITIONAL USES:

Burkina Faso: 3 bunches of leaves are boiled in 10 lit. water for 2 min., and used as a vapour bath twice daily for 2 days, drinking a glass of hot liquid each time to provoke diarrhoea. The leaves contain flavonoides, sterols, and essential oil of cineol: *de la Pradilla (1988) p.21*

E. Africa: an infusion of the leaves is used as a remedy for fever: *Kokwaro (1976) p. 165*

TOXICITY:

The oil is poisonous if swallowed, and can irritate the skin

Euclea crispa (Thunb.) Sond. ex Gurke [*E. lanceolata* E. Mey. ex A. DC.] (Ebenaceae)

DESCRIPTION:

A dense upright shrub 1–2m high, or a slender tree 8-20m high, found in bush clumps in grassland, open woodland and bushveld, among rocks, and at forest margins.

VERNACULAR NAMES:

E: Blue Guarri, Blue-leaved guarri; **Afr**: Bloughwarrie; **Ny**: mpata, mpukuso (9)

PLANT PARTS USED:

whole plant

TRADITIONAL USES:

Ethiopia: the fruit is used to make a remedy for malaria: *Geyid et al.(2005)*

Zambia: the plant is pounded and soaked, and the effusion drunk: *Vongo*

Euclea natalensis A. DC. [*E. multiflora* Hiern] (Ebenaceae)

DESCRIPTION:

Shrub or small tree up to 7m high, or medium tree up to 12m; variety of habitats from arid areas to bushveld, miombo woodland, riverind fringes, among rocks, and on koppies and anthills

VERNACULAR NAMES:

E: Hairy Guarri; **Afr**: Harige Ghwarrie; **B**: muntu-fita (3c, 23) **K**: muntu-kufita (3) **Ny**: manama, msumwa, mtana-wanjano (9), nhlangulane (22) **Tk**: mukula (8)

PLANT PARTS USED:

roots

TRADITIONAL USES:

E. Africa: the roots are pounded and boiled, and the decoction is drunk as a remedy for "ague": *Kokwaro (1976) p. 84*

Eulophia sp. (Orchidaceae)

VERNACULAR NAMES:

Not known

PLANT PARTS USED:

roots

TRADITIONAL USES:

Malawi: the root is pounded to make lather and mixed with water to wash the head of a malaria patient: *Williamson (1975) p. 111*

Euphorbia crotonoides Boiss. (Euphorbiaceae)

DESCRIPTION:

Much-branched annual herb 50 – 100cm high, with spreading white hairs and latex, lush ovate leaves. Found on sandy and stony soils in open grassy woodland or disturbed ground.

VERNACULAR NAMES:

Not known
PLANT PARTS USED:
latex
TRADITIONAL USES:
E. Africa: latex mixed with other ingredients is a remedy for malaria: *Kokwaro (1976) p. 91*

Euphorbia heterophylla L. [*E. geniculata* Cortega] (Euphorbiaceae)
DESCRIPTION:
Hardy invasive weed, from 30cm – 70cm high, upper leaves scarlet, contrasting with dark-green lower leaves, on roadsides and rural paths
VERNACULAR NAMES:
E: Cruel Plant, Painted Leaf, Fireplant, Desert Poinsettia; **Ny**: mpendo (3e), mpira, nama-finya (9)
PLANT PARTS USED:
roots, bark
TRADITIONAL USES:
Malaya: a decoction of the roots and bark has been used for ague: *Burkill (1935) p. 978; Burkill (1994) p. 69*
TOXICITY:
The latex and root are poisonous and have caused human death; the latex is toxic on contact with skin: *Watt et al.(1962) p. 408*

Euphorbia hirta L. [*Chamaesyce hirta* (L.) Millsp.] (Euphorbiaceae)
DESCRIPTION:
Hairy weed up to 70cm high, with numerous small flowers. Invades waste places and crops.
VERNACULAR NAMES:
E: Asthma Plant, Cat's Hair, Hairy Spurge, Pill-bearing Spurge, Snakeweed; **Ku**: nyama-toka (1) **Nm**: kiawa-ame (1a) **Ny**: cala-cankwale, naka-meso, naka-tobwa, nkholosa, pirilango (9) **To**: nakabele (5b)
PLANT PARTS USED:
whole plant
TRADITIONAL USES:
Zaire: the plant is decocted with other plants and used to wash infants with fever: *Bouquet (1969) p. 114,* quoted in *Burkill (1994) p. 71*
TOXICITY:
Tests of leaves on Brine shrimp showed LC_{50} 0.3253 and on Vero cell line 0.2208: *Brendler et al., (2010)*

Euphorbia thymifolia L.(Euphorbiaceae)
DESCRIPTION:
Annual prostrate herb up to 20cm, fibrous roots, many from base, with purplish leaves, red seeds.
VERNACULAR NAMES:
E: Chickenweed, Gulf Sandmat, Thyme-leaf Spurge
PLANT PARTS USED:
whole plant
TRADITIONAL USES:

Trinidad: a decoction of the plant is used as a fever remedy: *Wong p. 130,* quoted in *Burkill (1994) p. 78*
TOXICITY:
The latex has tumour-promoting properties and is implicated in the epidemiology of Burkitt's lymphoma. Its use in traditional medicine should be discouraged: *PROTA (2008) p. 301*

Evolvulus alsinoides (L.) L. (Convolvulaceae)
DESCRIPTION:
A bushy hairy perennial herb 20 – 30cm high, with spreading wiry branches arising from a small woody rootstock and bearing striking blue flowers.
VERNACULAR NAMES:
E: Dwarf Mornng-glory
PLANT PARTS USED:
leaves
TRADITIONAL USES:
General: the bitter leaves are widely used in preparing a tonic and febrifuge. For Ethiopia, see *Getahun (1975),* for India see *Sastri (1952) pp. 233-4,* for Nigeria see *Ainslie sp. no. 155,* for the Philippines see *Quisumbing (1951) pp. 756-7,* and for the Sudan see *Broun et al.(1929) pp. 322-3,* all quoted in *Burkill (1985) p. 530*
TOXICITY:
Acute oral toxicity studies on rats revealed the non-toxic nature of the roots: *Sathish et al. (2010)*

Faidherbia albida (Delile) A. Chev.[*Acacia albida* Delile] (Fabaceae, Mimosoideae)
DESCRIPTION:
A large tree, often reaching 20m. Comes into leaf at end of rains, and remains green throughout dry season. Widespread in hotter and drier areas, esp. on sandy alluvium on riverbanks, eg Kalahari Sands fringing Kafue Flats
VERNACULAR NAMES:
E: Apple-ring Acacia, White Thorn, Winter Thorn; **Afr:** Anaboom; **B:** mucesi (3c, 14, 23), munga-nunsyi (3a, 9a, 23) **I:** buungasiya, mungasiya, muunga-usiya (4), mulya-ŋombe, mungosia (19), mukoka (17) **I/ Le/ Lo/ Tk/ To:** muunga (2, 3a, 14, 16, 17, 19, 23, 22) **K:** mwiba (2, 3, 3a, 14) **Le:** mukoka (17) **Lo:** lunga-munene, mukwanga (3d, 14), mukwanga-banziba (3d), muunga (3, 14, 22), mungao-munsu (19) **Nm:** nanda (22) **Ny:** citonya, msangu-sangu, mtete (9), mtube-tube, nsangu-sangu (3e, 14), mutube-tube (20) **Ny/ To/ Tu:** nsangu (2, 3, 3e, 14), musangu (2, 3, 3a, 8, 9, 9a, 14, 19, 20, 23) **To:** mucakwe (19), mucangwe (3a, 9a), mujagwe (2, 3, 14), mucesi (2), musosa (8) **Tu:** ipogoro (7c)
PLANT PARTS USED:
bark, roots
TRADITIONAL USES:
Kenya:the Masai people use the plant to treat fevers: *Salawu et al.(2010)*
Nigeria: the bark is infused as a febrifuge: *Dalziel (1937) p. 202; Singha (1965),* quoted in *Burkill (1995) p. 235; Salawu et al.(2010)*
Senegal: an infusion of the crushed root is drunk morning and evening as a remedy for malaria and other fevers. Alternatively, an infusion of the bark is drunk, and used to bathe the patient morning and evening: *Kerharo et al. (1950) p. 102*
Tanzania: a bark decoction is taken as an emetic in fever: *Watt et al. (1962) p. 538; Wickens pp. 181-202,* quoted in *Burkill (1995) p. 235*

<u>Zambia</u>: a decoction of the bark is taken as an emetic in fever: *Nair (1967) p. 38*
TESTS:
In vivo tests on mice showed "remarkable dose-related chemo suppression in the various models used in the study comparable (sic) with chloroquine": *Salawu et al.(2010)*
TOXICITY:
The oral median lethal dose of ethanolic stem bark extract was greater than 5000mg/kg, suggesting that it is practically non-toxic: *Salawu et al.(2010)*

Ficus sur Forsk [*F.capensis* Thunb.; *F. mallotocarpa* Warb.] (Moraceae)
DESCRIPTION:
A tall tree up to 20m high with dense rounded crown. Widespread, especially near water, in open wooded grassland, and on termite mounds.
VERNACULAR NAMES:
E: Broom-cluster Fig, Bush Fig, Cape Fig, Fire Sticks; **Afr**: Besemtrosvy, Kooman, Wildevyeboom; **B**: muku (10), mukunyu (3c, 7a, 10, 14) **B/ K/ Ln/ Lo/ To**: mukuyu (3d, 7a, 10, 14) **B/ Ku**: nsambe, sambe (1) **I**: inkuzu (1c) **Lo**: kakeke, namonde (3d, 7a, 14) **Ny**: cikulu-mawele (3e, 7a), mkuyu (3e), mutunda, nkuyu (10)
PLANT PARTS USED:
leafy twigs
TRADITIONAL USES:
<u>Côte d'Ivoire</u>: a decoction of leafy twigs is used as a remedy for fever pains: *Kerharo et al. (1950) p. 131*
TESTS:
Organic and aqueous extracts of the plant were tested *in vitro* against chloroquine sensitive and resistant strains of *P.falciparum*, and showed activity value of $IC_{50}<30$ mg/ml: *Muregi et al. (2003)*

Flacourtia indica (Burm.f.) Merr. [*F. ramontchi* L'Her.; *F. hirtiuscula* Oliver] (Flacourtiaceae)
DESCRIPTION:
A small spiny shrub or tree up to 6m high. Widespread, especially in open woodland on poor soils.
VERNACULAR NAMES:
E: Governor's Plum; **Afr**: Goewerneurspruim; **B**: mubambwa-ngoma, mukulumbisya (3c), mupula-mpako (7a), mupulu-kuswa (3, 3c), mutumbwisya (3, 3c, 10, 20), mutumbusya, mwanga (20), mutumbwisyi (23), mutumbwizya (1) **B/ Ku**: ntumbuzya, ntunduvya (1) **I/ Lo/ Tk/ To**: mutumbulwa (3, 7a, 10) **I/ To**: mutimba-hula, mutimba-mvula (16, 17) **I/ Tk/ To**: mutumbula (4, 8, 10) **Ku/ Ny**: nkondo-nkonda (1) **Lb**: mukulu-ngufiya (10, 23) **Ln**: mumfumpu (6) **Ln/ Lo/ N**: mubulu-kusyu (3, 10) **Lo**: mubukusyu, tipa (3d), mubula-kusya (7a), mulula-lula (10), mutimbula (16) **Lv**: mufupu (6) **M**: canga (10, 20, 23) **Nm**: mlukuwa-mhuli (7c, 22) **Ny**: makoko-lono, mkondo-kondo, mtumbuzya, ntuza (3e), matyokolo, mtawa, mtema, mvunga-njati, ndawa, songoma (9), mkanda-mbazo (3e, 6), mtudzu (23), ntudza (3, 7a) **Tu**: mukondyo-ndyo (8)
PLANT PARTS USED:
leaves
TRADITIONAL USES:
<u>Tanzania</u>: the leaf sap is mixed with a root decoction as a malaria cure: *Burkill (1994) p. 156; Haerdi (1964) p. 71*

Philippines: sap from fresh leaves and tender stalks is useful in infantile fevers: *Quisumbing (1951) pp. 626-7,* quoted in *Burkill (1994) p. 156*

Flueggea virosa (Roxb. ex Willd.) Voigt subsp. virosa [*Securinega virosa* (Roxb. ex Willd.) Pax & K. Hoffm.] (Euphorbiaceae)
DESCRIPTION:
Many-stemmed, bushy shrub, 2 – 3m high, but sometimes a small spreading tree up to 4m. Common in deciduous woodland, at forest margins, on rocky outcrops, often on anthills.
VERNACULAR NAMES:
E: Whiteberry Bush, Snowberry Tree; **Afr**: Witbessiesbos; **B**: kasanso-bwanga (3, 3c, 23), mubwanga (3, 23) **B/ Ku/ Ny**: lukuswa-ula (1, 3e) **I/ Tk**: mufunda-baalu (4, 8, 17) **Lo**: katoma (3d) **Ny**: cilime, kali, kapira-pira, mdime, mkula-ngondo, mpo-nbona, msere-cete, ntanda-nyerere, waukulu (9), cipyelo, kapila-syila, kapyai-pyai (3e) **Ny/ Tu**: lumpunga (1, 3e, 8) **To**: mwilatuba (2, 3, 22) **Tu**: musanda-lima, muzanda-lima (8)
PLANT PARTS USED:
roots, leaves
TRADITIONAL USES:
Kenya: the roots are used in a remedy for malaria: *Bally (1937) p. 14*
Nigeria: a leaf decoction is drunk and used to bathe fever patients: *Burkill (1994) p. 139*
Tanzania, Zambia: an infusion of the roots is mixed into beef broth as a remedy for malaria: *Bally (1937) p. 14; (1938); Bally (1938),* quoted in *Watt et al. (1962) p. 417; Palgrave (2002) p. 472;*
TOXICITY:
Although used as fish-poison & ordeal poison in W.Africa, its toxicity seems low, and no human fatalities have been recorded: *PROTA (2008) p. 306*

Gardenia ternifolia Schum. & Thonn. subsp. jovis-tonantis (Wel.) Verdc.[*G. jovis-tonantis* (Welw.) Hiern; *G. asperula* Stapf & Hutch; *G. goetzei* Stapf. & Hutch] (Rubiaceae)
DESCRIPTION:
Savannah shrub, 5 – 15ft high, with fragrant white flowers opening for one night, and fibrous grey/green edible fruit.
VERNACULAR NAMES:
E: White/ Forest Gardenia; **B**: cikololo (3c, 14), manceba (3c, 14), mukololo (3, 3c, 7a, 9a) **I**: cikala-malanga (4) **K**: mwita-ngolwa (3, 14), mwitwa-ngolwa (7a) **Ln**: kasabeje (3, 14) **Lo**: cosombo (7a) **Ny**: likongono (9) **To**: mojolo (7a)
PLANT PARTS USED:
fruit, roots
TRADITIONAL USES:
E. Africa: a decoction of the fruit is taken as a remedy for malaria: *Kokwaro (1976) p. 187*
Tanzania: a root decoction is drunk as remedy for malaria: *Haerdi (1964) p. 139*
Zambia: an infusion of the fruit is taken twice daily as a remedy for malaria: *Haapala et al.(1994) p. 19*

Gloriosa superba L. [*G. simplex* L.; *G. virescens* Lindl.] (Colchicaceae)
DESCRIPTION:
The National Flower of Zimbabwe. A scandent plant, climbing by leaftip tendrils, with spectacular red/yellow flowers.
VERNACULAR NAMES:

E : Flame/ Fire/ Glory/ Superb/ Climbing/ Creeping/ Lily; **Ku**: kalume-kandiya (1) **Lo**: makuwa-kuwa (3d) **Nm**: jengaluwo-ngako (7c) **Ny**: citambala, tambala (9), msele (3e), nya-malokane (22) **Tk**: simuzingili (8) **Tu**: nyaka-jongwe (22), pamusele (8)

PLANT PARTS USED:

sap

TRADITIONAL USES:

<u>Tanzania</u>: the plant sap is drunk as a remedy for malaria: *Burkill (1995) p. 505; Haerdi (1964) p. 198; PROTA (2008) p. 310*

TOXICITY:

The plant contains colchicine, a toxic alkaloid. Severe reactions, including human deaths, have been reported: *PROTA (2008) p. 311*

Gnidia chrysantha (Solms ex Schweinf.) Gilg. [*Arthrosolen chrysantha* Solms-Laub. ex Shweinf.] (Thymelaeaceae)

DESCRIPTION:

Found in *miombo* woodland and vleis, a herb with conspicuous bright yellow flower-heads.

VERNACULAR NAMES:

Lo: simalembo (2, 3, 3d) **Ny**: katupe, swinya (3e)

PLANT PARTS USED:

leaves, roots

TRADITIONAL USES:

<u>Tanzania</u>: leaf sap and decocted roots are used as a body wash in fever: *Burkill (1995) p.188; Haerdi (1964) pp. 71-2*

TOXICITY :

Poisonous to stock : *Watt et al. (1962) p.1022*

Gomphocarpus fruticosus (L.) Ait. f. [*Asclepias fruticosa* L.] (Asclepiadaceae)

DESCRIPTION:

A herbaceous, perennial, multi-stemmed shrub or shrubby herb up to 2m high, with attractive pendulous creamy-yellow flowers. Found in disturbed ground at roadsides and abandoned fields.

VERNACULAR NAMES:

E: Milkweed, Wild Cotton; **Afr**: Gansie, Melkbos

PLANT PARTS USED:

roots

TRADITIONAL USES:

<u>Kenya</u>: the Tugen people use the roots to decoct a medicine for malaria: *Munguti (1994) p. 5*

TOXICIY:

All parts are poisonous to livestock: *Watt et al.(1962) pp.119-123*

Gossypium barbadense L. (Malvaceae)

DESCRIPTION:

Short-lived, sparsely-branched perennial shrub with yellow flowers, up to 3m high.

VERNACULAR NAMES:

E: Cotton; **B**: mutonge (9a, 23) **Lo**: mbwanda (16) **Ny**: tonje (3e)

PLANT PARTS USED:

leaves

TRADITIONAL USES:

Brazil: the leaves are macerated in water to make an antimalarial drunk 3x daily: *Di Stasis et al. (1994) p. 536; Milliken (1997) pp. 22, 25; Milliken (1997a) p. 223*

TOXICITY:

Tests on rats showed that "aqueous extract of cotton seed meal contains substances that can rapidly cause damage to testicular, liver, kidney & muscular tissue": *Thomas et al.(1991)*

Grewia mollis Juss. (Tiliaceae)

DESCRIPTION:

Shrub or small tree, up to 10.5m high. Found in semi-evergreen thickets and derived woodland in L. Mweru Basin.

VERNACULAR NAMES:

Not known

PLANT PARTS USED:

fruit

TRADITIONAL USES:

Nigeria: the fruit is used against fevers: *Ainslie sp. no. 171,* quoted in *Burkill (2000) p. 207*

TOXICITY:

Tests of stem bark extract on rats suggest that consumption of the plant material at high concentrations may elicit hypatotoxic effects in rats & possibly in humans: *Wilson et al.(2010)*

Gymnosporia senegalensis (Lam.) Loes. [*Maytenus senegalensis* (Lam.) Exell] (Celastraceae)

DESCRIPTION:

Much-branched shrub or small tree, up to 5m high, bearing profusion of white flowers. Widespread, in open woodland, often at edges of dambos.

VERNACULAR NAMES:

E: Confetti Spike-thorn; **Afr**: Rooipendoring; **B**: cungu (10, 20), ciyungu, munga-senge, mutu-mbwisya (3c), mutenda-nkwale (9a, 10) **B/ I/ K/ To**: mukuba (3, 4, 9a, 10) **Ku**: mutungwa (1) **Lb**: muwambwa-ngoma **Le**: kambolo (17) muwasa-nzolu (10) **Ln**: kabaŋakacina (6, 22) **Lo**: muoliba (3) **M**: cunga-linde (10) **N/ Tk**: muba- bala (10) **Nm**: luwenje, mcoma-fisi (7c), mwezya (22) **Ny**: cikumbelo, mlusye, mpavula (3e), ligoga, mpambulu, mtupa, namano-mano (9), mpelu (9, 10), msuka-mfuti (3, 3e), mtungba-bala (22), mupuru-puta (10) **Tk**: muthunga-phazi (?) **To**: muvunda (10, 17), muungula-mabele (?)

PLANT PARTS USED:

roots, bark

TRADITIONAL USES:

E. Africa: a root infusion is widely used as a remedy for fever: *Kokwaro (1976) p. 52*

Senegal: the bark is commonly used for infants with fever: *Burkill (1985) p.358; Kerharo (1964)*

Zimbabwe: the whole body, except the head, is washed with an infusion of the roots as a remedy for fever: *Gelfand et al.(1985) p. 176*

TESTS:

Aqueous and other extracts of the roots, leaves and stem bark were tested *in vitro* against P.falciparum, and showed strong antiplasmodial activity of $IC_{50}<10$mg/ml: *Gessler et al. (1994).* A methanolic extract of the plant tested against chloroquine resistant and sensitive strains of *P. falciparum* revealed an IC_{50} value of only 2.1 mg/ml: *El Tahir et al. (1999b)*

Hagenia abyssinica (Bruce) J. F. Gmel. (Rosaceae)
DESCRIPTION:
Slender tree up to 18m high. Found at edge of evergreen montane forest near souce of Shire River.
VERNACULAR NAMES:
E: African Redwood
PLANT PARTS USED:
roots
TRADITIONAL USES:
E. Africa: the roots are cooked with meat, the soup being drunk as a remedy for malaria: *Kokwaro (1976) p. 183*
TOXICITY:
The plant is widely used in Ethiopia as a medicine, yet it is toxic, & excessive consumption can cause blindness and death: *Getahun (1975)*

Hallea stipulosa (DC.) J.-F. Leroy [*Mitragyna stipulosa* (DC) Kuntze (Rubiaceae)
DESCRIPTION:
Timber tree up to 22m high, widespread in lowland evergreen swamp forest
VERNACULAR NAMES:
B/ M: mupa, muupa (3, 3b, 3c, 14, 20, 23) **Ln**: mutunta-mankwaji (3, 3b, 14), muzyu (19, 23)
PLANT PARTS USED:
leaves, bark
TRADITIONAL USES:
W. Africa & Zaire: the leaves and bark contain mitragynine, and are used against malaria: *Githens (1948) p. 96; Watt et al. (1962) p. 900*
TESTS:
Spencer tests: score 0 / 0: *Spenser et al. p. 16*
Tests on the bark have produced negative results: *Watt et al. (1962) p. 900*

Harrisonia abyssinica Oliv.(Simaroubaceae)
DESCRIPTION:
A scrambing shrub, or small/medium tree up to 6m high. Found only in Lake Mweru basin, in deciduous or evergreen thicket on deep alluvium fringing the lake
VERNACULAR NAMES:
Nm: kadhatula, msongwa, musoma-njaro, nsoma (22), mkussu (1a, 22) **Ny**: msangalasa (9)
PLANT PARTS USED:
roots, leaves
TRADITIONAL USES:
Tanzania: a decoction of the boiled roots with leaf sap is used as a remedy for malaria: *Burkill (2000) p.89; Haerdi (1964) p. 119; Hedberg et al.(1983) p. 249; Kokwaro (1976) p. 203; Watt et al. (1962) p. 941*
Tropical Africa: the plant is used throughout to treat malaria: *PROTA (2008) p.317*
TESTS:

Early tests proved negative (*Spencer et al: p. 170*) but in 1999 a methanolic extract of the plant was tested *in vitro* against *P. falciparum*, and showed activity, with IC$_{50}$ 10 mc/ml: *El Tahir et al., (1999b)*.

Further *in vitro* tests in 2006 confirmed these findings : *Kirira et al. (2006)*

An 80% MeOH extract of the root showed antiplasmodial activity when tested *in vitro* in Tanzania: *Maregesi et al., (2010)*

Methanolic root extracts have shown significant antiplasmodial activity against chloroquine-sensitive & chloroquine- resistant strains of *Plasmodium falciparum* : *PROTA(2008) p.317*

Harungana madagascariensis Lam. ex Poiret (Guttiferae)
DESCRIPTION:
Small evergreen tree up to 13m high, widespread as pioneer in *mutesyi* forests, sometimes in *Brachystegia* woodland in higher-rainfall areas.
VERNACULAR NAMES:
E : Orange-milk Tree ; **B**: katumbi (3), katumbe-tumbe(20, 23), mufifi (3, 7a, 9a), muswesya-namo (23), mutumbe (20) Ln: katunya (21); **Ny**: mbuluni, mpefu, msambitsa-mkanda, mtumu, ntungungu (9), msuwa-suwa (3e, 7a)
PLANT PARTS USED:
bark, leaves, roots
TRADITIONAL USES:
Sierra Leone: young leaves are boiled with fruits of *Solanum* sp. and drunk as needed: *Barnish et al. (1992)*
Tanzania: a decoction of the bark or the juice of the leaf and roots is drunk as a remedy for malaria: *Burkill (1994) p. 396; Haerdi (1964) p. 101; Kokwaro (1976) p. 103*
Madagascar: the leaves are used as a fever remedy: *Pernet et al. (1957)*; *Debray p. 75*, quoted in *Burkill (1994) p. 488; Githens (1948) p. 91; Watt et al. (1962) p. 495*.
The bark of the stems, branches and trunk is scraped, macerated and decocted to be drunk & used as an enema: *Wome (1985)*
TOXICITY:
Tests on rats altered metabolic activities, showing need for caution in administering: *Olagunju et al. (2004)*; tests of leaves on Brine shrimp showed LC$_{50}$ 0.748 and on Vero cell line 0.053:*Brendler (2010) pp. 138, 139, 286*

Heinsia crinita (Afzel.) G. Taylor subsp. crinita (Rubiaceae)
DESCRIPTION:
Shrub to small tree up to 7.5m high, found at low altitudes in riverine fringe forest and thickets
VERNACULAR NAMES:
E: Jasmine-gardenia; **Afr**: Jasmynkatjiepiering
PLANT PARTS USED:
roots
TRADITIONAL USES:
Tanzania: raw scrapings of the roots are eaten against malaria: *Haerdi (1964) p. 140*

Helianthus annuus L. (Asteraceae)
VERNACULAR NAMES:
E: Sunflower; **Ku/ Ny**: nyaka-zuba (1, 3e)

PLANT PARTS USED:
leaves, flowers
TRADITIONAL USES:
<u>Caucasus</u>: the leaves are used as a remedy for malaria: *Dalziel; Watt et al. (1962) p. 236*
<u>Italy</u>: the aerial parts are used as a remedy for malaria: *Watt et al. (1962) p. 236*
<u>Zaire</u> (?): a 10% tincture of the flowers with 70% alcohol is used as a febrifuge: *Watt et al. (1962) p. 236*
<u>Zambia</u>: A leaf infusion is used as a malaria remedy and as an insect killer: *Nair (1967) p. 95*
TESTS:
Tests in Poland have produced negative results: *Watt et al. (1962) p. 236*
TOXICITY:
The plant causes allergic contact dermatitis in sensitive individuals after contact with the sesquiterpene lactones contained in fragile, mutlicellular, capitate glandular hairs. Cattle have been poisoned in Europe after ingesting plants that did not have mature seeds. This is a result of nitrate toxicity, which has caused sickness and death: *(Cooper et al.(1984)*

Helichrysum nudifolium (L. f.) Less. var. nudifolium (Asteraceae)
DESCRIPTION:
A sun-loving perennial herb with pale yellow inflorescence and shiny light-green leaves, up to 1.5m high
VERNACULAR NAMES:
E: Kaffir-tea; **Afr**: Kaffertee; **Ny**: likango (9)
PLANT PARTS USED:
roots
TRADITIONAL USES:
<u>S. Africa</u>: to treat fever, the Southern Soto make a steam bath by pouring an infusion of the roots on to hot stones: *Burkill (1985) p. 478; Watt et al. (1962) p. 239*
TOXICITY:
Non-toxic

Helichrysum panduratum O. Hoffm. (Asteraceae)
DESCRIPTION:
Herbaceous perennial or shrub, up to 60 – 90cm.
VERNACULAR NAMES:
E: Strawflower; **Ny**: canje, canzi-campongo (9)
PLANT PARTS USED:
whole plant
TRADITIONAL USES:
<u>Malawi</u>: the plant is pounded and infused, the liquid being used as a bath for a fever patient: *Morris (1996) p. 263*
TOXICITY:
Non-toxic

Helinus integrifolius (Lam.) Kuntze [*H. scandens* (Eckl. & Zeyh.) A. Rich.; *H. ovatus* E. Mey.] (Rhamnaceae)
DESCRIPTION:
A climbing shrub, reaching up to 8m. high, with coiled tendrils and yellowish flowers. Found in deciduous woodland, often in rocky places and on quartzite hills on Copperbelt.

VERNACULAR NAMES:
E: Soap Creeper; **Lo**: kasese (3d), mulalawa (3, 3d, 19) **Tk**: mulalane (23)
PLANT PARTS USED:
roots
TRADITIONAL USES:
<u>Tanzania</u>: the roots, mashed and beaten to a foam, are drunk and rubbed on the body as a remedy for malarial convulsions in children: *Haerdi (1964) p. 111*
TOXICITY:
Non-toxic

Heliotropium indicum L. (Boraginaceae)
DESCRIPTION:
Annual, erect herb growing up to 15 – 20cm, with blue flowers. Found in low moist alluvial woodland, on muddy banks, gravel bars, and wasteland.
VERNACULAR NAMES:
E: Indian Heliotrope, Indian Turnsole; **Ny**: humbangayi (7c)
PLANT PARTS USED:
leaves, whole plant
TRADITIONAL USES:
<u>W.Africa</u>: the plant is infused to treat fever in children. The Igbo people use a leaf decoction to wash a feverish child. In Guinea, the whole plant is decocted as a febrifuge: *Dalziel (1937), Dawodu 3, Portères p. 17, Thomas p.1989*, quoted in *Burkill (1985) p. 294*
<u>Zaire</u>: a leaf infusion is used as a fever remedy: *Taton (1971) pp. 29-30,* quoted in *Burkill (1985) p. 294*
TOXICITY:
The plant can cause severe adverse reactions & deaths have been reported: *Globinmed*

Hemizygia bracteosa Briq. (Lamiaceae)
DESCRIPTION:
Erect greyish pubescent annual up to 3ft high, with pale pink flowers. Widespread, in marshy grassland.
VERNACULAR NAMES:
Ku: mkupa-imbu (1) **Lo**: lukambamwe (3), sinuke (3d)
PLANT PARTS USED:
whole plant
TRADITIONAL USES:
<u>Nigeria</u>: the plant is used to foment the body of a fever patient: *Burkill (1995) pp. 5, 638; Dalziel (1937) p. 463*

Heteromorpha arborescens *var.* abyssinica (Sprengel) Cham. & Schltdl. [*H. trifoliata* sensu Cufod; *H. collina* Eckl. & Zeyh.; *H. trifoliata* (H. L. Wendl.) Eckl. & Zeyh.]
(Umbelliferae)
DESCRIPTION:
Shrub or small tree up to 6m high. Found in gallery forest fringes, rainforest fringes at Victoria Falls, and in *miombo* woodland.
VERNACULAR NAMES:
E: Parsley Tree; **Afr**: Wildepietersieliebos; **Ny**: cikolola, kapoloni, mpamabana, mpoloni-

wamuna, msemfano, napose, ndodo-yafiti, nsyapita (9), cinkulu (3e, 7a) **Tk**: kanjoma (23), lulya-bazuba (8)
PLANT PARTS USED:
roots
TRADITIONAL USES:
Tanzania: a root decoction mixed with leaf juice is drunk as a malaria remedy: *Haerdi (1964) p. 163*
Zimbabwe: a root infusion is drunk as a remedy for fever: *Gelfand et al.(1985) p. 199*
TOXICITY:
In vitro bacterial & mammalian cell assays showed that the plant caused both DNA damage & chromosomal aberrations: *Fennell et al.(2004)*

Hibiscus micranthus Linn. f. (Malvaceae)
DESCRIPTION:
Shrub, 2.5m high. Found at Victoria Falls, in open *Combretum apiculatum, Acacia nigrescens, Colophospermum mopane* woodland on shallow soil overlying Karroo Basalt
VERNACULAR NAMES:
Nm: mburi (1a)
PLANT PARTS USED:
Whole plant
TRADITIONAL USES:
Zambia: the whole plant is used as a febrifuge: *Nair (1967) p. 58*

Hibiscus rosa-sinensis L. (Malvaceae)
DESCRIPTION:
Evergreen flowering shrub up to 2.5m high with large red scentless flowers
VERNACULAR NAMES:
E: Chinese Hibiscus, China Rose, Shoe Flower
PLANT PARTS USED:
leaves
TRADITIONAL USES:
The East: a leaf decoction is used as a lotion to cool a feverish patient: *Watt et al.(1962) p. 737*

Hibiscus sabdariffa L. (Malvaceae)
DESCRIPTION:
Annual or perennial herb or woody-based subshrub growing up to 2.5m high. Flowers white to pale yellow, with red spot at base of each petal
VERNACULAR NAMES:
E: African Mallow, Jamaica Sorrel, Natal Sorrel, Rosella; **B**: pupwe (20) **B/ M**: koloko-ndwe (20) **I/ To**: lukukwa (4, 6, 19), mukungu (4, 5, 6) **Lb**: bapupwe, ikombwe (18,19) **Le**: mulembwe (18, 19) **Ln**: ndambala (6) **Ln/ Lv**: mutete (18) **Lo**: lukuku, makuku (18, 19) **Lo/ To**: mundabwe, mundambi (18, 19) **Lv**: dikelenge (6) **Ny**: munzi (20) **Ny/ Tu**: lumanda (20)
PLANT PARTS USED:
leaves
TRADITIONAL USES:
Brazil: the leaves are decocted to make an antipyretic: *Di Stasis et al. (1994) p. 536*

TOXICITY:
Tests of calyx epicalyx on Brine shrimp showed LC$_{50}$ 0.0205 and on Vero cell line -1: *Brendler et al.(2010) pp 144, 286.* Showed as cytotoxic when tested on snails at 50-100 ppm: *Elsheikh et al.(1990)*

Holarrhena pubescens (Buch.-Ham.) Wall ex G. Don [*H. febrifuga* Klotzstch] (Apocynaceae)
DESCRIPTION:
Shrub flowering at about 1m. high, or small tree up to 8m high, with rough corky bark. Widespread, especially in drier areas and rocky places
VERNACULAR NAMES:
E: Fever Pod, Jasmine Tree; **Afr**: Koorspeulboom; **B**: mwenge-busyilu (3, 10) **I**: cilundine, (10) citundwe (4, 10) **K**: muteje (10, 23) **Ku**: mpala-mtowasilu, mpala-mwezamunko, mutowa-nsilu (1) **Ku/ Ny**: mweza-munko (1), mtowa-silu (1, 3e) **Ln**: mulya (3), mulya-walisyaka (23) **Nm**: mbele-bele (7c), msonga-lukuga, msonga-ti, mweri-weri, ŋweli-ŋweli (22) **Ny**: cikope, cipeta, ciwimbi, mfumba-mula, mkwale, mtombozi, njiriti (9), gowo-gowo, mbeza-munku, ndombo (3e), kacamba (9, 10), mkalancamba (10), mpondasilu (3, 3e), ndombozi-cipeta (6, 22) **Tk**: lusyalo (8) **Tu**: mubeza-munku (8)
PLANT PARTS USED:
bark, roots
TRADITIONAL USES:
Central Africa & Mozambique: the bark is used as a febrifuge: *Palgrave (1957) p. 29; Palgrave (1983) p. 786*
E. Africa: the roots are used in washing children with fever: *Kokwaro (1976) p. 27.* The bark is used as a febrifuge: *Githens (1948) p. 92; Watt et al. (1962) p. 84*
Malawi: the cortex of the root is used as a remedy for malaria: *Palgrave (1957) p. 29*
S. Africa: the bark has been used to treat fevers : *Palgrave (2002) p. 951*
TESTS :
Spencer tests: score 0 / 0 *Spencer et al: p. 153*
TOXICITY:
Aqueous extracts of the root bark, stem bark or leaves show relatively low toxicity, although containing the alkaloid *conessine*: *PROTA (2010) p. 333*

Hoslundia opposita Vahl. [*H. verticillata* Vahl] (Lamiaceae)
DESCRIPTION:
Shrub up to 5m high. Found at Lake Mweru, Kafulwe, Ndola
VERNACULAR NAMES:
B: katimbiti (9a) **Lo**: kanyange-nyange (3d) **Nm**: mhunga-mbu(22), mkalula, mswele, msyelele, munjinua (1a, 22) **Ny**:canzi, cinyeu, mpingu, tawinda-mazimu, uzambwiya (9), katand-ambo (3e) **To**: musombwani (3) **Tu**: katanda-imbo (8)
PLANT PARTS USED:
roots, leaves, whole plant
TRADITIONAL USES:
Côte d'Ivoire: an infusion of the plant is used as a febrifuge: *Bouquet (1974) p. 97,* quoted in *Burkill (1995) p. 6; Kerharo et al. (1950) p. 234*
Ethiopia: the leaves and stem are boiled in water, and the steam inhaled at bedtime: *Teklehaymanot, (2009)*

<u>Tanzania</u>: the leaves are infused and used as a cooling wash for feverish children. The roots and leaves are crushed and infused in water, and drunk in cases of fever: *Kokwaro (1976) p. 108.* The Syambala use the root as a fever remedy: *Bally (1937) p. 24; Watt et al. (1962) pp. 515-6.* A root decoction mixed with leaf juice is drunk as a remedy for malaria: *Burkill (1995) p. 6; Haerdi (1964) p. 188*

<u>Ubangi</u>: a leaf decoction is used to wash fever patients: *Vergiat (1970) p. 90,* quoted in *Burkill (1995) p. 6*

TESTS:

The hexane extract of the root bark showed good activity *in vitro* against *P.falciparum*, with an IC_{50} of 5.6mg/ml: *Schwikkard et al. (2002) p. 683*

TOXICITY:

Methanol extracts showed good toxicity profile with LD_{50} above 5000mg/kg when tested on rats: *Akah et al.(2010)*

Hugonia orientalis Engl. [*H. busseana* Engl.] (Linaceae)
DESCRIPTION:

A many-stemmed climbing shrub up to 6m high, with yellow flowers. Found in *Kirkia* woodland nr Beit Bridge, and in relict evergreen thickets near Kabwe.
VERNACULAR NAMES:

E: Ram's horn; **Afr**: Ramshoringbos
PLANT PARTS USED:

roots
TRADITIONAL USES:

<u>Mozambique</u>:The rootbark is decocted to treat malaria: *Mulhovo (1999)*

Hydrostachys polymorpha Klotzsch ex A.Braun (Hydrostachydaceae)
DESCRIPTION:

A perennial herb, attached to rocks in waterfalls or rapids by numerous thread-like roots from its compacted woody stem. Its flowers are borne on fleshy spikes above the surface.
VERNACULAR NAMES:

Not known
PLANT PARTS USED:

leaves
TRADITIONAL USES:

<u>Tanzania</u>: the leaf juice is drunk as a malaria remedy: *Haerdi (1964) p. 162*

Hymenocardia acida Tul. [*H. mollis* (Pax) Radcl.-Sm.] (Euphorbiaceae)
DESCRIPTION:

Small treeup to 10m high, smooth flaking bark; widespread in various types of woodland, especially on sandy soils.
VERNACULAR NAMES:

E: Large Red-heart; **Afr**: Grootrooihartboom; **B/ K/ Le/ Lo/ M**: kapempe (3, 3b, 3d, 7a, 14, 17, 20) **B/ Ln/ Lo/ M**: mupempe (3, 3b, 14, 16, 20) **I**: mpazu-pazu, mumpele-mpempe (4, 17, 24) **I/ To**: mumpempe, mupazu-pazu (4, 14, 17) **K/ Ku**: kapembe (1, 23) **Ku**: mwavi-kulu (1) **Ln**: kapepe (19), kapepi (21), kopepe (23) **Ln/ Lo/ Lv**: mupepe (3d, 6, 7a, 22) **Lo**: muyunda (3, 3b) **N**: mupembe (16) **Ny**: cikolo-wanga, cipundu, mkula-ngondo, mpempwe, mteru-teru, mtema (9), kabale (3e, 7a, 14), kalalitsi (3, 3b, 3e, 14) **Tk**: mpatugila (8) **To**: kapasu-pasu (23), mupasu-pasu (17), mupazu-pazu (14) **Tu**: kaputi-puti,

mpasu-pasu (8)
PLANT PARTS USED:
leaves
TRADITIONAL USES:
Côte d'Ivoire: the plant has a great reputation as a febrifuge. Leaf decoctions are drunk, and also used to bathe fever patients: *Kerharo et al. (1950) p. 77.* The leaf-sap is used in topical friction: *Bouquet (1974) p. 84,* quoted in *Burkill (1994) p. 86*
Mali: soft leaves are decocted to make a remedial drink: *Malgras (1992)*
Tanzania: The leaves are mashed in water and the liquid drunk, and pounded root-bark is eaten with porridge to cure malaria: *Burkill (1994) p. 86; Haerdi (1964) p. 94; PROTA(2008) p.340*

Hypoestes verticillaris (L.f.) Roem. & Schult. (Acanthaceae)
DESCRIPTION:
A polymorphic tufted herb up to 1m high, found in open places of wooded savannah and secondary and deciduous forest.
VERNACULAR NAMES:
Tk: sokwe (8) **Tu**: cikokoti (8)
PLANT PARTS USED:
whole plant
TRADITIONAL USES:
Tanzania: the whole plant is soaked in water, which is then used to bathe a feverish child: *Burkill (1985) p. 15*

Hyptis pectinata (L.) Poiret (Lamiaceae)
DESCRIPTION:
An invasive weed with erect stem, green hairy oval leaves, and blue flowers.
VERNACULAR NAMES:
E: Comb Hyptis, Comb Bushmint, Purple Top
PLANT PARTS USED:
leaves, whole plant
TRADITIONAL USES:
Côte d'Ivoire: a leaf infusion is drunk to relieve fever pains. An infusion of the plant is drunk by feverish infants: *Bouquet (1974) p. 97,* quoted in *Burkill (1995) p. 9; Kerharo et al. (1950) p. 235*
Madagascar & W.Africa: the leaf is used as a remedy for fever: *Githens (1948) p. 92; Watt et al. (1962) p. 516*
TESTS:
Spencer tests: score 0/0: *Spencer et al: p.162.* Tests have shown that the tops with flowers have no anti-malarial activity: *Watt et al. (1962) p. 516*

Hyptis spicigera Lam. (Lamiaceae)
DESCRIPTION:
An erect, aromatic herb up to 1m high, with spikes of very small white/ mauve flowers. Found on roadsides, wastes, and cultivated land, often on damp areas.
VERNACULAR NAMES:
E: Black Sesame, Black beni-seed, American Bushmint; **B/ Ku/ Ny**: mwinula-mponda (1, 3e) **Ku/ Ny**: jeli-jeli (1)

PLANT PARTS USED:
leaves
TRADITIONAL USES:
Burkina Faso: fresh leaves are rubbed between the hands, then lifted to the nostrils for inhalation. This should be done sitting down, and once only for each fever attack, since it causes a temporary malaise: *de la Pradilla (1988) p.22*
W. Africa: the whole plant, or the leaves, is infused and drunk hot to treat fevers. A leaf-mash in hot water is rubbed on the body to reduce the temperature: *Ainslie sp. 190, Bouquet (1974) p. 97,* quoted in *Burkill (1995) p. 11*

Imperata cylindrica (L.) P. Beauv. [*I. arundinacea* Cyr. var. *thunbergii* Hack.] (Poaceae)
DESCRIPTION:
A perennial rhisomatous grass up to 3m high with roots up to 1.2m deep, found in disturbed soils, roadsides, building sites.
VERNACULAR NAMES:
E: Kogongrass, Kunai Grass; **B**: ibamba (20) **B/ I/ Lb**: lubamba (4, 17, 19, 20) **I**: cilenje (4, 8, 19) **Le**: mukula-gisyika (19) **Lo**: silenje (19) **Ny**: cisisi, mtuku (3e), nyama-songole (3e, 9) **Tk**: cabweyo (8) **To**: kansonje (19)
PLANT PARTS USED:
roots, inflorescence
TRADITIONAL USES:
Burkina Faso: the roots are roasted, powdered, & mixed with karite butter to make an ointment, rubbed on to a child at night for 3 nights: *de la Pradilla (1988) p.23*
China: the root and inflorescence are used as a fever remedy, although tests have shown that the plant has no antipyretic effects: *Watt et al. (1962) p. 474*
Zambia: the roots and inflorescence are used as a fever remedy: *Amico (1977) p. 134*
TOXICITY:
The pollen can cause allergic reactions: *Khumar et al.(1998)*

Indigofera arrecta Hochst. ex A. Rich. (Fabaceae, Papilionoideae)
DESCRIPTION:
Shrub up to 3m high, but often smaller, with yellow flowers and chestnut-brown pods.
VERNACULAR NAMES:
E: Bengal Indigo, Natal Indigo
PLANT PARTS USED:
roots, leaves
TRADITIONAL USES:
W. Africa: the macerated roots and crushed leaves are drunk as a febrifuge: *Berhaut (1976) pp. 282-4,* quoted in *Burkill (1995) p. 365*

Indigofera tinctoria L. (Fabaceae, Papilionoideae)
DESCRIPTION:
Shrub up to 0.6m high in Zambia, with light green leaves and sheaves of violet/pink flowers. Collected west of Kazungulu.
VERNACULAR NAMES:
E: True Indigo
PLANT PARTS USED:
leaves

TRADITIONAL USES:

E. Africa: the leaves are used as an antipyretic: *Burkill (1995) p. 383; Watt et al. (1962) p. 613*

Ipomoea aquatica Forssk. [*I. reptans* (L.) Poir.] (Convolvulaceae)
DESCRIPTION:

Semi-aquatic invasive plant growing in water or moist soil. Stems 2 – 3m long, white trumpet-shaped flowers with mauve centres
VERNACULAR NAMES:

E: Water Spinach, Water Morning Glory, Water Convolvulus, Chinese Cabbage, Chinese Spinach, Swamp Cabbage; **Ku**: sanda (1) **Ku/ Ny**: kandambua (1)
PLANT PARTS USED:

whole plant
TRADITIONAL USES:

Indo-China: the plant is applied as a poultice in febrile delirium: *Burkill (1985) p. 533*

Ipomoea pes-caprae subsp. brasiliensis (L.) Oostr. [*Convolvulus brasiliensis* L.] (Convolvulaceae)
DESCRIPTION:

Creeping vine with sprawling runners from a woody rootstock.
VERNACULAR NAMES:

E: Beach Morning-glory, Goat's Foot, Railway Vine
PLANT PARTS USED:

leaves
TRADITIONAL USES:

Ecuador: a leaf decoction is given to treat fevers: *Coe et al. (1996) p. 99*
TOXICITY:

Tests of leaves on Brine shrimp showed LC_{50} 0.0087 and on Vero cell line 0.0394: *Brendler et al. (2010) pp. 160, 286*

Ipomoea purpurea (L.) Roth (Convolvulaceae)
DESCRIPTION:

Vine coiling round structures up to 2 – 3m high, with trumpet-shaped blue/ white/ purple flowers
VERNACULAR NAMES:

E: Common/ Purple/ Tall Morning-glory
PLANT PARTS USED:

stalks
TRADITIONAL USES:

S. Africa: a handful of stalks, about 6" long, is bruised in a pint of boiling water or fowl-broth, and the liquid drunk: *Bryant (1909) p. 44*
TOXICITY:

The entire plant is toxic, and can cause nausea & hallucinations: *Watt et al.(1962) p.309*

Jateorhiza palmata (Lam.) Miers [*J. bukobensis* Gilg.] (Menispermaceae)
DESCRIPTION:

A vigorous hairy climber with large palmate leaves and pendulous green/ cream flowers, found in moist woodland, forest and riverine fringes

VERNACULAR NAMES:
E: Calumba, Colomboa, Sarsaparilla; **Ku**: mpala-mabwe (1) **Ny**: kasana, mwana-wamfepo, njoka (9)
PLANT PARTS USED:
whole plant
TRADITIONAL USES:
India: the plant is used as an anti-pyretic: *Chopra (1933)*, cited by *Watt et al., (1962)*
TESTS:
Extracts of the plant have been tested against experimental malaria: *Watt et al. (1962) p. 758*

Jatropha curcas L. (Euphorbiaceae)
DESCRIPTION:
A poisonous, semi-evergreen shrub or small tree, up to 6m high, often planted as a living fence.
VERNACULAR NAMES:
E: Barbados Nut, Purging Nut, Physic Nut; **B**: citondo-mono (3c, 9a, 20) **Ln/ Lv**: monoza-nyemba, nyembe (2, 23) **Lo**: mufuma-ngoma, mulema-nanga (19) **Lv**: cimpulukwa, tumpo (19) **Ny**: civumulu (3e), masa-wasa, msapatonje, msatsimanga (9)
PLANT PARTS USED:
leaves, sap, roots
TRADITIONAL USES:
Benin: the leafy twigs are pounded in water, and one beerglass of the filtrate is drunk each day as a malaria remedy: *Adjanahoun et al. (1989)*, cited by *Neuwinger(1996)*
Brazil: the leaves are used to make a bath for fever patients: *Di Stasis et al. (1994)*
Ecuador: a leaf decoction is used to treat fever: *Coe et al. (1996)*
Madagascar: The roots are used to make a remedy: *Pernet et al.(1957)*; the leaves are boiled to make a bath: *Pernet (1957)*; the leaves and roots are decocted to make a malaria remedy: *Rasoanivao et al. (1992)*
Nigeria: a decoction of the young leaves is taken orally for fever: *Bhat (1990)*
W. Africa: the leaves and sap are used as a fever remedy. Hot or cold leaf infusions are drunk. The sap is given to a child with fever: *Ferry no. 96*, quoted in *Burkill (1994)*
Zambia: an infusion of the leaves is used as a bath to treat fever: *Nair (1967)*
TOXICITY:
Used as a fish and arrow poison. Oil from the nuts contains a toxalbumin, *curcin*, which is a drastic purgative: *Fanshawe et al.(1967)*. Seeds can produce a violent reaction, but recovery is usually rapid: *Greshoff (1900); Morris (1996); Palgrave (2002); Watt et al. (1927)*. The Pasteur Institute tested in 1919 and found at least 2 toxic principles, in the husk and kernel, but were unable to elucidate their chemical nature *(Anon, 1919)*. Leaves and fruit produce hydrocyanic acid: *Watt et al.(1962)*. Neuwinger reports extensive tests of seeds, roots and fruit, but no human fatalities: *Neuwinger (1996)*

Jatropha multifida L. (Euphorbiaceae)
DESCRIPTION:
Shrub or small tree, up to 6m high, with bright scarlet flowers and poisonous seeds. Cultivated in gardens at Livingstone as ornamental shrub
VERNACULAR NAMES:
E: Coral Bush, Coral Plant, Guatemala Rhubarb

PLANT PARTS USED:
Leaves, fruit
TRADITIONAL USES:
<u>Nigeria</u>: the leaves and fruit are boiled and drunk or used as a bath to treat fevers: *Ainslie, sp. no. 195,* quoted in *Burkill (1994) p. 93*
TOXICITY:
The fruit is attractive to children, but contains the toxalbumin ricin, which can cause vomiting, shock, renal & hepatic impairment & death: *Levin et al.(2000)*

Kalanchoe crenata (Andr.) Haw. (Crassulaceae)
DESCRIPTION:
A perennial succulent herb growing up to 2m high, with magenta, yellow or orange flowers
VERNACULAR NAMES:
E: Never-die
PLANT PARTS USED:
leaves, roots
TRADITIONAL USES:
<u>Burundi</u>: Leaves are heated by fire and the juice drunk: *Baerts et al. (1989)*
<u>Congo-Brazzaville</u>: leaves are warmed at fire, juiced, 2 drops into each nostril 2x daily to prvoke sneezing: *Diafouka (1997)*
<u>Côte d'Ivoire</u>: leaves are rubbed on feverish infants: *Bouquet p. 77,* quoted in *Burkill (1985) p. 558*
<u>Tanzania</u>: a root decoction mixed with leaf sap is drunk as an anti-malarial remedy: *Burkill (1985) p. 558; Haerdi (1964) p. 162*
TOXICITY:
Contains *bufadienolide* cardiac glycosides, which can cause cardiac poisoning in grazing animals

Keetia zanzibarica (Klotzsch) Bridson, subsp. cornelioides (De Wild.) Bridson [*Canthium zanzibaricum* Klotzsch] (Rubiaceae)
DESCRIPTION:
A scrambling shrub
VERNACULAR NAMES:
B: mubaba (3, 3c) **Ny**: zavuma (3e)
PLANT PARTS USED:
leaves, roots, bark
TRADITIONAL USES:
<u>Tanzania</u>: a mixture of leaf sap with decocted roots and bark is drunk to cure malaria: *Haerdi (1964) p. 137*

Khaya anthotheca (Welw.) C. DC [*K. nyasica* Stapf. ex Baker f.] (Meliaceae)
DESCRIPTION:
A tall evergreen timber tree up to 25m high, much-branched rounded crown, with smooth grey/brown scaling bark. Characteristic tree of fringing forest in eastern areas, also Lusaka and Copperbelt.
VERNACULAR NAMES:
E: Red Mahogany; **B**: musyikisyi-culu (3c) **B/ I/ K/ Ku/ Lb/ Le/ Lo/ Ny/ Tk/ To/ Tu**: mululu (1, 3, 3a, 3c, 10, 14, 19, 20, 22, 23) **Ku/ Ny/ To/ Tu**: muwawa (1, 20, 22) **Le**:

mongo (19) **Ln**: wululu (21) **Ny**: mlulu (3e, 14) **Ny/ Tu**: mbawa (3, 3a, 3e, 9, 10, 14, 20, 22) **Tu**: suwawa (10)

PLANT PARTS USED:

bark, roots

TRADITIONAL USES:

Angola: the bitter bark is used as a febrifuge: *Anon. (1939) p. 193,* quoted in *Watt et al. (1962) p. 744*

Côte d'Ivoire: a decoction of the bark is drunk as a febrifuge: *Kerharo et al. (1950) p. 156*

Tanzania: a root decoction is drunk against malaria: *Haerdi (1964) p. 121*

Kigelia africana (Lam.) Benth. [*K. pinnata* (Jacq.) DC.] (Bignoneaceae)

DESCRIPTION:

Tree up to 15m high, often hollow, grey flaking fissured bark. Distinctive fruit are sausage-shaped, up to 1m long, and weighing up to 10kg. Found by seasonal streams in fringing woodland, in hotter and drier areas.

VERNACULAR NAMES:

E: Sausage-tree; **Afr**: Worsboom; **B/ Ku**: civungula (1) **B/ Lo**: mufungu-fungu (1, 3, 3a, 3c, 3d, 7a, 10, 14, 20) **I**: ivunga, namutengwa, namuzungula (4, 10, 17, 23), muhungu (19), munzungula (4, 17) **I/ K/ Le/ Lo/ Tk/ To**: muzungula (3, 3a, 3d, 4, 8, 10, 14, 17, 23) **K**: kifungula (3, 14), lifungula (7a), mugungwela (10) **K/ Lb/ Le/ Ny**: mufungula (10, 17) **Ku/ Ny/ Tu**: muvungula (1, 8, 10, 20), mvungula (1, 3e, 10) **Ln**: ifungu-fungu (3, 14, 21), mufuno-funo (3, 7a, 10, 14) **Ln/ Lv/ N**: muvungo-vungo (6, 10, 22) **Lo**: munguli (3d, 14), mupungo-pungo, mupolota (10, 16), mupolata (19, 22) **M/ Tu**: muvee (10, 22) **Nm**: ŋwica (22) **Ny**: cizutu, mvula (3e, 7a, 14), galahungu (22), mvunguti (3, 3e, 7a, 9, 14), mvungutsi (6), mvungutwa (9), nvungula (1) **To**: kazungula, muntengwa (17), muveve, namutengwa (10) **Tu**: milungulu, mvungula (10)

PLANT PARTS USED:

leaves, bark

TRADITIONAL USES:

E. Africa: a decoction from the leaves is used against malaria: *Kokwaro (1976) p. 36*

Malawi: a decoction of the bark is used against malaria: *Morris (1996) p. 231*

TESTS:

Tests in 1996 and 2000 identified four comounds with anti-malarial activity, the most active ones being 2-(1-hydroxyethyl), naphthol[2,3-*b*], furan-4, 9-dione, and isopinnatal: *Schwikkard et al. (2002) p.679*

TOXICITY:

The green fruits are said to be poisonous, causing blistering if ingested: *Watt et al. (1962),* Tests of fruit on Brine shrimp showed LC_{50} >2 and on Vero cell line 0.149: *Brendler et al., (2010) pp. 165, 286, Adoum (2009) p.136, Kela et al (1989)*

Kirkia acuminata Oliver [*K. pubescens* Burtt] (Kirkiaceae)

DESCRIPTION:

A medium-sized deciduous tree, 12m high or more, with corky grey bark. Widespread in hotter and drier areas; abundant in Escarpment woodland at lower altitudes, often with *Julbernardia globiflora*

VERNACULAR NAMES:

E: White Kirkia; **Afr**: Witsering; **B/ Ku/ Ny/ Tu**: muzumba (1, 10, 20) **I**: mupumena (19) **I/ K/ Lo/ Tk/ To**: muzumina (3, 3d, 4, 10, 14, 19, 22, 23) **I/ To**: muzuminwa (10) **I/ K/**

To: musanta (3, 3a, 4, 10, 14, 17, 19, 22, 23) **Le**: lemba, mutaba-taba (17) **Le/ Lo/ N**: mulemba (3d, 10, 14) **Ln**: mwiyombo (10) **Lo**: mulemba-lemba (3a, 10, 22), muzunge (3d, 10, 14) **Lv**: muyombo (10) **Ny**: mtumbu (9, 10), ntunduwa, ntungundwa (9), mtumbwi (3, 3e, 14), muzumbu (3a), mzumba (3e, 14), ntumbwa (20) **Tk**: musatha (8) **To**: mutaba (17)
PLANT PARTS USED:
whole plant
TRADITIONAL USES:
<u>Zimbabwe</u>: a sacred tree, used against malaria: *Watt et al. (1962) p. 941*
TESTS:
Spencer tests: score 0 / 0: *Spencer et al. p. 162* ; extracts of the plants have proved to be ineffective against experimental malaria: *Watt et al. (1962) p. 941*

Lagenaria sphaerica (Sond.) Naud. (Cucurbitaceae)
DESCRIPTION:
Vigorous climber with robust stems up to 10m long, with pendent green spherical fruit the size of a cricket ball, and creamy fragrant flowers opening in the evening. Found on margins of evergreen and deciduous forest, riverine woodland, river banks and dry beds, dry scrub thickets.
VERNACULAR NAMES:
E: Wild Calabash; **Afr**: Wildekalbas; **Ku**: katenge-nene (1) **Lo**: malaka (3d) **Ny**: cipuzi, cisyalere, sopa (9)
PLANT PARTS USED:
leaves
TRADITIONAL USES:
<u>Tanzania</u>: the leaf sap is drunk against malaria: *Haerdi (1964) p. 81*

Laggera crispata (Vahl) Hepper & Wood [*Blumea alata* (D.Don) DC.; *B. crispata* (Vahl.) Merxm] (Asteraceae)
DESCRIPTION:
Annual or perennial herb up to 2.4m high, woody at base, several-stemmed, much branched. White/ pink/mauve/purple florets.
VERNACULAR NAMES:
Not known
PLANT PARTS USED:
leaves
<u>Malawi</u>: the leaves are used as an enema to treat fever: *Gelfand et al.(1985) p. 233*

Landolphia owariensis P. Beauv. (Apocynaceae)
DESCRIPTION:
Liana up to 15m high, or scrambling shrub, or small tree up to 10m high, with greeny-violet, yellow or grey fruit. Found in riparian woodland, savannah
VERNACULAR NAMES:
E: White Rubbervine, White-ball Rubber; **B**: Mubongo
PLANT PARTS USED:
roots, green fruits
TRADITIONAL USES:
<u>Côte d'Ivoire</u>: The roots or green fruits are decocted to make a steam bath to treat fever pains: *Bouquet (1969) p. 64,* quoted in *Burkill (1985) p. 163*

Lannea discolor (Sonder) Engl. (Anacardiaceae)
DESCRIPTION:
Small tree, up to 15m high, with rounded crown. Widespread; open woodland, especially on rocky slopes, or on sandy soil, often on anthills
VERNACULAR NAMES:
E: Live-long; **Afr**: Dikbas; **B**: babumba, kayimbe (22), nakabumbu (3, 3c, 7a, 14), nakaumba (23) **B/ Ku/ Ny**: kaumbu (1, 3b, 3c, 6, 10, 14) **B/ Lb/ Lo/ Ny**: kabumbu (3, 3b, 3c, 7a, 9a, 10, 14, 20) **I/ To**: mubombo (12, 17) **I/ Le/ Lo/ Tk**: musamba (10,17) **I/ Ln/ Lo/ N/ To**: mubumbu (3b, 3d, 4, 5, 7a,10, 12, 14, 22) **Ku**: syaumbu (1) **Lo**: mukako (3d, 14) **Lv**: muyombo (6) **N**: muuti (10) **Ny**: ciyumbu, mdya-kamba, mdya-tungu (9), kaumba (10), kawumbu (3e, 7a), mbale (3e, 14) **Tk**: mububwa, mubumbucuulu, sikalulumbu (8) **To**: cibu-mbu (10, 19, 23), muŋongwa (3, 3b, 14) **Tu**: tondo-tondo (8)
PLANT PARTS USED:
roots
TRADITIONAL USES:
Zambia and Zimbabwe: an infusion of the roots is used to treat children's fevers: *Palgrave (1957) p. 8*

Lannea edulis (Sonder) Engl. (Anacardiaceae)
DESCRIPTION:
Suffrutex about 0.3m high, with semi-procumbent, woody, perennial branchlets from large woody rootstock, red fruit Oct-Nov; in grassland, vleis, open woodland, esp. after fires.
VERNACULAR NAMES:
E: Wild grape **B**: luwimbwa (3, 3c, 9a, 20, 23), mumbulu-mbunye (1, 3c) **I/ To**: cibumbu (4, 17, 19) **Lb**: umulamba-lamba (22) **Lb/ Lo/ Tk**: kabumbu (23) **Le/ To**: mulamba (17) **Ln**: mulya-nsyinga (3) **Lo**: mufunge, sikabasya (3d), mulolo (3) **Ny**: cilembusya, magenge, mbulu-bunje (3), mlamba (3, 3e), mbulu-mbunje (3e), ndya-kamba, ufukula (9) **To**: kasiko-kamuumbu (17), mumbu (8)
PLANT PARTS USED:
root
TRADITIONAL USES:
Tanzania: a root decoction is drunk as a malaria cure: *Haerdi (1964) p. 126*

Lantana camara L. var. aculeata (L.) Moldenke (Verbenaceae)
DESCRIPTION:
A noxious and invasive weed, cultivated as an ornamental plant in S. Africa, but forming impenetrable thickets over 2m high in the wild: *Sharma et al. (2007) pp. 314, 315*
VERNACULAR NAMES:
E: Tickberry, Cherry Pie; **Ny**: Mfika, Mlunguzi
PLANT PARTS USED:
leaves, twigs, flowers, roots
TRADITIONAL USES:
Cameroon: a leaf decoction is used to treat malaria: *PROTA (2008)*
Madagascar: the leaves and roots are infused to make a febrifuge; *Rasoanivao et al. (1992);* leaves & stems are infused to make a drink, and boiled to make a steam bath: *Novy (1997)*
Mozambique: the leafy branches are used to treat malaria: *Mulhovo (1999)*
Rwanda: an infusion of pounded leaves is made to treat malaria: *Hakizamungu et al. (1988)*

Tanzania: the roots are used for malaria, and said to be effective in cases which are not responsive to quinine: *Hedberg et al.(1983); Pernet (1957), quoted in Burkill (2000)*

U.S.A.: This is an anti-pyretic and anti-spasmodic agent, included in the U.S. National Standard Dispensary (1926): *Dalziel (1937); Oliver-Bever (1960) and (1983); Schnell (1960), Wong (1976), quoted in Burkill (2000)*

TESTS:

Flower and twig extracts gave negative results in early antimalarial tests: *Spencer(1947); Watt et al.(1962)*

In vitro tests on the root bark in 1990 and 1991 showed significant antimalarial activity, with an IC_{50} value of 5-9 μg/ml: *Nkunya et al.(1991); Weenen et al. (1990), cited by Sharma et al. (1998)*

In vitro tests of leaves and twigs in 2004 found moderate activity (IC_{50} 11μg/ml): *Clarkson et al. (2004).*

In vivo tests on leaves showed insignificant inhibition of parasitism in mice: *Hakizamungu et al. (1988)*

In vitro tests of dichloromethane leaf extracts in 2008 showed moderate activity (IC_{50}< 14 mg./ ml): *Jonville et al.(2008)*

TOXICITY:

Fruit pleasant, and enjoyed by children, although poisonopus to livestock: *Morris (1996).* The leaves contain a quinine-like alkaloid, *lantanin*, which may sensitize an animal to ultra-violet light, producing dermatitis and intestinal haemorrhage: *Fanshawe et al. (1967); Watt et al. (1962).* Cases of renal failure were detected among Malagasy patients taking *L. camara*, as reported at the 7th Pan-African National Malaria Control Programmes meeting at Casablanca, Sept 2008: *Randrianarivelojosia (2008).* The green unripe fruit is reported as toxic to humans: *Sharma et al. (2007)*

Lantana trifolia L. [*L. salvifolia* Jacq.] (Verbenaceae)

DESCRIPTION:

Widespread woody herb or subshrub up to 2m high, sparsely branched, with pink/ mauve flowers, purple fruits.

VERNACULAR NAMES:

E : Lavender Popcorn, Shrub Verbena; **Lo**: likobeza (3d), likobeza-balisana (3) **Ny**: naka-sonde, naka-nunkhu (9)

PLANT PARTS USED:

leaves

TRADITIONAL USES:

Burundi: the leaves are decocted, alone or mixed with those of *Chenopodium ambrosioides* & *Crotalaria laburnifolia*, or mixed with *C. ambrosioides, Guizotia scabra* & *Ludwigia abyssinica,* to make a drink: *Baerts et al. (1989)*

Tanzania: small children with malarial rigor are given leaf juice to drink: *Burkill (2000) p.261; Haerdi (1964) p. 152*

TOXICITY:

An ethanol extract of *L.trifolia* was tested on brine shrimps, showing only mild toxicity at LC_{50} 32.3, and a dichloromethane extract none at LC_{50} 756 mg/ml: *Moshi et al.(2010)*

Laportea aestuans (L.) Chew (Urticaceae)

DESCRIPTION:

Annual herb up to 1m high, with green ovate leaves covered with stinging hairs, and white

flowers.
VERNACULAR NAMES:
E: West Indian Woodnettle; **Ku**: cipopo-lambwe, mpala-cipopo (1) **Ny**: kaba-bambwa (1)
PLANT PARTS USED:
leaves
TRADITIONAL USES:
Côte d'Ivoire: a leaf decoction is used to relieve fever: *Bouquet (1974) p. 172,* quoted in *Burkill (2000) p. 238*

Launaea cornuta (Hochst. ex Oliv. & Hiern) C. Jeffrey (Asteraceae)
DESCRIPTION:
Perennial herb up to 1.3m high with yellow flowers, on alluvial soils and disturbed ground along rivers and roadsides
VERNACULAR NAMES:
E: Bitter Lettuce
PLANT PARTS USED:
roots
TRADITIONAL USES:
E. Africa: the Giriama people of the coastal area of Kenya use it to prevent and cure malaria, except in cases of stomach ulcers: *Schippers (2004)*

Leea guineensis G.Don (Leeaceae)
DESCRIPTION:
Shrub 1m high, found in shade in *Syzygium cordatum* swamp forest; bright yellow/orange/red flowers, and red fruits turning black.
VERNACULAR NAMES:
E: Burgandy
PLANT PARTS USED:
leaves
TRADITIONAL USES:
Tanzania: leaf juice is used to bathe the body as a remedy for malarial convulsions: *Haerdi (1964) p. 114*
TOXICITY:
The berries have an irritating sap, but no idication that this has ever been a problem.

Leonotis leonurus (L.) Ait.f. (Lamiaceae)
DESCRIPTION:
Shrubby herb up to 4m high, with hairy stems. Found in grassland between *Brachystegia* covered rocky hills
VERNACULAR NAMES:
E: Lion's Tail, Wild Dagga
PLANT PARTS USED:
leaves, roots, fruit
TRADITIONAL USES:
Mauritius: the plant is used as a fever remedy: *Terrac (1947)*
S. Africa: the Zulu use a cold infusion of the leaves as a nasal douche to relieve headache in fevers. They use the root, mixed with the root of or green fruit of *Strychnos spinosa* and other plants, as a febrifuge: *Watt et al. (1962) pp. 517, 730.*

Leonotis mollissima Gurke (Lamiaceae)
DESCRIPTION:
Semi-woody herb up to 3m high. Branched below, flowering shoots simple, stiffly erect
VERNACULAR NAMES:
B: mwikala-sosa (2, 3c), nakafungu (23) **Nm**: muhasi (1a,   22) **Ny**: nlonga-ndundu (9)
PLANT PARTS USED:
roots
TRADITIONAL USES:
Tanzania: a root decoction mixed with leaf juice is drunk as a remedy for malaria: *Haerdi (1964) p. 189*

Leonotis nepetifolia (L.) Ait.f. (Lamiacae)
DESCRIPTION:
A widespread pantropical herbaceous plant, in some areas regarded as a pest.
VERNACULAR NAMES:
E: Lion's Ear; **Ku**: Jeli-jeli; **Lo**: Mutuhu; **Ny**: Cidya-tsongwe, Mdya-tsongwe, Mnya-balame, Nlonga-ntundu, Nyasuna
PLANT PARTS USED:
aerial parts, leaves, twigs, flowers, roots, whole plant
TRADITIONAL USES:
Brazil: an infusion of the whole plant is drunk against malaria: *Di Stasis et al. (1994)*
Central Africa: the plant is used as a remedy for fevers: *De Wildeman (1949)*, quoted in *Watt et al. (1962)*
Madagascar: the aerial parts are decocted to make a febrifuge: *Rasoanaivo et al. (1992)*
W. Africa: a decoction of the plant is used to steam the head to relieve fever, and a leaf infusion is drunk: *Burkill (1995); Dalziel (1937)*
W. Indies, E. Indies, Brazil: the juice of the leaf is used against malaria and typhoid: *Dragendorff (1898)*, cited by *Watt et al.(1962)*
TESTS:
In Puerto Rico, the plant has a reputation as an anti-malarial, which is not supported by experiments: *Asenjo (1945, 1948)*, quoted in *Watt et al. (1962)*
In vitro tests of the plant in South Africa found it to be only moderately effective (IC_{50} 15μg/ml): *Clarkson et al. (2004)*
TOXICITY:
Greshoff reported that it was considered poisonous: *Greshoff (1913), cited by Watt et al. (1962)* ; tests on mice showed the LD_{50} value to be 3807.9mg/kg: *Ayanwuyi et al.(2009)*

Leucas martinicensis R. Br. (Lamiaceae)
DESCRIPTION:
Erect annual herb, up to 1m high. Unbranched stems, finely hairy, small white flowers in dense axillary whorls. Found in disturbed places, and a weed of cultivation.
VERNACULAR NAMES:
E: Bobbin weed; **Ny**: canzi, mcalamira, mcansira, nlonga-ndundu, nyamula-katundu (9)
PLANT PARTS USED:
leaves, whole plant, ash
TRADITIONAL USES:

Burkina Faso: fresh leaves are rubbed on a sick child once a day for 3 days: *de la Pradilla (1988) p.24*

W. Africa: the plant ash is used to repel mosquitoes. The whole plant is made into an infusion, used as a wash or steam fumigation for fevers: *Brotherton, p. 136,* quoted in *Burkill (1995) p.16; Dalziel (1937) p. 461*

TOXICITY:

The berries are edible when red, but the unripe green berries are toxic.

Lippia javanica (Burm. f.) Spreng. [*L. asperifolia* Rich.; *L. whytei* Moldenke] (Verbenaceae)

DESCRIPTION:

A weedy shrub occurring in secondary vegetation.

VERNACULAR NAMES:

E: Fever-tea, Wild Tea; **Afr**: Beukebos, Maagbossie; **Lo**: Likobeza, Likobeza-balisana; **Ny**: Canzi, Mcanzi, Mcenjema, Naka-nunkhu, Vumba

PLANT PARTS USED:

leaves, whole plant.

TRADITIONAL USES:

E. Africa: a leaf decoction is drunk, and used to bathe the body: *Kokwaro (1976)*

Mozambique: reported to be efficient in treating malarial symptoms: *Bandeira et al. (2001)*

S. Africa: an infusion is made of this plant plus *Artemisia afra* as a remedy for malaria: *Smith (1888),* quoted in *Watt et al. (1962).* Its use as a mosquito repellent is reported: *Govere et al. (2000).*

S. W. Africa: the plant is used as a malaria remedy by the Nunquois bushmen: *Codd (1951),* quoted in *Watt et al. (1962)*

Zimbabwe: the leaves are decocted and sugar added as a remedy for malaria: *Dornan (1916),* quoted in *Watt et al.(1962); Gelfand et al. (1985)*

TESTS:

In vitro tests on the roots in 2004 revealed a high degree of antiplasmodial activity (IC_{50} 3.8); the stems tested at IC_{50} 4.5µg/ ml): *Clarkson et al. (2004); Pillay et al. (2008)*

TOXICITY:

Avoided by grazing animals (*Codd 1951*) but eaten to some extent *(Dyer 1937), quoted by Watt et al. (1962).* It contains *icterogenin* and is suspected as a cause of geeldikop: *Steyn (1949).* Ingestion by stock causes photosensitisation: *Watt et al. (1962).* No reports of toxic effects on humans.

Ludwigia erecta (L.) Hara (Onagraceae)

DESCRIPTION:

Annual herb up to 3m high, wth small yellow flowers. Found on riverbanks, moist wastelands and other wet localities.

VERNACULAR NAMES:

Not known

PLANT PARTS USED:

Whole plant

TRADITIONAL USES:

E. Africa: the whole plant is boiled, and the malaria patient bathes in the water: *Kokwaro (1976) p. 170*

TESTS:

In Kenya in 2007, *in vitro* tests of water and methanol extracts showed high antiplasmodial activity ($IC_{50}<5mg/ml$) against chloroquine sensitive (D6) and resistant (W2) *Plasmodium falciparum* clones: *Muthaura et al. (2007)*

TOXICITY:

The above tests suggested that there is potential to isolate active non-toxic principles.

Maclura africana (Bureau) Corner [*Cardiogyne africana* Bureau] (Moraceae)

DESCRIPTION:

A widely-scrambling spiny shrub, with dark-green leaves arranged spirally on short spine-tipped lateral branchlets Found in hot and dry riverine vegetation and along edges of pans

VERNACULAR NAMES:

E: Thorny Mulberry, African Osage-orange; **Nm**: mnembuwa (1a) **Ny**: mamba-huru (7c) **To**: mutala (3)

PLANT PARTS USED:

roots

TRADITIONAL USES:

Tanzania: A root decoction is drunk, and scrapings of the wood eaten, and the body bathed with a pulverised bark infusion: *Haerdi (1964) p. 67*

Maerua triphylla A.Rich var. pubescens (Klotszch) De Wolf [*M.pubescens* (Klotszch) Gilg.] (Capparaceae)

DESCRIPTION:

Much-branched bush, scrambler or small tree, up to 5m high, found in low-altitude, dry woodland, and often on anthills in *miombo* woodland

VERNACULAR NAMES:

E: Small Beadbean; **B**: kafumbe

PLANT PARTS USED:

Twigs

TRADITIONAL USES:

Mozambique: the twigs are used to treat malaria: *Mulhovo*

Maesa lanceolata Forsk. (Maesaceae)

DESCRIPTION:

Shrub or much-branched tree, up to 15m high. Widespread as a pioneer in gallery forest, swamp forest, and montane forest

VERNACULAR NAMES:

E: False Assegai; **Afr**: Valsassegai; **B**: musangula, mutendefwa (3c) **Ny**: cikundu, kabalisi (3e), ciwombosi, cingu-ngu, manga-cule, mdenjele, mdyatso-ngwe, mkakama, mtanda-mpira (9), msanji-pila (3, 3e, 9)

PLANT PARTS USED:

fruit

TRADITIONAL USES:

Ethiopia, E. and S. Africa: The fruit is used by the Karanga people around Morgenster as a febrifuge, and elsewhere as a purgative in fever: *Bally (1937) p. 21; Bally (1938), Greenway (1941), Greshoff (1900), Sim (1907), quoted in Watt et al. (1962), p. 787; Dragendorff (1898) p. 515*

Mangifera indica L. (Anacardiaceae)
DESCRIPTION:
Widely planted as a fruit tree
VERNACULAR NAMES:
E: Mango; **B**: mawembe, miyembe (20), mwembe (3c) **B/ Ku/ Ny**: manga (1, 3e), yembe (7a) **M**: mangonye (20) **Ny**: mango (9)
PLANT PARTS USED:
bark, leaves
TRADITIONAL USES:
DRC Kinshasa: the fruits are used to make a malaria cure: *Tshiamuene et al.(1995)*
Ecuador: a decoction of bark and leaves is used against fevers: *Coe et al. (1996)* p. 96
Madagascar: the fruits or leaves are used to make malaria cures: *Pernet (1957)* a leaf decoction is used as a febrifuge: *Rasoanaivo et al. (1992) p.119*
W. Africa: a leaf-decoction is used as a febrifuge in Côte d'Ivoire and Nigeria: *Ainslie, sp. no. 221, Bouquet (1974) p. 16,* quoted in *Burkill (1985) p. 83*
TOXICITY:
The sap, leaves and green fruit of mangoes can cause dermatitis; cattle deaths have been recorded: *Watt et al.(1962)* p. 48

Manihot esculenta Crantz. [*M. utilissima* Pohl.] (Euphorbiaceae)
DESCRIPTION:
Laxly-branched shrub up to 4.5m high, with large tuberous roots, for which it is much cultivated, especially in Barotseland and L. Bangweulu Basin
VERNACULAR NAMES:
E: cassava, ayucca, manioc; **B**: kalundu (20) **B/ Ku**: kalundwe (1, 3c, 9a, 20) **B/ Ku/Lb/ Le/ Ny**: tute (1, 3c, 3e, 9a, 17, 19, 20) **I/ K/ Le/Ln/ N/ To**: makamba (4, 17, 19) **Le**: cindongo, sikamba(17) **Lo**: katetele, linangwa, lingunga, malose, mufuta (19), likamba (3d) **Lo/ Lv/ To**: mwanja (17, 19) **Lv**: lupa (19) **M**: kaliya (20) **M/ Tu**: mayawo (20) **Ny**: cikawu, nyimbula (3e), nyumbula (20) **Ny/ Tu**: cinangwa (3e, 20, vinangwa (20) **To**: mwanja (17) **Tu**: muhogo, vikau (20)
PLANT PARTS USED:
leaves, roots
TRADITIONAL USES:
Ecuador: a decoction of leaves and roots is used to treat fever: *Coe et al. (1996) p. 100*
TOXICITY:
The raw tubers contain highly toxic glycosides: *Watt et al.(1962) p. 423*

Manilkara mochisia (Baker) Dubard [*M. macaulayae* Hutch. & Corbishley; *M. umbraculigera* Hutch. & Corbishley; *Mimusops mochisia* Baker] (Sapotaceae)
DESCRIPTION:
Deciduous tree, up to 20 m high; sometimes with several trunks from common base, and root suckers; found in Central and Southern Provinces, and Zambezi Valley and adjoining plateau, often on termite mounds
VERNACULAR NAMES:
E: Lowveld Milkberry; **Afr**: Laeveldmelkbessie; **Ku/ Ny**: cipuwe (1, 3e), mugonambwa (1) **Lo**: mucisa (3, 3d, 23), muhumu (3d) **Ny**: mkabwa (3e) **Tk**: muuse (3, 3d, 23), muhumu (3d) **Ny**: mkabwa (3e) **Tk**: muuse (8) **To**: mupusyo (3) **Tu**: muugabwa (8)

PLANT PARTS USED:
whole plant
TRADITIONAL USES:
<u>Kenya</u>: a decoction is used by the Tugen people as a fever remedy: *Beentje (1994) p. 454*

Maprounea africana Muell. Arg. [*M. vaccinioides* Pax]
DESCRIPTION:
Small deciduous tree with spreading branches, 4-8m high, in open woodland, sandy soils, riverine fringe forest
VERNACULAR NAMES:
E: Magic nut; **Afr**: Towerneut; **B**: kafula-mume (3c, 14, 20, 23), mutumbwa (3c, 4), mupasa (3, 3c, 14, 20) **B/ Ln/ Lv**: kavula-mume (3, 6, 14, 22) **I**: kamwaya (4, 5, 23) **Ln**: kavulawuni (21) **Lo**: lumwamwa (3d, 14), mumwaa (3, 3d, 14, 16, 23) **Ny**: cigaga (3e, 14) **Tk**: mumwaya (8)
PLANT PARTS USED:
roots
TRADITIONAL USES:
<u>Mozambique</u>: the rootbark is used to treat malaria :*Mulhovo (1999); PROTA (2008) p.376*
TOXICITY:
The rootbark is lethal if taken in large amounts. The acetone extract showed high toxicity to mice:*PROTA (2008) p.375*

Margaritaria discoedia (Baill.) G. L. Webster [*Phyllanthus flacourtioides* Hutch.; *P. discoides* (Baill.) Muell. Arg.] (Euphorbiaceae)
DESCRIPTION:
Shrub or small tree up to 8m high in medium/low altitudes in wooded grassland or bushveld; in forests it can reach 20m or more
VERNACULAR NAMES:
E: Peacock-berry; **Afr**: Poubessie; **I**: katuba-tuba (4, 23) **Ln**: ?citakaci (21), kasangala (3) **Nm**: msenga, usenga (22) **Ny**: mcele-cete, mnawa, mpicila-nyambo, mtundu-waŋono, nakatiko, napala-pala, ntanda-nyerere (9) **To**: mule-nkanga (22), mulya-nkanga (3)
PLANT PARTS USED:
roots, leaves
TRADITIONAL USES:
<u>Tanzania</u>: root decoction mixed with leaf sap is used against malaria: *Burkill (1994) p. 115; Haerdi (1964) p. 95*
TESTS:
A study in Tanzania found that the root bark of this plant was active, with the active constituent being *securinine*: *Schwikkard et al. (2002) p. 682*

Microglossa pyrifolia (Lam.) O. Kuntze (Asteraceae)
DESCRIPTION:
Woody climber up to 10m high; sometimes flowering as a herb. Widespread, in evergreen forest, *cipya* woodland, and on kopjes
VERNACULAR NAMES:
B: matongo (3, 3c)
PLANT PARTS USED:
leaves

TRADITIONAL USES:

E. Africa: an infusion of leaves is drunk as a remedy for malaria. It causes vomiting as a side effect, being very bitter: *Beentje (1994) p. 558; Kokwaro (1976) p. 68*

W. Africa: A leaf infusion or decoction is drunk to treat malaria. A lotion is used as a diaphoretic. An enema made from the plant is used to treat fever in an infant: *Burkill (1985) p. 485; Dalziel (1937), quoted in Watt et al. (1962) p. 251*

TESTS:

A dichloromethane leaf extract was tested *in vitro* against chloroquine-resistant *P.falciparum* strain (W2). Activity of IC_{50} <15 mg/ ml was shown: *Muganga et al. (2010)*

TOXICITY:

The extract was also tested against human foetal lung fibroblasts (W1-38) *Muganga et al. (2010)*

Mikania chenopodiifolia Willd. [*M. cordata* (Burm. f.) B. L. Robins.] (Asteraceae)

DESCRIPTION:

A fast-growing creeping and twining perennial, up to 5m long, often flowering as a herb, which smothers its host plant. Found at forest margins and edges of swamps

VERNACULAR NAMES:

E: American Rope, Bitter Vine, Chinese Creeper, Climbing Hempweed, Mile-a-minute; **Ny**: civumulo, lyundumula, matholisa (9), kayalika (3e)

PLANT PARTS USED:

whole plant, leaves

TRADITIONAL USES:

Senegal: friction with a plant-decoction is administered to relieve fever pain: *Burkill (1985) p. 487; Kerharo (1974) p. 288*

Tanzania: the leaf-sap is drunk against malaria: *Burkill (1985) p. 487; Haerdi (1964) p. 170*

Zaire: vapour baths of a decoction of the whole plant, followed by friction using the lees, are used as a febrifuge: *Bouquet (1969) p. 95, quoted in Burkill (1985) p. 487*

Milicia excelsa (Welw.) C.C.Berg. [*Chlorophora excelsa* (Welw.) Benth.; *Maclura excelsa* (Welw.) Bureau] (Moraceae)

DESCRIPTION:

Large deciduous tree up to 50m high, found in low-altitude evergreen forest and wet savannah

VERNACULAR NAMES:

E: African Teak

PLANT PARTS USED:

leaves

TRADITIONAL USES:

Central Africa: a leaf decoction is used as an anti-febrile: *De Wildeman (1949) p. 1,* quoted in *Watt et al. (1962) p. 772*

Zambia: the leaves are used to make a febrifuge: *Amico (1977) p. 118*

TOXICITY:

The sawdust can be irritant to woodworkers

Mimulus gracilis R.Br. (Scrophulariaceae)

DESCRIPTION:

Erect/ascending stiffly erect perennial/ annual herb 4 – 35cm high with white/ mauve

flowers; found in upland savanna
VERNACULAR NAMES:
E: Monkey Flower
PLANT PARTS USED:
whole plant
TRADITIONAL USES:
Lesotho:the plant makes a lotion to bathe feverish patients: *Burkill (2000) p. 76; Watt et al. (1962) p. 938*

Mollugo nudicaulis Lam. (Molluginaceae)
DESCRIPTION:
Annual herb with a rosette of prostrate leaves and a flowering stem up to 25cm high, found in grassy moist savanna
VERNACULAR NAMES:
E: Naked-stem Carpetweed
PLANT PARTS USED:
whole plant
TRADITIONAL USES:
Burkina Faso: a small bunch is crushed and steeped for up to 5 hours to make a wash, used on infants twice daily for 3 days: *de la Pradilla (1988) p.25*

Momordica balsamina L. (Cucurbitaceae)
DESCRIPTION:
A vine of woodlands, wooded grasslands, riverine fringes, and sandy soils up to 1600m
VERNACULAR NAMES:
E: African cucumber, Balsam apple, Balsamina, African pear; **Ny**: Mwalisaka; **To**: Nkaka.
PLANT PARTS USED:
whole plant, fruit
TRADITIONAL USES:
Mozambique: frequently used to treat malaria, being effective in controlling vomiting: *Bandeira et al., (2001)*
W. Africa: the plant is much used as a wash for fever: *Burkill (1985); Kerharo (1973); Watt et al. (1962)*
TESTS:
In vitro tests on the stem found a high degree of antiplasmodial activity (IC$_{50}$ µg/ml=5.3): *Clarkson et al.(2004).* Further tests in 2006 found it to be effective in combination with two other plants: *Benoit-Vical et al. (2006)*
TOXICITY:
Tests in 2008 of stem and flower extracts on Chang liver and adipose cells "raise concern for chronic use" with a toxicity score -2: *Van de Venter et al. (2008)*

Momordica charantia L. (Cucurbitaceae)
DESCRIPTION:
A climber with bitter, rugged courgette-like fruits.
VERNACULAR NAMES:
E: Bahama pear, Bitter gourd, Bitter melon, Carilla plant, Wild cucumber; **Afr**: Bitterappel; **Ny**: Kari, Likulupsya.
PLANT PARTS USED:

aerial parts, leaves, twigs, stem, whole plant.
TRADITIONAL USES:
Brazil: a leaf decoction or infusion is used throughout the Amazon area to treat malaria, being drunk 3x daily until the patient is cured: *Brandão et al. (1992); Di Stasis et al. (1994); Milliken (1997)*
Burkina Faso: 3 bunches of leafy twigs are boiled for 2 min. in 10 lit. water, used as a wash twice daily for 2 days: *de la Pradilla (1988)*
Colombia: a leaf decoction or infusion is used to treat malaria, being drunk 3x daily until the patient is cured: *Garcia-Barriga (1992),* cited in *Milliken (1997)*
Ecuador: a decoction of leaves and stem is given to treat malaria: *Coe et al.(1996)*
Guyana: a leaf decoction or infusion is used to treat malaria, being drunk 3x daily until the patient is cured: *Lachman-White et al. (1987),* cited in *Milliken (1997)*
Madagascar: the aerial parts are decocted to treat malaria: *Rasoanivao et al. (1992)*
Nigeria:the leaves are crushed and the juice diluted with water, & drunk with onion and salt: *Adjanohoun et al. (1991)*
W. Africa: the plant is commonly used as a febrifuge, either as a wash or mixed with palm wine or drinking water: *Ainslie, sp. no. 232, Irvine (1930),* quoted in *Burkill (1985); Kerharo (1973)*
W. Indies: a leaf decoction or infusion is used to treat malaria, being drunk 3x daily until the patient is cured: *Ayensu (1981),* cited in *Milliken (1997)*
Zaire: the leaves are crushed in the hands and massaged into the body for fever pains: *Bouquet (1969),* quoted in *Burkill (1985)*
TESTS:
In vivo tests of the leaves revealed no antimalarial activity*: Amorim et al. (1991)*
In vivo tests on mice revealed no antimalarial acivity during a 5-day test: *Ueno et al. (1996)*
In vivo tests showed that *M. charantia* was ineffective in a 4-day suppressive test in mice (ED70=904mg/kg/day): *Munoz et al. (2000)*
An 80% methanol extract of the whole plant was tested in vitro against the chloroquine-sensitive Ghanaian strain of *P. falciparum: Mesia et al. (2007)*
TOXICITY:
Administration of *M. charantia* extract up to 800mg/kg body weight is safe...& tolerated by the body: *Abalaka et al.(2009)*
For BST tests on stem bark, see *Adoum (2009) p. 137*

Momordica foetida Schum. & Thonn. (Cucurbitaceae)
DESCRIPTION:
Perennial herb with woody rootstock, and stems trailing or climbing up to 4.5m. Smells awful when crushed. Bright orange fruit with red seeds. Found in forest edges, woodland, wooded grassland and riverine fringes, and in disturbed land.
VERNACULAR NAMES:
Ny: cikhaka, cisyalere (?), kacitose, kamoto, liku-lupsya, mkungunjila, namgoneka, tungwi (9)
PLANT PARTS USED:
roots
TRADITIONAL USES:
Ethiopia: the people of Dek Island, L.Tana, boil leaves and stems in water, & inhale the steam at bedtime: *Teklehaymanot (2009)*

<u>Togo</u>: stems and leaves infused for several hours at 100g/ lit then drunk, or the intertwined vines used as sponges to bathe skin with the infusion: *Beloin et al. (2005)*

<u>Paraguay</u>: the root is used as a remedy for fever: *Watt et al. (1962) p. 364*

TOXICITY :

Grazed by cattle in Sudan, but avoided as poisonous in Kenya: PROTA (database 2007)

Monanthotaxis buchananii (Engl.) Verdc. (Annonaceae)

DESCRIPTION:

Shrub, liana, or small tree 1.5 – 7.5m high, found in gulley forest and groundwater forest

VERNACULAR NAMES:

Not known

PLANT PARTS USED:

roots

TRADITIONAL USES:

<u>Tanzania</u>**:** a root decoction is drunk as a malaria remedy: *Haerdi (1964) p. 38*

Morinda lucida Benth. (Rubiaceae)

DESCRIPTION:

Tree (rarely a shrub) 2.4 – 18m high, sometimes planted round villages

VERNACULAR NAMES:

E: Brimstone Tree

PLANT PARTS USED:

Aerial parts, stem bark, root bark

TRADITIONAL USES:

<u>Congo-Brazzaville</u>: bark from stem, branches & trunk is decocted to make a steam bath to treat mataria patients: *Bouquet (1969)*

<u>W.Africa</u>: Used to treat malaria and other diseases: *Schwikkard et al. (2002) p. 686*

TESTS:

Leaf extracts were found to inhibit more than 60% of *P.falciparum* when tested *in vitro*: *Tona et al.(1999) pp. 193-201.*

Both *Koumaglo et al.* and *Sittie et al.* isolated active *anthraquinones* from *M. lucida,* the most active being *damnacanthal: Schwikkard et al. (2002) p. 686*

TOXICITY:

No danger reported: *Oduola et al.(2010)*

Moringa oleifera Lam. [*M. pterygosperma* Gaertn.] (Moringaceae)

DESCRIPTION:

Deciduous, laxly-branched tree up to 10m. high, whose foliage appears whitish from a distance; found in Gwembe Valley

VERNACULAR NAMES:

E: Horse-radish Tree; **To**: zakalanda (23)

PLANT PARTS USED:

root, bark, leaf, flower, seed & gum

TRADITIONAL USES:

<u>Africa & Madagascar</u>: the exudate, the roots and the bark are used as a remedy for fever: *Githens (1948) p. 97; Watt et al. (1962) p. 781*

<u>India</u>: the root, bark, leaf, flower, seed & gum are used to treat fever.

TESTS:

Extracts have given negative results against experimental malaria: *Watt et al. (1962) p. 781*
TOXICITY:
Tests of seed extract on mice showed LD$_{50}$ 20-50: *Gruenwald et al. (2000), cited by Brendler (2010) p.*

Mucuna poggei Taubert (Fabaceae , Papilionoideae)
DESCRIPTION:
Liana with woody stems 15 – 18m high, found in evergreen forest. Pods densely rusty-tomentose
VERNACULAR NAMES:
E: Buffalo Bean, Itchy Bean, Hellfire Bean; **B**: sepe (3, 3c) **I**: lubabangwe (4) **Ny**: citedze, citeya, likwanya, namalopa (9)
PLANT PARTS USED:
roots
TRADITIONAL USES:
Tanzania: root shavings are decocted and drunk as a cure for malaria: *Burkill (1995) p. 406; Haerdi (1964) p. 61*
TOXICITY:
Any potentially harmful alkaloids can be eliminated or made safe by proper processing: *Tuleun et al.(2008)*

Mucuna pruriens (L.) DC. (Fabaceae , Papilionoideae)
DESCRIPTION:
Twining plant, up to 11m high, densely covered with long irritant hairs; found at Lake Tanganyika, nr. Crocodile Island, and in the Gwembe Valley
VERNACULAR NAMES:
E: Buffalo Bean, Itchy Bean, Hellfire Bean; **B**: kaenya (3c), kaminda (3, 3c) **B/ Ku/ Ny**: bukasi, nkhasi (1) **B/ Lb**: sepe (3c, 23) **Ku/ Ny**: mpezi, nkhaze (1) **Lo**: kayuwe (3d), kayuye (23) **Lo/ To**: muyuyu (3, 3d) **Ny**: citedze (3, 9), likwanya, namalopa (9) **Tk/ To/ Tu**: mpesi (8) **To**: mutimbya-ndavu (17)
PLANT PARTS USED:
leaves
TRADITIONAL USES:
W. Africa: the leaf sap is used as a febrifuge: *Berhaut (1976) pp. 432-4, Bouquet (1974) pp. 136, 138,* quoted in *Burkill (1995) p. 407; Kerharo et al. (1950) p. 122*
TOXICITY:
See under *Mucuna poggei*

Mukia maderaspatana (L.) M. J. Roem. (Cucurbitaceae)
DESCRIPTION:
Hairy climber or trailer up to 2.5m from perennial woody rootstock, spreading over vegetation in damp woodland, grassland, and riverine margins
VERNACULAR NAMES:
Ku: kazimbili-nkonde, njolole (1) **Ku/Ny**: kazewe-lezya
PLANT PARTS USED:
seeds, whole plant
TRADITIONAL USES:

<u>Burkina Faso</u>: 2 bunches of the plant are boiled for 10 min. in 5 lit. water, and used twice daily for 3 days to wash sick infants: *de la Pradilla p.26*

<u>Nigeria and India</u>: the seeds are chewed or decocted to cause sweating: *Ainslie sp. no. 62, Sastri (1952) p.336,* quoted in *Burkill (1985) p. 601*

TOXICITY:

Information from Dept.Animal Health, Zambia, states that it is deadly poisonous to grazing cattle: *Trapnell CRS 400, RBG Kew*

Myrothamnus flabellifolius (Sond) Welw. (Myrothamnaceae)

DESCRIPTION:

Aromatic, prostrate, ascending or erect shrub up to 2m high, with much-branched twigs. Locally abundant on dry, rocky hills up to 2000m

VERNACULAR NAMES:

E: Resurrection Bush; **B**: camba-lupili (3, 3c) **Ny**: canasa, cisoni (9)

PLANT PARTS USED:

leaves

TRADITIONAL USES:

<u>Malawi</u>: a leaf infusion is used to treat fevers: *Morris (1996) p. 416*

Neorautanenia mitis(A. Rich.) Verdc. (Fabaceae , Papilionoideae)

DESCRIPTION:

Erect, scrambling or climbing shrubby perennial with a large tuberous root and blue/violet flowers. Found in woodland, grassland and old lands, often in rocky places and vlei margins

VERNACULAR NAMES:

Lo: tinde (3d) **Ny**: nakangunde, nankobwe (9)

PLANT PARTS USED:

tuber

TRADITIONAL USES:

<u>Zimbabwe</u>: the fever patient sits in an infusion of the tuber: *Gelfand et al.(1985) p. 150*

TESTS:

In studies carried out in 2003, extracts from tubers of *N. mitis* exhibited activity aganst larvae of *Anopheles gambiae* and *Culex quinquefaciatus* mosquitoes. The activity of the crude extracts was comparable to those of the standard mosquitocides *deltamethrin* and *alphcypermethrin,* and the findings support the use of the plant for the control of mosquitoes, esp. at breeding sites: *Joseph et al. (2004)*

Nymphaea lotus L. [*N. zenkeri* Gilg.] (Nymphaeaceae)

DESCRIPTION:

A bulb plant producing submerged and floating leaves, with striking white flowers. Found in stagnant water from lakes to small pools

VERNACULAR NAMES:

E: Tiger Lotus, White Lotus, Egyptian White Water-lily; **Tk**: makwangala (8)

PLANT PARTS USED:

rhizome

TRADITIONAL USES:

<u>Philippines</u>: the rhizome is infused to make a refreshing drink as a remedy for fevers: *De Wildeman (1948) p. 5,* quoted in *Watt et al. (1962) p. 802*

Ochna macrocalyx Oliv. (Ochnaceae)
DESCRIPTION:
Usually a rhyzomatous shrublet, with only leaves and orange/yellow flowers showing above ground
VERNACULAR NAMES:
E: Dwarf Ochna; **Ny**: mlanda-mpete, mlungamu, mpelele, mpeuma, ngunda-nguluwe (9)
PLANT PARTS USED:
roots
TRADITIONAL USES:
Tanzania: a root decoction mixed with leaf sap is drunk: *Haerdi (1964) p. 100*

Ochna schweinfurthiana F. Hoffm. (Ochnaceae)
DESCRIPTION:
Small tree found in open deciduous woodland
VERNACULAR NAMES:
E: Brick-red Ochna; **B**: coni (9a), iconi (3c), munawe (1, 23) **Lb**: munyango, mwiyenzela (23) **Ln**: musengu (9a) **Nm**: kawa-tundwe (7c) **Ny**: ciwombola, kanzoki, mlungamu, msimpa, napose, ngunda-ndanguluwe (9), mconi, mlanga-mpete, mtanta-mpete, patwe (3e) **To**: ngulamabele (23)
PLANT PARTS USED:
twigs, bark
TRADITIONAL USES:
Burkina Faso: 2 bunches of leafy twigs are boiled for 5 minutes in 10 lit. water, to be used as a wash twice daily for 3 days: *de la Pradilla p.26*
Tanzania: pap made from dried pulverised bark is given for malaria: *Haerdi (1964) p. 101*

Ocimum americanum Sims [*Ocimum canum* L.] (Lamiaceae)
DESCRIPTION:
An erect, branching herb up to 50cm, common on cultivated land, wastelands, and open *Brachystegia* woodland
VERNACULAR NAMES:
E: Hairy Basil; **B/Ku**: lwena, mununka; **Ku**: mukupa-imbu; **Nm**: msumba-mpungu; **Ny**: canzi, kafavumba, leyani, manunka, mpungabwe, msinyani, nunkamani
PARTS USED:
plant, aerial shoots, flowers, leaves, sap, stems
TRADITIONAL USES:
Brazil: the plant, leaves, and flowers are used as a diaphoretic: *Dragendorff (1898)*, quoted in *Watt et al. (1962); Di Stasis et al. (1994)*
Ghana: a leaf decoction is drunk, sap squeezed into the patient's eyes, and the plant is put into bathwater. The leaves are rubbed on the body: *Irvine (1930)*, quoted in *Burkill (1995)*
India: A decoction of the leaves mixed with black pepper is given to pregnant women with malaria: *Singh et al. (1993)*
Malawi: an infusion of the leaves is used as a remedy for fevers. The plant is burnt in a hut to repel mosquitoes: *Morris (1996); Williamson (1975)*
Madagascar: the stems and seeds are decocted to make a malaria remedy: *Rasoanaivo et al. (1992)*
Nigeria: the leaves are chopped up and eaten as a febrifuge: *Ainslie, sp. no. 248*, quoted in *Burkill (1995)*

<u>Rwanda</u>: the leaves are used to treat malaria: *Van Puyvelde et al.(1975)*
<u>Zambia</u>: a leaf infusion is used for vapour treatment: *Vongo (1999)*
TESTS:
In vitro tests on the whole plant showed a high antiplasmodial activity of $IC_{50}=4.2\mu g/ml$:
Clarkson et al. (2004); Pillay et al. (2008)
TOXICITY:
No reports of toxic effects on humans

Ocimum gratissimum L subsp. gratissimum [*O.gratissimum var. suave* (Willd.) Hook f.; *O. urticifolium* Roth].
DESCRIPTION
Herb or shrub up to 4m.
VERNACULAR NAMES
E: African Basil; **B**:biswa; Lo: litandamwe, lwenye, mukubamwe, saminami; **Lv**: mofatela-khethee
PLANT PARTS USED
Leaves
TRADITIONAL USES
<u>Angola</u>: a leaf infusion is used as a febrifuge and sudorific: *Watt et al. (1962)*
<u>Ghana</u>: a leaf infusion is taken by draught or enema to treat fevers: *Irvine (1961),* quoted in *Burkill (1995)*
<u>Côte d'Ivoire</u>: a leaf decoction is taken in baths and draughts for coughs and fever: *Burkill (1995); Kerharo et al. (1950)*
<u>Nigeria</u>: 26% of rural practioners in Edo State used a leaf (?) infusion in a steam bath to treat malaria, making this plant by far the most widely used there: *Obuekwe (n.d.)*
Leaves are powdered with salt & gunpowder(!) to make a decoction for drinking: *Adjanohoun (1991)* cited in *Prelude med.plants database (2011)*
<u>Rwanda</u>: the leaves are used to treat malaria: *Van Puyvelde et al.(1975)*
<u>West Africa</u>: hot infusions of the leaves are used to treat fevers: *Ainslie (1937), Irvine (1930), 37, (1961), Iwu (1986), Oliver-Bever (1960); Walker (1953),* quoted in *Burkill (1995); Burtt Davy et al. (1937); Dalziel (1937)*
TESTS:
Spencer tests: 0 / 0: *Spencer et al. p. 163*
TOXICITY:
No reports of toxic effects on humans

Ocotea usambarensis Engl. (Lauraceae)
DESCRIPTION:
Large tree up to 22m, with stout unbranched 10m bole. Rare and local in evergreen seepage forest
VERNACULAR NAMES:
E: East African Camphor
PLANT PARTS USED:
roots
TRADITIONAL USES:
<u>E. Africa</u>: the roots are pounded and soaked in water, the infusion being drunk as a remedy for malaria: *Kokwaro (1976) p. 115*

Olea europaea L. subsp. africana (Mill.) P. S. Green [*O. africana* Mill.; *O. chrysophylla* Lam.] (Oleaceae)
DESCRIPTION:
Small, much-branched evergreen tree up to 20m. Only at Victoria Falls in forest on gorge sides
VERNACULAR NAMES:
E: Olive; **Ny**: nakatimba (9)
PLANT PARTS USED:
roots, leaves, bark
TRADITIONAL USES:
Kenya: the Wanderobo and Kipsigis people use a root or bark decoction as a malaria remedy: *Beentje (1994) p. 472*
S. Africa: the leaves and bark have been used as a remedy for "intermittent fever". The resin has been used as a febrifuge: *Watt et al. (1962) p. 808*

Oncoba spinosa Forsk. (Flacourtiaceae)
DESCRIPTION:
Much branched spinous shrub or small tree up to 13m high; widespread, often in fringing forest.
VERNACULAR NAMES:
E: Snuff-box Tree, Fried-egg Flower; **Afr**: Snuifkalbassie; **B**: musombo (3c, 14), nzombo (1) **B/ Ku**: muzombo (1) **K**: musakalala (23) **Ku**: matanta, mutanta (1) **Lo**: kakoma (3, 3d, 14), munze-neze (14), mukoma (19), mulangu (3, 3d, 14, 19, 23), munze-nze (3d) **Ny**: daza, dzorba, matseka, msece, msewe, mtawa (9), mlaza (3, 3e, 14), mpamalaza, msese, mtanta (3e, 14), ncowana, umthu-mgwa (22) **Tk**: musangu-sangu (8, 23) **To**: mukumbuzu (3, 14), munsangwa, musakalala (14) **Tu**: musese (8)
PLANT PARTS USED:
seeds, leaves
TRADITIONAL USES:
Nigeria: the seed-oil is drunk as a fever remedy: *Ainslie sp. 253,* quoted in *Burkill (1994) p. 161*
Tanzania: the leaf sap is drunk as a malaria cure: *Haerdi (1964) p. 71*

Opilia amentalea Roxb. [*O. celtidifolia* (Guill. & Perr.) Endl. Ex Walp.; *O. tomentella* (Oliv.) Engl.] (Opiliaceae)
DESCRIPTION:
Scandent evergreen shrub, widespread in higher-rainfall areas, mostly in *mutesyi* forest
VERNACULAR NAMES:
B: itawa-tawa (3c), mulele (3, 3c) **Ny**: kalizakulu (3e)
PLANT PARTS USED:
Roots, leaves
TRADITIONAL USES:
E. Africa: the roots are infused or decocted as a remedy for fever: *Kokwaro (1976) p. 170*
Mali: a leaf decoction is drunk 2x daily and applied as wash for 3 days; leaves are decocted with those of *Trichilia emetica* and a teacup drunk 2x daily for 3 days, unless side-effects caused; the leaves are boiled with the fruit-juice of *Tamarindus indica* and filrate drunk until healed:the leaves are crushed in water to make a bath to treat neuralgic malaria: *Togola et al. 2005)* cited in *Prelude med. plants database (2011)*; a leaf decoction is drunk; another

recipe prescribes the decoction for a body wash, warning against drinking, as causing diarrhoea:*Grøenhaug et al., (2008)* cited in *Prelude med. plants database (2011)*
TOXICITY:
Used as a fish poison: *Adjanohoun et al.(1979)*. Tests on molluscs showed cytotoxic activity: *Kela et al. (1989)*. See also *Grøenhaug et al.(2008)*

Ottelia ulvifolia (Planch.) Walp. (Hydrocharitaceae)
DESCRIPTION:
Submerged aquatic in muddy pools, yellow/white flowers just above water
VERNACULAR NAMES:
Not known
PLANT PARTS USED:
leaves
TRADITIONAL USES:
Central African Republic: the leaves are mashed and made into an embrocation against fever: *Vergiat (1977) p. 88,* quoted in *Burkill (1994) p. 412*

Ozoroa reticulata (Baker f.) R. & A. Fern. [*Heeria reticulata* (Baker f.) Engl.; *O. insignis* Delile subsp. *reticulata* (Baker f.) J. B. Gillett)] (Anacardiaceae)
DESCRIPTION:
Much-branched tree up to 13m high, with dense rounded crown, found in *cipya* and *miombo* woodland
VERNACULAR NAMES:
E: Raisin Bush, Tar Berry Resin-Tree; **Afr**: Teerbessieharpuisboom; **B**: munembwa (9a, 10, 14), musombo (10, 23), nama-bele, namu-wale (9a, 10) **B/ Lb**: mabele (10, 14, 23) **I/ Ln/ Lo/ Tk/ To**: mulilila (3, 4, 8, 10, 14, 16, 19) **K**: camakupa (10) **Ku**: cifiko (1) **Ln**: nya-monga (14) **Ln/ Lo/ Tk**: mulilela (3, 8, 10, 16) **M/ Ny**: masimya (9, 10) **Nm**: kala-kala, mkala-kala, mlago (22), mkala (1a, 22) **Ny**: cifiko, cifitye, mlunga, mnembwa (3e, 14), kanyera, mbewe, nulambwe (9), mtuku-mpuko (3), nama-sira (9, 10) **To**: cafica (22), cimpempe (14), muambwe (23), musambwe (10, 14) **Tu**: msimbiti, mtuka-mbako (10)
PLANT PARTS USED:
roots, root-bark
TRADITIONAL USES:
Tanzania: to treat malaria, the roots are eaten raw, or the powdered root-bark is infused and drunk like tea: *Burkill (1985) p. 84; Haerdi (1964) p.125*
TOXICITY:
Brine shrimp tests showed toxicity at LC_{50} 2.21mg/ml: *Moshi et al.(2004)*

Panicum maximum Jacq. (Poaceae)
DESCRIPTION:
Perennial tufted grass with short, creeping rhizome, up to 2m high. Likes shady damp areas.
VERNACULAR NAMES:
Ku: mapuya, nkwinde (1) **I**: namununke (4, 19) **I/ Tk**: kafuwa-kaswi (8) **I/ Tk/ To**: civuse (4, 8) **Le**: namuncaca (19) **Lo**: nandu-ndu (3d), simuhupu (19) **Tu**: navwale (8)
PLANT PARTS USED:
seeds
TRADITIONAL USES:
Zambia: the seeds are used as a febrifuge: *Amico (1977), p. 134*

Paralepistemon shirensis (Oliv.) Lejoly & Lisowski [*Turbina shirensis* (Oliv.) A. Meeuse] (Convolvulaceae)
DESCRIPTION:
Robust climber reaching up to 15m, with woody hairy stems and white flowers. In woodland, dry riverine forest, grassland, roadsides, sandy soils
VERNACULAR NAMES:
To: civavani; **Tu**: cikuli
PLANT PARTS USED:
roots
TRADITIONAL USES:
Mozambique: the root is pounded and mixed with porridge: *Mulhovo (1999)*

Parinari curatellifolia Planchon ex Benth. [*P. mobola* Oliver](Chrysobalanaceae)
DESCRIPTION:
Medium-sized tree, us. c.9m high, with dense rounded crown. Widespread in many types of woodland, often left standing in African villages as shade and fruit tree
VERNACULAR NAMES:
E: Mobola Plum; **Afr**: Grysappel; **B/ K/ Lb/ Le/ Ln/ Ny/ Tu**: mupundu (3, 3c, 9a, 10, 14, 19, 20, 23) **B/ Ku**: mpundu (1) **I**: ibula (4), tufukusa (17) **I/ Lo/ M/ Nm/ Tk/ To**: mubula (1c, 3, 3a, 3d, 4, 9a, 10, 14, 16, 19, 20, 22) **Ln/ Lv**: muca (3, 9a, 10, 14, 22) **Le**: tukupu-twambula (17) **Ln**: muhota-hota, mwica (6, 22) **Lv**: muhota (6, 22) **M/ Ny/ To/ Tu**: muula (3, 3e, 9, 9a, 10, 14, 20) **N**: mukolwa (10) **Nm**: mnaziya-polini, mwula (22) **Nm/ Ny/ To**: mbula (3e, 22) **Ny**: cakate, mpembu (9), mpundu (3e, 14)
PLANT PARTS USED:
leaves, bark
TRADITIONAL USES:
Côte d'Ivoire: the leaves are decocted and used both as a drink and a lotion to ameliorate fever: *Burkill (1985) p. 383; Kerharo et al. (1950) p. 90*
Malawi: an infusion of bark is used as a remedy for malaria: *Morris (1996) p. 246*
Tanzania: a bark decoction is used as a remedy for malaria: *Brenan et al.(1949), De Wildeman (1949) p. 1*, quoted in *Watt et al. (1962) p. 890; Burkill (1985) p. 383; Smith (2004) p.71*
Zambia: a decoction of bark is used as a remedy for malaria: *Nair (1967) p. 34*
TOXICITY:
"The high LD_{50} value (7.27gm/kg) obtained was a clear indication that *P. curatellifolia* herbal preparations could be safe for use" : *Ogbonnia et al.(2009)*

Parkia filicoidea Welw. ex Oliver [*P. bussei* mwensa (20) (Fabaceae, Mimosoideae)
DESCRIPTION:
Large tree up to 16m high, buttressed at base. Found in evergreen fringing forest at Lake Mweru
VERNACULAR NAMES:
B: musepa (3c, 14, 20, 23) **Ny**: mkundi (2, 3, 3e, 9, 14), mpeza, msenya (3e, 14), mtanga-tanga (9) **Ny/ Tu**: mpesa (20)
PLANT PARTS USED:
bark
TRADITIONAL USES:

Tanzania: a bark decoction is drunk against malaria: *Haerdi (1964) p. 50*
TOXICITY:
Poor storage can lead to production of toxic mould.

Paullinia pinnata L. (Sapindaceae)
DESCRIPTION:
A vine with white flowers and green/ red fruit, widespread in evergreen and mixed forest
VERNACULAR NAMES:
K: Kasisi; **Ku**: Cikumba; **Ln**: Cifui; **Nm**: Mgogote; **Ny**: Cika, Mapirano, Mkanda-nkhuku, Mkanda-nyalugwe, Mwana-mwamfepo.
PLANT PARTS USED:
whole plant, leaves, leaf sap, twigs, roots, root bark
TRADITIONAL USES:
Burkina Faso: 3 bunches of leafy twigs are boiled for 15 min. in 10 lit. water; half a glass drunk, and used as a wash, twice daily for 3 days: *de la Pradilla (1988)*
Côte d'Ivoire: a plant decoction is drunk, used as a lotion, and steamed as a vapour bath, to reduce fever pains: *Burkill (2000); Kerharo et al. (1950)*
E. Indies: roots and root bark are applied as a rubeficient over the liver to relieve induration after malaria: *Dragendorff (1898), Greshoff (1913), quoted in Watt et al.(1962)*
Malawi: an infusion of the leaves is drunk as a remedy for malaria: *Morris (1996); Oliver-Bever (1986), quoted in Burkill (2000)*
Mali: an infusion or macerate of the leaf or leaf sap is used to treat fever: *Irvine (1961), Oliver-Bever (1983), quoted in Burkill (2000); Dalziel (1937); Kerharo (1963,1974)*
Senegal: a preparation of leafy twigs is taken as a febrifuge: *Burkill (2000); Kerharo (1963)*
Tanzania: leaf sap is drunk as a malaria remedy: *Burkill (2000) p. 29; Haerdi (1964)*
Congo: the leaves boiled in bathwater are prescribed for feverish children: *Bouquet (1969), quoted in Burkill (2000)*
TOXICITY:
Widely used as a fish poison, but no reports of toxic effects on humans in spite of widespread and prolonged use

Pavetta crassipes K. Schum. (Rubiaceae)
DESCRIPTION:
Small sparsely-branched tree up to 6m high, found in open woodland on shallow stony soils
VERNACULAR NAMES:
Ku: mcoka (1) **Lb**: musyasya (23) **Le**: tinana-na (23) **Ny**: likaka-lyaŋombe, lilime-lyaŋombe, anjatali, mpewu, mteru-teru (9), matuwa-ngoma, mdya-nsefu (3e), mtangoma (3) **To**: mufundu (23)
PLANT PARTS USED:
roots, leaves
TRADITIONAL USES:
Malawi: a root decoction is used as a fever remedy: *Morris (1996) p. 461*
Tanzania: the pulverised leaves are mixed with sap to treat malaria: *Haerdi (1964) p. 142*
TESTS:
Alkaloid extracts were tested *in vitro* against chloroquine resistant and sensitive strains of *P.falciparum*, and produced significant inhibitory activity of IC_{50} 1.23 mg/ml: *Sanon et al.(2003)*

TOXICITY:
The above test also demonstrated weak cytotoxicity against three human cell lines.

Pavetta sp.
DESCRIPTION:
VERNACULAR NAMES:
not known
PLANT PARTS USED:
roots
TRADITIONAL USES:
Tanzania: a root decoction is drunk against malaria: *Haerdi (1964) p. 143*

Peltophorum africanum Sond. (Fabaceae, Caesalpinioidea)
DESCRIPTION:
Small graceful tree with spreading crown, 5 – 10m high; widespread in central areas, often in open high grass woodland on rejuvenated soils
VERNACULAR NAMES:
E: African Wattle; **Afr**: Huilboom; **B**: mwenza (10) **B/ K**: mwikala-nkanga (3, 3a, 3c, 10, 14, 17, 23) **B/ Ku**: citeta (1) **I**: mukasala (10) **I/ Le/ Lo/ Tk/ To**: munyeele (3, 3a, 4, 8, 10, 17, 22, 23) **I/ To**: mungele (19) **K**: mponjo, syimwakasala (10) **Ku**: citumbula-mala (1) **Lb**: mbali-mbali, ipunga-ŋombe (22) **Le**: muyeele (17) **Ln**: mwezengele (10), mwezenyele (3, 14) **Lo**: mulungwa, musyande (3d, 14), musenu (10) **M**: nyaka-mbalilo (10) **N**: masyande (10) **Ny**: mnyele (3, 14), mteta (3e, 14), muteta (3a), nyele (3e) **To**: muzenze-nze (3, 14) **Tu**: punyele (22), utete (10)
PLANT PARTS USED:
roots
TRADITIONAL USES:
Zimbabwe: a root infusion is drunk as a sudorific: *Gelfand et al.(1985) p. 136.*
TOXICITY:
In vitro tests of leaf, bark & root extracts on brine shrimps & Vero monkey kidney cells showed no toxicity: *Bizimenyera et al.(2007)*

Pentanisia prunelloides (Klotsch) ex Eckl. & Zeyh.) Walp. subsp. latifolia (Hochst.) Verdc. (Rubiaceae)
DESCRIPTION:
Beautiful blue grassland flower, stout hairy stems from woody rootstock.
VERNACULAR NAMES:
E: Wild Verbena, Broad-leaved Pentanisia; **Lb:** insanki (22)
PLANT PARTS USED:
roots
TRADITIONAL USES:
S. Africa: the Zulus use a hot decoction of the root as a fever remedy: *Watt et al. (1962) p. 902*

Pentas purpurea Oliv. (Rubiaceae)
DESCRIPTION: erect perennial herb <1.3m, in grassland & open woodland, purple/blue flowers
VERNACULAR NAMES:

Not known
PLANT PARTS USED:
juice
TRADITIONAL USES:
E. Africa: the juice is used as a febrifuge: *Bally (1937) p. 22*
TOXICITY:
Reported to be poisonous to sheep: Watt et al.(1962)

Pergularia daemia (Forssk.) Chiov (Apocynaceae)
DESCRIPTION:
Perennial twining herb with milky sap, up to 4m. Found in woodland and riverine forest fringes
VERNACULAR NAMES:
E: Trellis-vine
PLANT PARTS USED:
whole plant
TRADITIONAL USES:
Nigeria: the plant is used in combination with others to treat fever: *Dawodu,* quoted in *Burkill (1985) p. 235*
TOXICITY:
Highly toxic in all its parts, due to the presence of numerous *cardenolides* & *cardenolide glycosides*: *PROTA (2010)*

Phragmites australis (Cav.) Steud. [*P.communis* Trin.] (Poaceae)
DESCRIPTION:
Perennial grass with short, creeping rhizomes, widespread and locally dominant on sandy alluvium fringing seasonal watercourses
VERNACULAR NAMES:
B: lutete (20) **B/K/Ny/To/Tu**: matete (3, 19, 20) **I**: ibu (4), lubu (4, 19) **K**: sjitete (19) **Lo**: litaka (19), mutaka (3, 19)
PLANT PARTS USED:
roots
TRADITIONAL USES:
W.Africa: the roots are used as a fever remedy: *Burkill (1994) p. 488*

Phyllanthus fraternus G.L.Webster (Euphorbiaceae)
DESCRIPTION:
Herbaceous weed of roadsides and wastes
VERNACULAR NAMES:
E: Gulf Leaf flower
PLANT PARTS USED:
leaves
TRADITIONAL USES:
Côte d'Ivoire: the pounded leaves are used to relieve fever pains: *Bouquet (1974) p. 85,* quoted in *Burkill (1994) p. 121.*
W. Indies: the plant has a reputation as a malaria remedy: *Burkill (1994) p. 121; Dalziel (1937) p. 157.* However, tests have given negative results.
TESTS:

Negative results have been recorded in early experiments: *Spencer et al. (1947); Kerharo (1974) pp. 427-8*

Sittie et al. isolated from this plant two alkamides, *(E,E) –octa-2,4-dienamide*, and *(E,Z) – deca-2,4-dienamide*, with moderate antiplasmodial activity: *Schwikkard et al. (2002) p. 681*

TOXICITY:

Used as fish-poison – see reference above

Phyllanthus muellerianus (O.Kuntze) Exell (Euphorbiaceae)

DESCRIPTION:

Glabrous shrub or woody climber, widespread in tropical Africa

VERNACULAR NAMES:

B: kapele (3), mupetwa-lupe (3, 3c, 7a, 23), musamba-mfwa, mweyema (3c) **Ln/ Lv**: mulya-sefu (6) **Ny**: kapika-nduzi (3e, 7, 7a), lungwisyi, mkuza-ndola, mpika (3e) **To**: mununka-calisya (3, 7a)

PLANT PARTS USED:

leaves, roots

TRADITIONAL USES:

<u>Côte d'Ivoire</u>: leaf sap is used as a wash for fevers: *Burkill (1994) p. 122; Dalziel (1937) pp. 157-8*

<u>Nigeria</u>: a root decoction is used as a febrifuge: *Ainslie sp. 272,* quoted in *Burkill (1994) p. 122*

<u>W. Africa</u>: the leaves and roots are used as a fever remedy: *Burkill (1994) p. 488*

<u>Zambia</u>: an infusion of leaves is used to bathe the body in cases of fever: *Haapala et al.(1994) p. 22*

TESTS:

An ethanolic extract of the plant was tested *in vitro* against the chloroquine-resistant FcB1 strain of *P. falciparum*, and showed good antiplasmodial activity together with weak cytotoxicity (IC_{50} 2.3 – 13.7 mc/ml): *Zirihi et al (2005)*

TOXICITY:

 See above.

Phyllanthus pentandrus Schumach. & Thonn. (Euphorbiaceae)

DESCRIPTION:

Common weed of cultivated sandy soil

VERNACULAR NAMES:

Ku: kauluzi (1) **Le**: mbwebe (22)

PLANT PARTS USED:

leaves, roots

TRADITIONAL USES:

<u>Nigeria</u>: the leaves and roots are decocted and drunk at frequent intervals as a fever remedy: *Ainslie sp. 273,* quoted in *Burkill (1994) p. 124; Watt et al. (1962) p. 427*

Phyllanthus reticulatus Poir. (Euphorbiaceae)

DESCRIPTION:

Weak-stemmed shrub or scrambler up to 6m, widespread, and often forming thickets

VERNACULAR NAMES:

B/ Ku/ Ny: kapulula-mbuzi (1, 3e) **Ku/ Ny**: kapululu (1, 3e) **Lo**: cizeze, pisape (3d) **Ny**: kabaka-bala, lungwe, mdya-pumbwa, ntanda-nyere (9), ntanta-nyelele, ntanta-yelele (1) **To**: mwicecele (3)
PLANT PARTS USED:
leaves
TRADITIONAL USES:
Tanzania: the mashed leaves are rubbed on the body as a malaria remedy: *Burkill (1994) p. 125; Haerdi (1964) p. 96*
TESTS:
Leaf extracts were tested *in vitro* against the K76 chloroquine- sensitive and ENT36 chloroquine-resistant strains of *P. falciparum*, and proved active, at $IC_{50}<10$ mc/ml: *Omulokoli et al. (1997)*
TOXICITY:
The root & fruit were used in Tanganyika for "criminal poisoning": *Brenan et al.(1949)*

Phyllocosmus lemairianus(De Wild. & Th. Dur.) Th. & H. Dur. [*Ochthocosmus lemaireanus* De Wild. & Th. Durand] (Ixonanthaceae)
DESCRIPTION:
Small, much-branched evergreen tree up to 8m high, with corky bark. Widespread in higher-rainfall areas, us. in *miombo* woodland
VERNACULAR NAMES:
B: cilumbwe-lumbwe (3c), kampolo-pombwe (3, 3b, 3c), kampombwe (3, 3b, 3c, 14, 23), mulumbwe (20), mumpombwe (20), mwembembe (3c, 14) **B/ K/ Ln/ Lo**: mulumbwe-lumbwe (3, 3b, 3c, 14, 19, 20) **B/ Ku**: kapolo-pombwe (1) **B/ Lb**: mulumbwe (20, 23) **B/ Le**: musenga-meno (14, 17) **I**: ilumba-lumba, mulumba-lumba (4, 5) **I/ To**: mulya-pwele (4, 14, 17, 19, 23), mwiinya-meenzyi (17) **Ku**: kasokopya (1) **Ln/ Lv**: cikuku, mutuku-mwisi (6, 14) **To**: munozya-meenda, musozi-wabeembela (17) **Ny**: kapulula (14)
PLANT PARTS USED:
leaves, roots
TRADITIONAL USES:
Tanzania: leaf sap and a root decoction are drunk as a malaria cure: *Haerdi (1964) p. 90*

Physalis angulata L.(Solanaceae)
DESCRIPTION:
Annual herb up to 4ft high with oval green/violet leaves and yellow edible fruits
VERNACULAR NAMES:
E: Cutleaf Ground-cherry, Cape Gooseberry, Winter Cherry, Wild Tomato
PLANT PARTS USED:
leaves, whole plants
TRADITIONAL USES:
Brazil and Peru: plant infusions are used to treat fever: *Di Stasi (1994), Milliken (1997) pp. 22, 25*
Nicaragua: a leaf infusion is taken to treat fever and malaria: *Coe et al. (1996) p. 105*
Nigeria: a leaf infusion is taken as a remedy for fever: *Ainslie sp.no. 274,* quoted in *Burkill (2000) p.116*
TESTS:
In vitro tests on aqueous and methanolic leaf extracts of the plant against *P. falciparum* produced a very high activity (IC_{50} <3mg/ml). The extracts were also active *in vivo* when

tested on mice infected by *P. berghei*: *Lusakibanza et al. (2010)*
TOXICITY :
Tests on animals have revealed no potential dangers for humans

Phytolacca dodecandra L'Herit. (Phytolaccaceae)
DESCRIPTION:
Scandent shrub up to 7m high with poisonous orange/red fruit. Found in *miombo* forest
VERNACULAR NAMES:
E: Gopo Berry, African soapberry; **B**: ipoko (7a) **Nm**: hoko (1a) **Ny**: mpombwe (9)
PLANT PARTS USED:
leaves, roots
TRADITIONAL USES:
Ethiopia: the plant is used to treat malaria: *Tadeg et al. (2005)* cited in *Prelude med. plants database (2011)*
S. Africa: the Zulu make an emetic from the leaves and root as a fever remedy: *Watt et al. (1962) p. 837*
TOXICITY:
Used as poison to control fresh-water snails: *Molgaard et al. (2000), Karoonamoorthi et al.(2008).* Tests on berry extracts showed no toxic effects except for irritation to the eyes: *Lambert et al.(1991)* Showed as cytotoxic when tested on snails at 50-100 ppm: *Elsheikh et al.(1990)*

Piliostigma thonningii (Schumach.) Milne- Redh. [*Bauhinia thonningii* Schumach]
(Fabaceae, Mimosoideae)
DESCRIPTION:
Small/medium tree up to 10m high, with spreading branches and pendulous fruits. Widespread: often in grassy woodland on rejuvenated soils
VERNACULAR NAMES:
E: Camels-foot, Monkey-bread; **Afr**: Kameelspoor; **B**: cibumbe (20), mumfumbe (3, 3b) **B/ Ku/ Ny**: citimbe (1, 3c, 3e, 9, 10, 20, 23) **B/ Lb/ M**: mufumbe (3, 3b, 3c, 10, 14, 20, 23) **B/ M**: cifumbe (3c, 20) **I/ Ln/ Lo/ Ny/ Tk/ To/ Tu**: musekese (1c, 3, 3b, 4, 5, 10, 17, 19, 20, 22, 23) **K**: kafumbe (23), kifumbe (3, 3b, 10, 14) **Ln**: kanungi (3, 3b, 14), mulolo-waunene-wacikulu (10) **Lo**: mubaba (3, 3b, 10, 14, 22), musyeke-sye, mwababa (3d) **N**: mulamata (10) **Nm**: mfundwa-mboga, mtinda-mbogo, mtindwa-mbogo (22), msyinda-mbogo (1a, 22) **Ny**: mfumbe (3e), msyowa (10, 20), mtukutu (3e, 23) **Ny/ Tk/ Tu**: msekese (3, 3b, 3e, 8, 9, 14, 20) **Tu**: mutukutu (10)
PLANT PARTS USED:
bark, leaves, roots, fruit
TRADITIONAL USES:
E. Africa: the bark is macerated and infused to treat malaria: *Aubreville p. 215, Baumer p. 110,* quoted in *Burkill (1995) p.148*

Mali: the leaves are infused or the tips of flower-buds are decocted to remedy fevers and malaria: *Malgras (1992)*

Nigeria: roots and fruits in proportion of 2:1 are decocted in water and boiled to concentrate. The liquid is drunk to treat fever: *Bhat (1990), p. 384*

Zambia: the bark and leaf are used as a febrifuge: *Amico (1977) p. 115*

Pistia stratiotes L. (Araceae)
DESCRIPTION:
Invasive floating plant often forming large dense mats
VERNACULAR NAMES:
E: Water-cabbage, Water-lettuce; **B**: bombwe (3c) **B/ Ku**: tena-tena (1) **Ku/ Ny**: kambu-mbu (1) **Lv**: lungwe (6, 22) **Ny**: kabu (3e)
PLANT PARTS USED:
whole plant
TRADITIONAL USES:
Zambia: a decoction of the plant is used to treat fever: *Amico (1977) p. 132*
TOXICITY:
Only toxic if large quantities eaten: *Russell et al. (1997)*

Pittosporum viridiflorum Sims subsp. ripicola (J. Leonard) Cufod. [*P. malsanum* Bak.] (Pittosporaceae)
DESCRIPTION:
A small tree or large shrub, up to 10m. high, found in riverine or evergreen mountain forest
VERNACULAR NAMES:
E: Cheesewood, White Cape beech; **Afr**: Bosbeukenhout, Bosboekenhout, Kaasuur, Kersuurboom, Stinkbas, Witboekenhout, Zeepbas; **Ny**: mkunguni, mkunguti, mscera.
PLANT PARTS USED:
bark
TRADITIONAL USES:
E. Africa: the bark is decocted and the liquid stirred into soup as a remedy for malaria and other fevers. The liquid is bitter and induces vomiting: *Kokwaro (1976)*
S. Africa: the Zulu and Xhosa use a bark decoction or infusion as a remedy for feverish conditions: *Githens (1948), Watt et al. (1962)* . A piece of bark, about 3" by 2", is pounded and steeped in a pint of boiling water. This is drunk, and then sufficient plain water to excite vomiting is drunk. Sometimes a double quantity of the bark is used to make an enema: *Bryant (1909); Grace et al. (2003).* The stem bark is used against chest complaints and malaria: *Iwu (1993)*
TESTS, ANTI-MALARIAL:
In vitro tests on the whole plant revealed a very high degree of antiplasmodial activity IC_{50} = 3:mg/ml *Clarkson et al. (2004)*
TOXICITY:
No reports of toxic effects on humans, apart from the use of the bark as an emetic above. Methanol and water extracts were cytotoxixc (SI, 0.96 – 2.51) on Vero E6 cells: *Muthaura et al., (2007)*

Pityrogramma calomelanos (L.) Link var. aureoflava (Hook.) Weath. ex F. M. Bailey (Pteridaceae)
DESCRIPTION:

Fern having fronds with golden undersides
VERNACULAR NAMES:
E: Silver Fern, Gold Fern, Dixie Silverback Fern
PLANT PARTS USED:
fronds
TRADITIONAL USES:
Trinidad: the infused fronds are used to make a tea, drunk as a fever remedy: *Wong p. 109,* quoted in *Burkill (2000) p. 462*
TOXICOLOGY:
As an arsenic hyperaccumulating fern, contamination is possible: *Francesconi et al., 2002*

Platostoma africanum P. Beauv. (Lamiaceae)
DESCRIPTION:
Slender prostrate/erect branched annual herb up to 60cm tall, found in damp savanna and waste places
VERNACULAR NAMES:
Not known
PLANT PARTS USED:
leaves
TRADITIONAL USES:
Ghana: the leaf sap is instilled into the eyes for fever: *Irvine, pp. 345-6,* quoted in *Burkill (1995) p. 28; Dalziel (1937) p. 463*
Nigeria: it is used in the same manner as *Ocimum* species: *Burkill (1995) p. 28; Dalziel (1937) p.463.*

Platycerium elephantotis Desv. (Polypodaceae)
DESCRIPTION:
Staghorn fern with dark green fronds up to 1.5ft long
VERNACULAR NAMES:
E: Cabbage Fern, Elephant Ear
PLANT PARTS USED:
sap
TRADITIONAL USES:
Tropical Africa: the sap is drunk as a fever remedy: *Burkill (2000) p. 560*

Plectranthus cylindraceus Hochst. ex Benth. (Lamiaceae)
DESCRIPTION:
Erect woody herb with light green leaves and pale mauve flowers, with strong menthol scent
VERNACULAR NAMES:
E: Vick's Plant, Mentholatum Plant; **Tk**: lumole (8)
PLANT PARTS USED:
leaves
TRADITIONAL USES:
E. Africa: a decoction of leaves is drunk, together with an infusion from *Microglossa parvifolia,* as a remedy for fever: *Kokwaro (1976) p. 113*

Plectranthus laxiflorus Benth. (Lamiaceae)
DESCRIPTION:

Coarse scrambling/erect glandular herb up to 10'; copious loose panicles of violet flowers; smells strongly of lemons when crushed; in openings in montane forest & scrub
VERNACULAR NAMES:
E: White Spur Flower
PLANT PARTS USED:
leaves
TRADITIONAL USES:
<u>S. Africa</u>: the Zulu use the powdered leaves as an enema for the relief of feverishness: *Watt et al. (1962) p. 525*

Plicosepalus kalachariensis (Schinz) Danser [*Loranthus kalachariensis*: Schinz] (Loranthaceae)
DESCRIPTION:
Aerial perennial succulent parasitic shrub with bright red fruit reaching 1m
VERNACULAR NAMES:
E: mistletoe
PLANT PARTS USED:
leaves
TRADITIONAL USES:
<u>Tanzania</u>: the leaf sap is drunk against malaria: *Haerdi (1964) p. 109*

Plumbago zeylanica L. (Plumbaginaceae)
DESCRIPTION:
A small straggly shrub with white flowers, found in deciduous thicket, anthills or mopane woodland.
VERNACULAR NAMES:
E: Ceylon leadwort, Leadwort, White-flowered leadwort; **Afr**: Swartwaterbossie; **I, Tk**: kalutenta; **Ku/ Ny**: sikalilo; **Lo**: sikalutenta; **Ny**: mtendasiwa; **Tu**: thenda-sipa
PLANT PARTS USED:
whole plant
TRADITIONAL USES:
<u>India</u>: the herb is used as a sudorific: *Watt et al.(1962)*
<u>Nigeria</u>: it is used as a fever remedy: *Gelfand et al.(1985)*; the roots are pounded & mixed in water with black soap & a cobra's head to drink: *Adjanahoun et al. (1991)* cited in *Prelude med. plants database (2011)*
TESTS, ANTI-MALARIAL:
In vitro tests of the leaves in 2001 and 2004 revealed a high level of antiplasmodial activity (IC_{50}= 3 µg/ml) *Clarkson et al. (2004); Simonsen et al. (2001); Pillay et al. (2008)*
TOXICITY:
The long succulent roots contain an irritant poisonous juice, *plumbagin*: *Watt et al. (1962), Fanshawe et al. (1967)*. Brine shrimp tests showed no toxicity in an ethanol extract at LC_{50} 232.3 mg/ml: *Moshi et al. (2010)*

Pollichia campestris Aiton (Caryophyllaceae)
DESCRIPTION:
Small, much-branched sprawling shrublet, found in open woodland, often on sandy soils.
VERNACULAR NAMES:
E: Waxberry; **Afr**: Suikerteebossie; **To**: kafuwa-kanswi

PLANT PARTS USED:
roots
TRADITIONAL USES:
E. Africa: a decoction of roots is drunk as a remedy for malaria, vomiting and diarrhoea being side effects: *Kokwaro (1976) p. 51*
TOXICITY:
See above

Polycarpaea corymbosa (L.) Lam. (Caryophyllaceae)
DESCRIPTION:
Erect annual herb, usually unbranched, found on sandy soils in open woodland and grassland, sometimes as a weed of cultivated ground.
VERNACULAR NAMES:
E: Old Man's Cap; **Lo**: mwamfu (3d)
PLANT PARTS USED:
whole plant
TRADITIONAL USES:
Guinea: an infusion is used to wash down fever patients: *Burkill (1985) p. 344*

Polygala persicariifolia DC. (Polygalaceae)
DESCRIPTION:
Annual slender-stemmed herb up to 60cm high, found in forests, grassland, roadsides on hillslopes
VERNACULAR NAMES:
E: Knotweed-leaved Milkwort
PLANT PARTS USED:
whole plant
TRADITIONAL USES:
Tanzania: the plant is macerated and mixed with *Biophytum helenae* pulp; the mixture is drunk and also used to bathe the body to relieve malarial spasms: *Haerdi (1964) p. 78*

Portulaca oleracea L. (Portulacaceae)
DESCRIPTION:
A weed of roadsides, gardens, railway lines
VERNACULAR NAMES:
E: Common Purslane, Garden Purslane, Pigweed; **Afr**: Misbredie, Porselein; **Ny**: nyele-nyele
PLANT PARTS USED:
Aerial shoots; whole plant
TRADITIONAL USES:
Brazil: the aerial shoots are decocted to make an antipyretic: *Di Stasis et al. (1994)*
Colombia: the plant is used to treat malaria: *Blair (1991),* cited in *Milliken (1997)*
North Africa: purslane has been used as a sudorific: *Watt et al. (1962)*
Zambia: the plant is used as a sudorific: *Amico (1977)*
TOXICITY:
Moderately toxic, with LD_{50} values of 1853.5 to 1875.5mg/kg: *Musa et al. (2007)*

Protea madiensis Oliv. subsp. madiensis (Proteaceae)
DESCRIPTION:

Shrub or small tree up to 5m high; found on Nyika Plateau, and Ika hills
VERNACULAR NAMES:
B: musoso (23) **I**: mulemu (4) **Lo**: muzungi (8) **Nm**: mzaza (22) **Tu**: ciyele (8)
PLANT PARTS USED:
roots
TRADITIONAL USES:
<u>Tanzania</u>: a root decoction mixed with leaf sap is drunk as a remedy for malaria and headache: *Haerdi (1964) p. 73*

Pseudolachnostylis maprouneifolia Pax [*P. dekindtii* Pax] (Euphorbiaceae)
DESCRIPTION:
Small deciduous tree, up to 13m high, with dense rounded crown; widespread in open woodland, esp. on sandy soil or rocky slopes
VERNACULAR NAMES:
E: Kuduberry, Duikerberry; **Afr**: Koedoebessie; **B/ M**: musangati (3, 3b, 3c, 10, 14, 20, 23) **B/ Ny/ Tu**: musolo (1, 20) **K**: musole (3, 3b, 14) **K/ Ln**: kabala-bala (3, 3b, 6, 23) **Ln/ Lv**: musalya (6) Lo: mangile, mokungu (22), mubu (3d), mukunu (16) Lo/ Tk: mukunku (8/ 19) **Lo/ To**: mukunyu (3, 3b, 3d, 14) **Nm**: mtunguru (22) **Ny**: condi, likulu-psya, mbwa-nyanya, mjenjete (9) **Ny/ Tu**: msolo (3, 3b, 3e, 9, 10, 14, 20) **Tk**: mukunthu (8) **Tu**: iputila-mpanda (7c), lubebya (10, 20)
PLANT PARTS USED:
leaves
TRADITIONAL USES:
<u>Zimbabwe</u>: a leaf decoction is drunk as a remedy for fever: *Gelfand et al.(1985) p. 167*

Psidium guajava L. (Myrtaceae)
DESCRIPTION:
A shrub or small tree, up to 10m. Widely used as fruit and medicine.
VERNACULAR NAMES:
E: Guava; **Afr**: Koejawel; **B**: Mapela; **Ku/ Ny**: Guaba; **Ny**: Makepera
PLANT PARTS USED:
leaves, resin
TRADITIONAL USES:
<u>Colombia</u>: an infusion of the leaves or "resin" is taken for malaria: *Milliken (1997a)*
<u>Nicaragua</u>: a decoction or infusion of the fruit and leaf is used to treat malaria: *Coe et al.(1996)*
<u>Java</u>: the "resin" is used as a febrifuge: *Watt et al.(1962)*
<u>Malawi</u>: a leaf decoction is used against malaria: *Morris (1996)*
<u>South Africa</u>: Interviews with traditional healers in Kwa-Zulu Natal showed this plant to be one of the most frequently used in treating malaria: *Nundkumar et al. (2000)*
<u>Zimbabwe</u>: a leaf decoction is drunk against fever: *Gelfand et al. (1985)*
TESTS:
The antimalarial activity of a 2% extract of the resin has been demonstrated by *Blair (1991)* and *Garcia-Barrega (1992),* cited by *Milliken (1997a)*
In vitro tests on the leaves in 1994 revealed antimalarial activity (IC_{50} 10 μg/ml): *Gessler et al. (1994)*

In vitro tests on stem-bark extract in 2002 showed IC$_{50}$ values of 10-20 μg/ml. Phytochemical analysis revealed the presence of flavonoids, terpenoids, anthraquinones and seccoirridoides: *Nundkumar et al. (2002); Pillay et al. (2008)*
TOXICITY:
The fruit, seeds, leaves and root contain hydrocyanic acid: *Watt et al. (1962), p. 1145*
Tests in 2008 of leaf and root extracts on Chang liver and adipose cells "raise concern for chronic use" with a toxicity score -2: *Van de Venter et al. (2008) p. 84*

Psorospermum febrifugum Spach (Guttiferae)
DESCRIPTION:
Shrub or small tree up to 9m high, scattered through open woodland over wide range of altitudes
VERNACULAR NAMES:
E: Christmas Berry; **B/ Ku/ Lb**: katumbi (1/ 10) **Ku/ Tu**: kavundula (1, 8) **Lb**: kapota (10, 23), mufilu (23) **Ln**: muhota-hota (6, 22), **Lo**: mubemu (3d) **Lv**: muhota (6, 22) **M**: mfifi (10) **Ny**: ciwungu, mfele, msilanyama, mkosomola, mtsiloti (9), mdima (9, 10), mziloti (3e) **Tk**: kabuye (8) **Tu**: mulosio (22)
PLANT PARTS USED:
Whole plant, roots
TRADITIONAL USES:
Angola: the trunk bark is pounded & decocted to make tea: *Bossard (1996)* cited in *Prelude med. plants database (2011)*
Burundi: the leaves are decocted in banana beer: *Polygenis-Bigendako, M.-J. (1990)*, cited in *Prelude med. plants database (2011)*
E. Africa: the plant is "apparently febrifuge": *De Wildeman (1949) p. 2,* quoted in *Watt et al. (1962) p. 498.* The roots and exuded latex are used to treat fever: *Githens (1948) p. 101*
Ivory Coast: the plant is used to treat fever: *Bouquet et al.(1974)* cited in *Prelude med. plants database (2011)*
Malawi: A decoction of the roots is used as a remedy for fever: *Morris (1996) p. 320*
W. Africa: the plant is used as a fever remedy: *Dalziel (1937) p. 88; Oliver-Bever (1960) pp. 35, 79, Walker p. 206,* quoted in *Burkill (1994) p. 404*
TESTS :
Spencer tests: score 0 / 0: *Spencer et al. p.162*
TOXICITY :
Brine shrimp tests showed toxicity LC$_{50}$ 12.69mg/mg/ml: *Moshi et al. (2006)*

Psychotria kirkii Hiern [*P. petroxenox* K. Schum.] (Rubiaceae)
DESCRIPTION:
Small shrub with white/ greenish flowers and red/orange fruit. In woodland savanna, often near rocks or anthills
VERNACULAR NAMES:
B: citapa-tapa, namulilo-wakasyika (3c), mumbwe, namulilo (3) **Tk**: musozya-wabasibombe (8)
PLANT PARTS USED:
roots
TRADITIONAL USES:
Tanzania: a root decoction is drunk against malaria: *Haerdi (1964) p. 144*

Psychotria zombamontana (Kuntze) E. M. A. Petit [*P. meridio-montana* E. M. A. Petit; *Grumilea kirkii* Hiern] (Rubiaceae)
DESCRIPTION:
Shrub or small tree, with white/ creamy yellow flowers and yellow/ red fruit. In understorey of evergreen forest and along margins
VERNACULAR NAMES:
E: Red Bird-berry; **Ny**: cipeta, kanyole, mkolongo, mpimbi-nyolo, msasaula
PLANT PARTS USED:
bark
TRADITIONAL USES:
Malawi: a bark infusion is used for fever: *Morris (1996) p. 462*

Pterocarpus angolensis DC. (Fabaceae, Papilionoideae)
DESCRIPTION:
Small-medium deciduous tree up to 20m high; widespread in different woodland types
VERNACULAR NAMES:
E: Mukwa, Bloodwood; **Afr**: Kiaat; **B**: mupyka-kulu (3c) **B/ Ku/ Lo/ Ny/ Tu**: mulombwa (1, 2, 3, 3a, 3c, 9a, 10, 14, 20, 23) **I**: ibanga, umbanga (4, 5, 17) **I/ Ku/ Lb/ Le/ Lo/ N/ Ny/ Tk/ Tu**: mulombe (3a, 3d, 8, 10, 16, 17, 19, 20) **I/ Lo**: mukwa (2, 3, 3a, 3d, 4, 5, 9a, 10, 23) **K**: mukula-kula (2, 14), ndombe (19) **K/ Ln/ Lv/ To**: mukula (2, 3, 3a, 6, 9a, 10, 19, 21, 23) **Ku**: mbangozi, mulome (1) **Lb**: tulemba (19) **Le/ To**: muzwa-mulowa (3a, 9a, 10, 17) Lv: mukala (22) M: mtumbati (10) M/ Nm: muva-malopa (10, 22) N: mulumbe (10) **Ny**: kasakula, mbangozi, ndombwa (3e), mlombe (3e, 14), mlombwa (3e, 9, 10, 14), mtumbali (9), mulomba (9a)
PLANT PARTS USED:
roots
TRADITIONAL USES:
Zambia, Zimbabwe:
A root decoction is used against malaria: *Palgrave (1957) p. 332; (2002) p. 390*

Pterolobium exosum (J.F.Gmel.) Baker f. [*P.stellatum* (Forsk.) Brenan; *Cantuffa exosa* J.F. Gmel.] (Fabaceae, Caesalpinioideae)
DESCRIPTION:
Many-stemmed climbing shrub up to 15m, with thorny stems, often forming dense thickets; in forest margins, riverine bush, bushveld, anthills
VERNACULAR NAMES:
E: Redwing; **Afr**: Rooivlerk; **Tu**: kombora-kombora (8)
PLANT PARTS USED:
leaves
TRADITIONAL USES:
E. Africa: the leaves are used as a febrifuge: *Bally (1937) p. 16*
Tanzania: the Chagga eat the leaves, fresh or dry, with butter, as a fever remedy: *Watt et al. (1962) p. 642*

Pupalia lappacea (L.) A. Juss. (Amaranthaceae)
DESCRIPTION:
Annual/ perennial herb, erect or sprawling, found in sandy soils
VERNACULAR NAMES:

Lo: lilamatwa, mukakabwa (3), luwele (3d)
PLANT PARTS USED:
leaves
TRADITIONAL USES:
Tanzania: leaves are used as a febrifuge: *Irvine, pp. 359-60,* quoted in *Burkill (1985) p. 62*

Ranunculus multifidus Forssk. (Ranunculaceae)
DESCRIPTION:
A perennial herb up to 2m. tall, widespread in tropical and southern Africa, in damp places of submontane grassland.
VERNACULAR NAMES:
E: Buttercup; **Afr**: Botterblom, Kankerblare, Rhenoster; **To**: Kapupuba; **Tu**: Sansamwa.
PLANT PARTS USED:
bark, whole plant
TRADITIONAL USES:
S. Africa: Europeans in the Louis Trichardt district mixed the ashes of the plant with beef or mutton fat, and rubbed the ointment on the fontanelle of a feverish baby: *Watt et al. (1962)*
TESTS:
In vitro tests on the whole plant revealed a very high level of antiplasmodial activity (IC $_{50}$=2.3µg/ml): *Clarkson et al. (2004)*
TOXICITY:
The plant contains *anemonol* (*Githens 1948*) which blisters the skin and is considered toxic and dangerous to sheep, although it is widely drunk as a medicine to treat a variety of diseases: *Steyn (1934),* cited by *Watt et al.(1962); Fanshawe et al. (1967)*

Rauvolfia caffra Sonder [*R. natalensis* Sonder] (Apocynaceae)
DESCRIPTION:
Tree up to 25m high with long straight corky bole, in swamp and fringing forest
VERNACULAR NAMES:
E : Quinine Tree; **Afr**: Kinaboom; **B/ I/ K/ To**: mubimbi (3, 3c, 4, 5, 14, 23) **B/ Ny**: mwimbe (3c, 3e, 9, 14) **K/ Ln**: mutoto (14) **Ln**: mutoci (14) **Ln**: utoto (3) **Ny**: ciwimbi, mvumba-mvula, ngwimbi (9), mnyesani, msabuwa, mvumba-mvula (3e, 14), mpamba-mvula (3, 3, 14), mvuba-mvula (3), nyesani (3e)
PLANT PARTS USED:
bark
TRADITIONAL USES:
E.Africa:a bark or root decoction is taken as an astringent, purgative or emetic to treat fever: *PROTA (2008) p.480-1*
Malawi: The bitter bark is used as a remedy for malaria: *Morris (1996) p. 219.*
TESTS:
Storrs casts doubt upon its efficacy (*Storrs 1979 p. 40),* but it was not among the *Rauvolfia* spp. tested by *Spencer et al.* with negative results.
TOXICITY:
Contains the toxic alkaloid *ajmaline,* and several countries prohibit its use in medicine: *PROTA (2008) p.481*

Reissantia indica (Willd.) N. Halle [*Hippocratea indica* Willd.] (Celastraceae)

DESCRIPTION:
Scrambling shrub, liana, or small tree about 5m high, with drooping branches, in dry woodland, rocky outcrops, gorges, or riverine fringe forests
VERNACULAR NAMES:
E: Ribbed Paddle-pod
PLANT PARTS USED:
leaves, roots
TRADITIONAL USES:
Tanzania: the leaf sap and root decoction are drunk as a cure for malaria: *Haerdi (1964) p. 107*

Rhamnus prinioides L'Herit. (Rhamnaceae)
DESCRIPTION:
Shrub or small tree up to 7m high; in montane forest, and fringing forest on Plateau
VERNACULAR NAMES:
E: Shiny-leaf; **Afr**: Seepblinkblaar
PLANT PARTS USED:
roots
TRADITIONAL USES:
Kenya: a root decoction is used against malaria by the Kipsigi people: *Beentje (1994) p. 358*
TESTS:
Organic and aqueous extracts of the plant were tested *in vitro* against chloroquine sensitive and resistant strains of *P.falciparum*, and showed activity value of $IC_{50} < 30$ mg/ml: *Muregi et al. (2003)*
TOXICITY:
In vitro bacterial & mammalian cell assays showed that the plant caused both DNA damage & chromosomal aberrations: *Fennell et al.(2004)*

Rhus longipes Engl. var. **longipes** (Anacardiaceae)
DESCRIPTION:
Weak-stemmed shrub or small tree up to 5m highg, sometimes scandent up to 11m. Widespread at edges of fringing and swamp forest up to 2200m
VERNACULAR NAMES:
E: Large-leaved Rhus; **B**: kasalasya (3c), mula-lusya (3c, 23), namula-lusya (3, 3c) **Ku**: cisitu (1) **Ln**: mukanda-cina (21) **M**: musangula (23) **Ny**: cimwa-manzi, mtatu (3), kamwazi, matsutula, mdima, mpila-kukuru, mtatatu (9)
PLANT PARTS USED:
roots
TRADITIONAL USES:
Tanzania: a root decoction is drunk as a malaria remedy: *Burkill (1985) p. 86; Haerdi (1964) p. 126*

Ricinus communis L. (Euphorbiaceae)
DESCRIPTION:
A shrub or small tree with palmate leaves, widely cultivated throughout the tropics. Common on waste land and near villages.
VERNACULAR NAMES:

E: Castor bean, Castor-oil plant/ tree, Palma Christi; **Afr**: Kasterolieboom; **B**: kasekelele (3), kaselele, mubalika (2, 3c), lumono, mbono (20), mumono (2, 3c, 20) **B/ K/ Lb/ Lo/ Ny/ Tu**: mono (1, 2, 3, 3c, 3d, 3e, 19, 20) **I**: ibono-ntelemba, intelemba, mubona-ntelemba (4, 5) **I/ Lb/ Le/ To**: mabono (4, 5, 19) **Ln/ Lv**: imono, katete, (19, 20), zimono (19) **Lo**: libono-bono, mabono-bono, mbonwa, monwa (19), tomo (3d) **Lv**: lilinji, linjinje, malindi (19) **M**: impulya (20) **Ny**: mbalika, msatsi (9), muzinga (23), nsatsi (3e, 20), nsyafuta (20)
PLANT PARTS USED:
bark, leaves, stems, whole plant, oil
TRADITIONAL USES:
Côte d'Ivoire: the leaves are used as a remedy for fever: *Bouquet (1974)*, quoted in *Burkill (1994); Gelfand et al.(1985); Kerharo et al. (1950)*
India: the dried root is used as a febrifuge: *Watt et al. (1962)*. The leaves are warmed in the oil and applied to the head of malaria patients before shivering starts: *Singh et al. (1993)*. The leaves, smeared with Ricinus Oil, are tied over head, palms and feet before shivering starts: *Anis et al.(1986)*
Malawi: a leaf decoction is drunk against fever: *Morris (1996)*
Vietnam: the plant is used as a diaphoretic: *Watt et al. (1962)*
W. Africa: A lotion is used against fever: *Gelfand et al. (1985)*. The leaves are used as a fever remedy: *Burkill (1994)*. The oil is added to paraffin-based insecticidal sprays as an anti-malaria agent: *Burkill (1935); Burkill (1994); Dalziel (1937)*
TESTS:
In vitro tests on the stems revealed anti-plasmodial activity (IC $_{50}$=8µg/ml): *Clarkson et al. (2004); Pillay et al. (2008)*
TOXICITY:
Highly toxic. *In vitro* bacterial & mammalian cell assays showed that the plant caused both DNA damage & chromosomal aberrations: *Fennell et al.(2004)*
The oil is well known as a purgative. The seeds contain a protein, *ricin*, which is extremely poisonous, producing after a lapse of several days diarrhoea and vomiting, with death from depression of the respiratory centre: *Fanshawe et al.(1967)*. In Malawi many cases have been reported of the murder of unwanted children by mixing seeds with their food: *Williamson (1975)*. As few as two seeds can be fatal: *Watt et al. (1962)*.

Rinorea welwitschii (Oliv.) Kuntze (Violaceae)
DESCRIPTION:
Shrub or slender tree 3 – 4m high; yellow flowers visited by hornets; in forest fringing Zambesi
VERNACULAR NAMES:
Not known
PLANT PARTS USED:
leaves
TRADITIONAL USES:
Zaire: a leaf decoction is used in a vapour bath to relieve fever pains: *Bouquet (1969) p. 243*, quoted in *Burkill (2000) p.285*

Rotheca myricoides (Hochst.) Steane & Mabberley [*Clerodendrum myricoides* (Hochst.) Vatke] (Lamiaceae)
DESCRIPTION:
Shrub 2.4 – 3m high, with two-tone blue flowers, smells bad when crushed; found in rocky

thickets, along streams, at at edges of evergreen forest
VERNACULAR NAMES:
E: Butterfly Bush, Blue Glory Bower, Blue Wings, Blue/ Smelly Cat's Whiskers; **Ku**: mapolwa (1) **Lo**: mutume (3, 3d, 9a) **Nm**: mnindi, mnindini, mnindini-ndi (22) **Ny**: cipaupa, kafupa, kawinga-zimu, msuka-ana (9)
PLANT PARTS USED:
roots
TRADITIONAL USES:
E. Africa: the roots are chewed or pounded, and water added. The infusion is drunk as a remedy for malaria: *Kokwaro (1976) p. 220*
Kenya: a root decoction is used against malaria: *Beentje (1994) p. 615*
TOXICITY:
The plant is poisonous if ingested: *Beentje p. 615*

Rothmannia manganjae (Hiern.) Keay [*Gardenia manganjae* Hiern.] (Rubiaceae)
DESCRIPTION:
Shrub or small tree, up to 12m high, with scented white flowers. Found in low/ mid-altitude forests and associated woodland
VERNACULAR NAMES:
E: Scented Bells
PLANT PARTS USED:
roots
TRADITIONAL USES:
Tanzania: a root decoction mixed with leaf juice is drunk as a remedy for malaria: *Haerdi (1964) p. 145*

Rourea orientalis Baill. [*Byrsocarpus orientalis* (Baill.) Baker; *B. tomentosus* Schell.] (Connaraceae)
DESCRIPTION:
Deciduous shrub, woody climber, or small tree, up to 4m high; widespread, esp. in thickets and *mutesyi* forest; persisting as a suffrutex in fire-climax grassland
VERNACULAR NAMES:
E: Short-pod; **Afr**: Kortpeul; **B**: kapakati (3, 3c, 23), nacisungu (3c) **B/ Ku/ Ny**: kapulula-mbusyi (1, 3c, 3e, 20, 23) **I**: mutunduti (4, 8) **Lb**: nakakwete (23) **Ln**: kakuhu (3) **Lo**: kazongwe (3d), mukolwe (3, 3d), mutengu-musyamba (16) **M**: kapulu-mbuzi (23) **Ny**: kozi, mkula-nsinga (3e), msanla-njazi, msitoti, msuka-cuma, ntanda-nyerere (9) **Tk**: mulebayisya (8) **To**: kasingini (23), kazingini (3), muca (?) **Tu**: unaha-mbalala (7c)
PLANT PARTS USED:
leaves
TRADITIONAL USES:
Malawi: a leaf infusion is used against fevers: *Morris (1996) p. 276*
TOXICITY:
The plant contains *methionine sulfoximine*, a neurotoxic amino acid. It is this substance in a leaf decoction of *Cnestis ferruginea* which has caused over 50 human deaths in Senegal: *Garon et al.(2007)*

Ruspolia seticalyx (C.B.Clarke) Milne-redh. (Acanthaceae)
DESCRIPTION:

Flowering herb up to 1m high, with tubular red flowers in spikes
VERNACULAR NAMES:
Not known
PLANT PARTS USED:
roots
TRADITIONAL USES:
Tanzania: a root decoction is drunk against malaria: *Haerdi (1964) p. 184*

Salix mucronata subsp. mucronata Thunb. [*S. subserrata* Willd.; *S. safsaf* Trauty.]
(Salicaceae)
DESCRIPTION:
Much-branched shrub or small tree up to 7m high; widespread, esp. in south, in fringing
bush on sandy alluvium, sometimes among rocks
VERNACULAR NAMES:
E: Safsaf Willow, Wild Willow; **Afr**: Safsafwilger; **B**: fisaka (3c, 14), mulunga-wiwa (10),
mupula (3c, 10, 14) **I/ K/ To**: musompe (10, 14, 23) **K/ Ln**: kambulafita (6, 14, 23) **Ku/
Ny**: musondozi, musuci (1) **Lb**: kalungawiba (23), mutesu (10, 23) **Lo**: lyandumaka (3, 3b),
lyandumuka, musangu-sangu (3d, 14) mutoya (10) **Lo/ Tk**: muvule (10, 23) **Ny**:
mpungabwe (9), msondozi (3, 3b, 3e, 9, 10, 14), mtundu (14) **Tk**: lilongo (8) **Tk/ To**:
lilongwe (8, 14) To: cibumbya, mvule (14)
PLANT PARTS USED:
roots, leaves
TRADITIONAL USES:
Botswana & Zimbabwe: the bitter leaves are widely used as a remedy for fevers: *Palgrave
(2002) p. 122*
Central Africa: the roots are used against fevers: *Palgrave (1957) p. 402*
E. Africa: the roots are used in fever cases: *Kokwaro (1976) p. 197*
S. Africa: a leaf decoction is used as a remedy for fever: *Watt et al. (1962) p. 925*
Zambia: the roots are used to treat fever: *Storrs (1979) p. 288*

Salvadora persica L. (Salvadoraceae)
DESCRIPTION:
Evergreen, tangled shrub, up to 7m, forming thickets, giving off acrid smell in hot sun;
found in Zambezi Valley
VERNACULAR NAMES:
E: Toothbrush Tree, Mustard Tree; **Afr**: Mosterdboom; **Nm**: mswake, muce (22) **To**:
cengeno (3)
PLANT PARTS USED:
roots
TRADITIONAL USES:
W. Africa: the powdered root bark is made into a paste with water and applied to the head
in fever cases, probable as a counter-irritant: *Burkill (2000) p. 3*
TOXICITY:
Tests on mice indicate that the plant extract has adverse effects on male & female
reproductive systems & fertility: *Damani et al.(2003).* LD$_{50}$ dose >1200mg/ml: *I.J.Ex.Biol
1969 7 p.250*

Sapium ellipticum (Hochst. ex C. Krauss) Pax [*S. mannianum* (Muell. Arg.) Benth.]

(Euphorbiaceae)
DESCRIPTION:
Evergreen tree up to 7m high, in evergreen forest, often on anthills, esp. near dambos
VERNACULAR NAMES:
E: Jumping-seed Tree; **Afr**: Springsaadbom; **B**: mukondo-kondo, munka-nsyimba, mutanta-nsange(3c) **Ln**: mutundu (6, 22) **Lv**: mumwameme (6, 22)
PLANT PARTS USED:
roots
TRADITIONAL USES:
Tanzania: a root-concoction is drunk for malaria: *Burkill (1994) p. 137; Haerdi (1964) p. 98*

Schoenoplectus senegalensis (Steud.) J. Raynal [*Scirpus articulatus* L.] (Cyperaceae)
DESCRIPTION:
Rush-like annual plant with spongy rounded stems, in or near standing water
VERNACULAR NAMES:
E: Bulrush
PLANT PARTS USED:
whole plant
TRADITIONAL USES:
E. Africa: the whole plant is used to treat fever: *Kokwaro (1976) p. 234*

Sclerocarya birrea (A.Rich.) Hochst. subsp. caffra (Sond.) Kokwaro [*S. caffra* Sonder] (Anacardiacerae)
DESCRIPTION:
Tree up to 18m high ; widespread in hotter and drier areas
VERNACULAR NAMES:
E: Marula; **Afr**: Maroela; **B**: musewe (20) **B/ Ku/ Ny**: musebe (3c, 10, 14, 20, 23) **I**: mutaba-nzovu (19) **I/ K/ Lo/ Tk/ To**: muwongo (3, 3a, 3d, 10, 14, 19, 23) **Lo**: mulala (3a), mulula (3, 3d, 10, 14, 19, 23), muyombo (3d, 14) **M**: musele (10) **N**: mubongo (10) **Nm**: mgongho, mngongho (22) **Ny**: mgamu (3e, 14), msebe (3a), mtondwoko (9), tsua, tsula, umganu (22) **Ny/ Tu**: msewe (3, 3e, 8, 14) **Tk/ To**: mongwe (8, 10) **To**: mugongo (10), mungongo, munongo (19)
PLANT PARTS USED:
bark
TRADITIONAL USES:
Burkina Faso: 4 bark scraping 3 cm wide are boiled for 20 min. to produce one glass of liquid, used to bathe a sick infant twice daily for 3 days: *de la Pradilla p. 28*
N.E.Transvaal: the bark is gathered just before the leaves appear, and taken as a prophylactic. A brandy tincture of the bark or powdered bark in teaspoon doses is taken against malaria: *Palgrave (2002) p. 540; Storrs (1979) p. 184; Watt et al. (1962) p. 53.*
Zambia: the bark is used to treat malaria: *Amico (1977) p. 122*
TESTS:
Methanolic and aqueous extracts were tested *in vitro* against D6 and W2 strains of *P. falciparum*, but showed only moderate antiplasmodial activity. However, in combination with *Turraea robusta* or *Boscia salicifolia*, marked synergistic and additive interactions were observed: *Gathirwa et al.,(2007)*

TOXICITY :
Eating the fruit can cause enteric and malaria (*sic*): *Pijper,* quoted in *Watt et al. (1962) p. 54.* Tests on stem bark showed LD$_{50}$ 37.1 : *Adoum (2009) p. 136.* See also *Kela et al.(1989)*

Scoparia dulcis L. (Scrophulariaceae)
DESCRIPTION:
Erect annual herb growing up to 2m high, with serrated leaves and many small white flowers
VERNACULAR NAMES:
E: Bitterbroom, Broomweed, Sweet Broom, Licorice Weed, Riceweed
PLANT PARTS USED:
leaves, roots, aerial parts
TRADITIONAL USES:
Ecuador: a leaf and root decoction is used to treat malaria: *Coe et al. (1996) p. 104*
Gambia: a lotion made from the leaves is used in fever: *Rosevear (1961),* quoted in *Burkill (2000) p. 78;*
Madagascar: the aerial parts are decocted to make a febrifuge: *Rasoanaivo et al. (1992) p. 124*

Securidaca longepedunculata Fresen. (Polygalaceae)
DESCRIPTION:
Small tree, up to 6m high, occurring in various types of woodland and bush
VERNACULAR NAMES:
E: Violet Tree; **Afr**: Krinkhout; **B**: muluka, mwimba-finoka (3c), mupapi (2, 3, 3b, 3c, 10, 14, 20) **I/ K/ Tk/ To**: mufufuma (4, 5, 10, 14, 17, 19) **K/ Lb**: lupape (10), mucaca (14) **Ku/ Ny/ To**: bwazi (3, 3b, 8, 9, 10, 14, 20) **Ku/ Ny/ Tu**: kapapi (1, 8) **Lb/ Le**: ulupapi (22) **Le**: mupapala (17) **Ln**: ndembo (22) **Ln/ Lv**: mutata (3, 3b, 6, 8, 10, 14, 19, 21) **Lo/ Ny/ Tu**: mwinda (3, 3b, 3d, 8, 10, 14, 16, 19, 22, 23) **Lv**: lundi (22), lutata, mwisi (22), muyise (6) **M**: muwuluka (10) **Nm**: mteeyu, muteju (22), nengo-nengo, nteyo (1a, 22) **Ny**: ciguluka, cosi, masace, muraka (9), cisanje (3e), mpuluka (2, 3, 3b, 3e, 14), mukuluka (10), mwinda (3e, 14), nakabwazi (10) **Ny/ Tu**: kapapi (3e, 8) **Tk/ To**:muwaama (8, 10, 17) **To**: njefu, (10), mwama (10, 23), mufufu (10, 17) **Tu**: bwaze, muruka (10), kitwa-nkumbi (22)
PLANT PARTS USED:
leaves, roots, seeds
TRADITIONAL USES:
Côte d'Ivoire: the crushed roots are used to relieve fever pains: *Kerharo et al. (1950) p. 32*
Malawi: the roots are used against malaria: *Morris (1996) p. 444; Williamson (1975) p. 212*
Mozambique: a leaf infusion is used to wash a fever patient: *Gelfand et al.(1985) p. 161*
Nigeria: Roots and seeds are used to treat fevers: *Etkin et al.(1991) p. 238*
Tanzania: A root decoction is used to treat malaria in Eastern Tanzania: *Chhabra et al. (1991) p. 145*
Zambia: in Nuanetsi district the root is used as a fever remedy: *Watt et al. (1962) p. 853.*
Zimbabwe: the infused root is used as a Shona and Ndebele remedy: *Pardy p. 1674,* quoted in *Watt et al. (1962) p. 855.* The powdered root is mixed with porridge as an antipyretic: *Gelfand et al.(1985) p. 161*
TESTS:
A dichloromethane extract of the leaf when tested *in vitro* against *P. falciparum* showed an activity of IC$_{50}$ 7 mg/ ml: *Bah et al. (2007)*

TOXICITY:
Ethanolic extracts of root bark tested against brine shrimp showed LD$_{50}$ value of 0.32mg/ml: *Kamba et al.(2010), Kela et al.(1989)*

Senna didymobotrya (Fresen.) Irwin & Barneby [*Cassia didymobotrya* Fresen.] (Fabaceae, Caesalpinioideae)
DESCRIPTION: A rapidly-growing shrub, up to 5m high, planted as hedge-plant on Copperbelt
VERNACULAR NAMES:
E: Peanut Senna
PLANT PARTS USED:
roots, leaves
TRADITIONAL USES:
Burundi:
Leaves decocted to make drink, & mixed with decocted leaves & bark of *Alangium chinense* for enema: *Baerts et al. (1989)*
E. Africa: a decoction of roots and leaves, varying from two to five glasses, is drunk. Vomiting is caused, and the fever clears. If necessary, the vomiting is induced by taking warm water, since an overdose may be fatal: *Kokwaro (1976) p. 118* ; *Kareru et al.,(2007); Jeruto et al. (2008)* ; a root decoction is drunk to treat malaria: *PROTA (2008) p. 506*
TESTS:
In vitro tests on twigs found anti-plasmodial activity of IC $_{50}$ = 9µg/ml): *Clarkson et al. (2004)*
TOXICITY:
The leaves & roots are used as fish poison. Dangerous to humans when ingested: *Kokwaro (1976) p. 118* ; decoctions from all parts are poisonous & can cause violent vomiting & diarrhoea, & may be fatal. But under carefully controlled manufacturing conditions it is potentially useful: *PROTA(2008) p.508.*

Senna obtusifolia (L.) Irwin & Barneby [*Cassia obtusifolia* L.] (Fabaceae, Caesalpinioideae)
DESCRIPTION:
Herbaceous or shrubby, up to 1.5m high, common in Southern Province
VERNACULAR NAMES:
E: Chinese Senna, Sicklepod; **Ku:** mpala-ntanga (1) **Ln/ Lo:** cangu (3, 3d) **Lo:** mubungwe (3d) **Lv:** kajiha-musongo (6)
PLANT PARTS USED:
roots, leaves, whole plant
TRADITIONAL USES:
Brazil: a decoction of the roots or an infusion of the whole plant is drunk by the Macuxi and Wapixana and by Luso-Brazilians to treat malaria: *Milliken (1997a) p. 222*
W. Africa: the leaf is used to treat fevers: *Burkill (1995) p. 159*
TOXICITY:

Toxic to cattle & poultry. Rats fed for 30 days on diets containing *S.obtusifolia* seed suffered diarrhoea, weight loss etc.: *Voss et al.(1991)*

Senna occidentalis (L.) Link [*Cassia occidentalis* L.] (Fabaceae, Caesalpinioideae)
DESCRIPTION:
An annual invasive weed 1 – 1.5m. high, common in waste places, roadsides, and croplands
VERNACULAR NAMES:
E: Septic Weed, Negro Coffee, Stinking Weed, Wild Coffee, Coffee Weed; **Ku**: khesya, maluwa (1); **Lo**: katima, museke-seke (3, 3d, 23); **Lv**: cinyalala, luweni (6); **Nm**: segese (7c)
PLANT PARTS USED:
whole plant, aerial parts, leaves, roots, twigs
TRADITIONAL USES:
Angola: A seed infusion is taken for malaria: *Batalha et al., (1979)*
Benin: Leaves and roots are decocted as a malaria remedy: *Adjanohoun (1989); Natabou Degbe (1991)*
Brazil: a decoction or infusion of the plant is drunk 3x daily until the patient is well: *Brandao (1992), Di Stasi (1994), Lorenzi (1991),* cited in *Milliken (1997)*
Burkina Faso: a) 5 bunches of leafy twigs are boiled for 15 min. in 10 lit. of water. Adults drink and wash in the liquid, twice daily for 4 days; infants are washed in it. b) a small bunch of leafy twigs is boiled in 1 lit. water for 10 min., then the fire is put out and a small bunch of *Cymbopogon citratus* is steeped in the liquid for 10 min. Two litres a day to be drunk between meals, for 4 days: *de la Pradilla (1988)*
Cameroon: the leaves are decocted to treat Malaria: *Mapi (1988)*
Colombia: a decoction or infusion of the plant is drunk 3x daily until the patient is well: Garcia-Barriga (1992), cited in *Milliken (1997)*
Congo: The plant is decocted with aerial pts of *Acanthospermum hispidum, Chenopodium ambrosioides, Ocimum sp., Hyptis suaveolens, Sesamum indicum & Aframomum sp.* to make a bath for child patients: *Bouquet (1969)*; the leaves are infused to treat malaria: *Tona et al. (1999)*
Congo-Brazzaville: roasted seeds are infused to make drinks: *Koechlin (1951)*; the roots are decoctedand drunk, 100ml adults, 50ml chidren, 3x daily; the roots or leaves are decocted with various combinations of *Lantana camara, Carica papaya, Annona senegalensis, Cymbopogon citratus,* or *Mangifera indica,* to be drunk 2x or 3x daily: *Diafouka (1997)*
Congo-Kinshasa: stems & leaves are decocted, 1 glass daily for 5 days: *Kasuku et al.(1999)*
Côte d'Ivoire: the whole plant is used as a febrifuge: *Kerharo et al. (1950)*
East Africa: the leaves are rubbed on the body as a remedy for fever. The leaves are boiled in water, and the steam used as a vapour bath: *Kokwaro (1976)*
Ecuador: a decoction of the leaves and roots is used to treat fever: *Coe et al. (1996)*
Ethiopia:at Bahir Dar Awraja, the plant is used to treat malaria: *Tadesse (1947)*
Gabon and W. Indies: the roots are used as a quinine substitute: *Walker (1953),* quoted in *Burkill (1995); Berk (1930),* quoted in *Watt et al. (1962)*; the leaves are infused to make a drink to treat malaria: *Raponda-Walker et al.(1961)*
Guinea Conakry: twigs, leaves & roots are used to make a drink: *Pobeguin (1911)*
Kenya:plant is mixed with soup or water as remedy: *Heine et al.(1988)*
Indonesia: the "coffee" is used as a febrifuge: *Watt et al. (1962)*
Madagascar: the plant is used as a remedy: *Pernet et al. (1957), Ratsimamanga (1979)* ; the leaves & seeds are used to treat malaria:*Terrac (1947)*; the aerial pts.are decocted:

Rasoanaivo et al. (1992); a root decoction is drunk 3x daily:*Novy (1997); Rahantamalala (2000)*

Mali:fresh powdered roots are decocted to treat malaria:*Malgras (1992)*

Mauritius:in Rodriguez Island, roots are infused to treat malaria: *Gurib-Fakim et al. (1994)*

Mozambique: reported to be efficient in treating malarial symptoms: *Bandeira et al. (2001)*

Nicobar Islands: A fresh leaf aqueous extract is given to children in fever: *Dagar et al., (1991)*

Nigeria: a leaf infusion is mixed with lime juice to treat malarial diarrhoea: *Dalziel (1937)*. The Hausa use a leaf fumigation, and rub the decoction on the body until palpitations cease. The Yoruba mix the decoction with palm-oil to rub on children with convulsions. The Ijo add salt to the leaf sap for malaria: *Ainslie sp. no. 79,* quoted in *Burkill (1995); Dalziel (1937); Watt et al. (1927)*.

Senegal: patients are laid on a bed of leaves under a cloth to produce sweating, or leaf decoction is drunk: *Kerharo (1964b), Kerharo et al. (1974)*; to treat a child, leaves are infused in water to drink or bathe: *Van der Steur (1994)*

Sierra Leone:the leaves are decocted to treat malaria: *Sawyer (1983)*

S. Africa: the leaves are used as a purgative in cases of fever and malaria: *Githens (1948)*. A tincture of the seeds has been used as a febrifuge: *Watt et al. (1962)*

Sudan: a root decoction is preferred to the leaves: *Aubreville (1950),* quoted in *Burkill (1995).*

Tanzania: a leaf infusion is taken orally as febrifuge: *Gessler et al. (1995).*

Ubangi: patients are placed in vapour baths to which leaves have been added, inhalation of the steam causing strong sweating: *Kerharo (1964b)*

Uganda:in Sango Bay area & Mbarara, leaves are decocted to treat malaria: *Ssegawa et al.(2007); Katuura et al. (2007)*

Venezuela: a decoction or infusion of the plant is drunk 3x daily until the patient is well: *Pittier (1978),* cited in *Milliken (1997)*

W. Africa: the leaves, roots, and seeds are used to treat fevers: *Burkill (1995).*

Zambia: the leaves are used as an antimalarial and febrifuge: *Amico (1977)*

TESTS:

In vitro tests on leaf extracts in 1999 produced more than 60% inhibition of the parasite *Plasmodium falciparum*, (IC_{50} 6μg/ml): *Tona et al., (1999).*

In vivo tests of the root bark in 2001 showed marked antimalarial activity, reducing parasitaemia by >50% when tested at an oral dose of 200mg/kg/day: *Tona et al. (2001).* When tested as a mosquito repellent, smouldering plants proved ineffective: *Palsson et al. (1998)*

TOXICITY:

The literature is too extensive to be quoted here. See Appendix 3

Senna petersiana (Bolle) Lock [*Cassia petersiana* Bolle] (Fabaceae, Caesalpinioideae)
DESCRIPTION:

Shrub or small deciduous tree, found in riverine fringes and wooded grassland
VERNACULAR NAMES:

E: Monkey-pod; **B**: kafungu-nasya (3c, 7a), mukupa-ciwa, musamba-mfwa (3c) **Ku**: mbangozi (1) **Ny**: kusandore (23), mtanta-nyelele (3e, 7a), mbweba-nyani, mdya-pumwa, mpatsa-cokolo, mpika-maunga, mtele-vele, mwandu-zulala, ntanda-nyerere, ntewelewe (9), nembe-nembe (22)
PLANT PARTS USED:

bark, leaves
TRADITIONAL USES:
Malawi: the bitter bark is used against malaria: *Morris (1996) p. 340*
S. Africa & tropical Africa: the leaves, said to contain *anthraquinone* & *tannin*, are used as a febrifuge: *Watt et al. (1962) p. 573.* Various parts of the tree are used as a purgative and to treat fevers: *Palgrave (2002) p. 342*
Zambia: the leaves are used as a febrifuge: *Amico (1977) p. 113*
TESTS:
Aqueous and organic fractions of roots and leaves were tested *in vitro* against the multi-drug resistant Vietnam-Smith strain of *P. falciparum* (VI/S). High activity at $IC_{50}<3mg/ml$ was seen: *Connelly et al.(1996)*
TOXICITY:
Harmless to humans.

Senna siamea (Lam.) Irwin & Barneby [*Cassia siamea* Lam.] (Fabaceae, Caesalpinioideae)
DESCRIPTION:
Evergreen rapidly-growing tree up to 15m high, planted for shelter belts and fuel plantations
VERNACULAR NAMES:
E: Siamese Cassia; **Ku**: maluwa (1)
PLANT PARTS USED:
leaves
TRADITIONAL USES:
Burkina Faso: a bunch of leaves is boiled for 20 min. in a litre of water. To be drunk with lemon juice throughout the day. The leaves contain alkaloids: *de la Pradilla (1988) p. 14*
Guinea: the leaves are used to treat malaria: *Sugiyama et al. (1992)*
Nigeria: the plant is used for treating malaria: *Odugbemi et al. (2007)*
Sierra Leone: roots are decocted to drink: *MacFoy et al. (1983), (Kruger et al. (1985)*; leaves decocted until yellow are used to make drinks, poultices and to bathe malaria patients: *(Kruger et al. (1985)*; leaves & roots are decocted for 2 hrs to treat malaria: *Sawyerr (1983)*
Togo: a root decoction is drunk to treat malaria: *Adjanohoun et al. (1986)*; the leaves are infused to treat malaria: *Gbeassor et al. (1989)*
TOXICITY:
Analysis of the leaves revealed toxicants tannin, oxalate & phytate, showing the need for processing before consumption: *Smith (2009)*

Senna singueana (Delile) Lock [*Cassia singueana* Delile; *C. singueana* Delile var. *glabra* (Baker f.) Brenan] (Fabaceae, Caesalpinioideae)
DESCRIPTION:
Shrub or small tree up to 5m high; Found in woodland and wooded grassland; on anthills on Plateau.
VERNACULAR NAMES:
E: Scrambled Egg, Sticky Pod, Winter Cassia, Winter-flowering Senna; **B**: kafungu-nasya (3), lubangeni (1), mufungu-nasya (10), mukupaciwa (10, 20), munsoka-nsoka (10), munumka-nsimba (10, 23) **I**: kasalala (4, 10, 23) **K**: pwamalonda (10) **Ku**: mkuzya-ndola, mukuza-ndola, mutanta-nyelele (1) **Ku/ Tk**: kalisabwe (1, 8) **Lb**: kafunga (10), kafungu (23) **Ln**: munina-mpuku (3) **Lo/ To**: mululwe (3) **Nm**: mhumba, msambi, msambi-sambi, mtungulu, musambilya (22), msambirya (7c) **Ny**: kalusapwe (14), mtanta-nyerere (3, 6, 9), mtawe-tawe (3e, 10), ndia-pumbwa (10), kamoto, mdya-pumbwa, mpatsa-cokolo, mtere-vere,

muluza, namagara, napala-pala, ntanda-nyerere, ntewelewe, waŋono (9) **Ny/ Tu**: ntanta-nyelele (10) **Tu**: kalibabwe (8), punda-mbuzi (22)
PLANT PARTS USED:
leaves, fruit
TRADITIONAL USES:
Burkina Faso: 4 bunches of leaves are boiled for 10 min. in 10 lit. of water. Used to wash patients twice daily for 3 days: *de la Pradilla (1988) p. 16*
Cameroun: the pods & leaves are infused to treat malaria: *Aubreville(1950)*
Kenya:the ground-up stem bark is decocted & drunk hot: *Nanyingi et al.(2008)*
Tanzania: leaf sap is drunk against malaria: *Haerdi (1964) p. 45* ; a root decoction is drunk: *Chhabra (1987)*
Nigeria: the leaves, leaf sap, and fruit pods are infused & used both internally & externally against fevers: *Burkill (1995) p. 639; Dalziel (1937) p. 180; Kudi et al.(1999)*; leaves & pods are boiled in water: *Quimby et al.(1964)*
Zimbabwe: the plant is used to treat malaria: *Chinemana(1985)*
TOXICITY:
Stem bark & leaves contain tannins, but digestibility is quite good. More research needed into leaves: *PROTA (2008) pp. 513,4*

Sesamum indicum L. (Pedaliaceae)
DESCRIPTION:
Annual erect flowering plant up to 1m high, with white/ purple flowers
VERNACULAR NAMES:
E: sesame; **B**: busosyi (20) **Nm**: mtunda-tunda (22) **To**: bwengo (20)
PLANT PARTS USED:
whole plant
TRADITIONAL USES:
Transvaal: the Soto people use a decoction for malaria: *Beyer (1927),* quoted in *Watt et al. (1962) p. 832*

Sesbania sesban (L.) Merr. (Fabaceae, Papilionoideae)
DESCRIPTION:
A shrub/ tree up to 8m, widespreading branchlets with racemes carrying yellow/ purple flowers.In reedswamps & flooded areas, eg L. Mweru, Luangwa R., Zambesi, Beit Bridge
VERNACULAR NAMES:
E: Egyptian/West Indian Pea, Corkwood Tree; **B**: misekesi (3c); **B/ Ku**: soyo (1); **I**: muhubu (4); **I/To**: mufubu (19, 4, 23); **Ln**: kanseki (3) kuŋ-nayi (6, 22); **Lo**: kamfule, musyeke-syeke (3d), lilongo (2, 3, 3d, 23); **Lv**: ciseks(6, 22); **Ny**: cigoma, msala-sese (3e) **Tk**: kanzyomwa (8); **To**: mbele-mbele, musebe-sebe (2, 3); **Tu**: musala-sese (8); **U**: kanzeki (2)
PLANT PARTS USED:
Leaves, roots, juice
TRADITIONAL USES:
West Africa: the plant is used as a fever remedy: *Githens (1948)*
TESTS:
An 80% MeOH extract of *S. sesban* root was tested *in vitro* against *P. falciparum*, and antiplasmodic activity observed: *Maregesi et al., (2010)*
TOXICITY:

Toxicity factor when tested on mice = LD50 45mg/ml: *J. Ex. Biol. 1980:18 p.594*

Setaria longiseta P. Beauv. (Poaceae)
DESCRIPTION:
Loosely-tufted perennial grass
VERNACULAR NAMES:
E: Bristle-grass
PLANT PARTS USED:
whole plant
TRADITIONAL USES:
W. Africa: the plant is used to make a prophylactic drink for children in the dry season: *Burkill (1994) p. 341*

Setaria megaphylla (Steud.) T. Durand & Schinz [S. Chevalieri Stapf] (Poaceae)
DESCRIPTION:
Robust, tufted perennial grass, often growing in clumps; up to 3m high. Common in woodland shade, forest glades, and streamsides, from sea level up to 1800m.
VERNACULAR NAMES:
E: Ribbon Grass, Broad-leaved Bristle-grass, Buffalo Grass; **Afr**: Buffelgras
PLANT PARTS USED:
Roots
TRADITIONAL USES:
W. Africa: the roots are used against fever: *Burkill (1994) p. 488*
TESTS:
In vitro tests on the whole plant in 2004 revealed a very high level of antiplasmodial activity (IC $_{50}$= 4.5µg/ml): *Clarkson et al. (2004)*
TOXICITY:
The young, wilted grass may contain hydrocyanic acid. The seed may be toxic to small birds: *Watt et al. (1962).* When the ethanolic root extract was tested on rats, it was found to be slightly toxic, with LD_{50} value 2074.8mg/kg: *Okokon et al.(2007)* No reports of toxic effects on humans.

Sida acuta Burm. f. (Malvaceae)
DESCRIPTION:
Small erect much-branched perennial shrub or herb, from 30cm to 100cm high, with yellow flowers; found in wastelands, roadsides, disturbed soils.
VERNACULAR NAMES:
E: Common Wireweed; **Ny**: masace, mpempwe
PLANT PARTS USED:
leaves
TRADITIONAL USES:
Mexico: the Zapotecs use the leaves to bathe fever patients: *Frei (1998) pp. 159, 162*
TESTS:
Aqueous and ethanol extracts of the powdered plant were tested *in vitro* against chloroquine-resistant and chloroquine-sensitive strains of *P. falciparum,* revealing activity of IC_{50} 3.9 – 5.4mc/ml: *Banzouzi et al. (2004)*
TOXICITY:

Poisonous to goats, causing a lysostomal storage disease. Brine shrimp tests revealed that the plant is toxic, with LD$_{50}$ value of 99.4mg/ml: *Orech et al. (2005)*

Sida rhombifolia L. (Malvaceae)
DESCRIPTION:
Erect or sprawling perennial or annual herb, 50 – 120cm high, with creamy/yellow flowers; found in waste ground, roadsides, or rocky areas
VERNACULAR NAMES:
E: Arrowleaf Sida, Jelly Leaf, Cuban Jute
PLANT PARTS USED:
leaves
TRADITIONAL USES:
Ecuador: a leaf decoction is taken to treat fever: *Coe et al. (1996) p. 102*
TOXICITY:
In vitro tests showed the plant to be slightly toxic: *Assam et al. (2010)*

Sikotamukwa
DESCRIPTION:
A kind of bush
VERNACULAR NAMES:
I: sikotamukwa (5)
PLANT PARTS USED:
roots
TRADITIONAL USES:
Zambia: the roots are burnt and children suffering from fever are "smoked" in the fumes: *Fowler (2000) pp. 622, 833*

Siphonochilus kirkii (Hook.f.) B. L. Burtt [*Kaempferia kirkii* (Hook f.) Wittm. & Perring; *K.rosea* Schweinf. ex Benth. & Hook f.] (Zingiberaceae)
DESCRIPTION:
Perennial herb growing from a short rhizome, with showy pink flowers; found in semi-shade in woodland, often near anthills.
VERNACULAR NAMES:
E: Rose Ginger; **Ny:** cikasu, kurri, manjanu, mbiricila (9) **Ny/ Tu:** nyama-kalapi (3e, 8)
PLANT PARTS USED:
rhizome
TRADITIONAL USES:
Zambia: the rhizome is used to treat malaria: *Amico (1977) p. 130*

Smilax anceps Willd. [*S. kraussiana* Meisn.] (Smilacaceae)
DESCRIPTION:
Climbing or scrambling shrub with greenish-white flowers, and red fruit; often found in secondary forest
VERNACULAR NAMES:
E: West African Sarsparilla; **B:** cikowa-nkanga (22), mukolo (3c), nkololo (3, 3c), nkololwe (3, 3c, 23) **B/ Ln:** mukolo-bolo (3, 3c) **Ln:** muhalapoli (3) **Lo:** tukona (3d) **Ny:** kakwazi, mkwandula (9), mpama (3) **Tu:** intuntu (7c)
PLANT PARTS USED:

roots
TRADITIONAL USES:
<u>Ghana</u>: a root decoction is used to make a vapour bath: *Burkill (2000) p. 95; Dalziel (1937) p. 479*
<u>S. Africa</u>: the root is used as a febrifuge: *Githens (1948) p. 104; Watt et al. (1962) p. 716*

Solanum incanum L. [*S. bojeri* Dunal; *S. iodes* Dammer; *S. sodomeum* L.] (Solanaceae)
DESCRIPTION:
Prickly shrub up to 3m high, with hairy white underleaves, round green/orange fruit ; found on roadsides, disturbed land, riverine and forest margins
VERNACULAR NAMES:
E: Nightshade, Bitter/ Poison/Snake/Sodom Apple; **Nm**: matungusa, mtula (22), mtungu-jamita, mtungusa (1a, 22), mutuntula (9a) **Ny**: mbwanya-nyanya, mthika,mthula, mtungwisa, ntundu-were (9)
PLANT PARTS USED:
roots
TRADITIONAL USES:
<u>E. Africa</u>: a root decoction is used against fever: *Kokwaro (1976) p. 205*
TOXICITY:
Goats fed on the unripe dried fruits for 9 weeks developed pathological symptoms leading to death: *Thaiyah et al.(2010)*

Sorghum spp. (Poaceae)
DESCRIPTION:
Shrub up to 3m high, widely cultivated as cereal crop
VERNACULAR NAMES:
E: Sorghum; **B**: cinganda, cipapebwe, isumbu, kafubu, kasabwe, kasyokwe, katete, lumpampa-lunono, masali, musoma, mususulu (20) **B/ Ny**: kasela, njasyi (20) **B/ Tu**: kancebele, mapemba, mapira-tutile, pande, pande-kanondo (20) **I**: impunga (1, 4), inkoti, katumwe, kolwe, lubele, maceme, matuba, musale, musonde, nakasingo, nkolwe, sipewa-musazinoko (4) **I/ Le/ To**: maase (1, 4, 17) **I/ To**: mabele (17), muceme (4, 19) **K**: luvuma (19) **Ku**: kamkoba, kandodo, mapila (1) **Ku/ Tu**: kandondo (20) **Le**: cilunda (19), mawo (17) **Lo**: lukesyu, makonga, mulwa, munanana, muswe (19) **M**: nkumba, sonkwe (20) **Ny**: ka-emena (1), msonde (3e) **To**: ingase, inzembwe (17), lusili (19) **Tu**: cipamba, citana-wazyija, ciwose, kamcila, kanama, kandondo-mutaba, kangolo, kwanje, ndumba-luango, nkoce, ntente, nyaka-panda, pate (20)
PLANT PARTS USED:
roots
TRADITIONAL USES:
<u>E. Zimbabwe</u>: the roots are used as a remedy: *Quin (1954)*, quoted in *Watt et al. (1962) p. 488*

Sphaeranthus suaveolens (Forssk.) DC. (Asteraceae)
DESCRIPTION:
Erect or trailing perennial herb up to 1m high
VERNACULAR NAMES:
Not known
PLANT PARTS USED:

leaves, inflorescence
TRADITIONAL USES:
E. Africa: the leaves are decocted, and the liquid rubbed on to the body as a remedy for malaria: *Kokwaro (1976) p. 71*
Zambia: the leaves and inflorescence are used as a febrifuge: *Amico (1977) p. 126*

Sphenostylis marginata subsp. erecta (Baker f.) Verde [*S. erecta* (Baker f.) Hutch. ex Baker f.] (Fabaceae, Papilionoideae)
DESCRIPTION:
Trailing or semiscandent suffrutescent herb up to 1.5m long from woody rootstock, with pink/ purple flowers
VERNACULAR NAMES:
B: cimamba, mumamba (3c) **I**: kapululu (4) **Ku**: mbamba-sanga (1) **N**: cindele (8) **Ny**: kawowo, msulu (3e), mfutsa, ngunga (9), nkunga (3e, 9) **Tk**: mukululu, nkululu (8)
PLANT PARTS USED:
roots
TRADITIONAL USES:
Zimbabwe: the powdered root is used as an enema for fever: *Gelfand et al.(1985) p. 154*

Steganotaenia araliacea Hochst. (Apiaceae)
DESCRIPTION:
Unbranched or sparsely branched shrub up to 2m high, or small tree up to 8m, found in low-altitude woodland or rocky outcrops
VERNACULAR NAMES:
E: Carrot-tree, Popgun Tree; **Afr**: Geewortelboom; **B**: citebi-tebi, kapolo-polo, mutebi-tebi (3c, 14) **B/ Le/ Tu**: mutebe-tebe (3, 8, 9a, 17) **I**: cipolo-polo (23), mumpilip-ifwa (17) **I/ Le/ To**: mukanda-ŋombe (14, 17) **I/ To**: mumpele-mpemfwa (17) **K**: kipolo-polo (3, 14), lumonokyulu (3, 14) **Lo**: muketu (3, 3d, 14, 22), mukunya-nsefu, sindoke (3d, 14) **Nm**: mbotyola (22) **Ny**: biyo, cipombo, mtebe-tebe, nyongoloka (3e, 14), cifyopola (3, 3e, 14), mpanda-njovu, nakauti, ngombo-uti (9), mpoloni (3, 3e, 9, 14) **Tk**: mufeka (8) **To**: mupelewa (3, 14) mufuti, mutobolo (14), mumpelefwa (17) **Tu**: ntewe-tewe (8)
PLANT PARTS USED:
bark, roots
TRADITIONAL USES:
Kenya: the bark is chewed by the Turkana people as a fever remedy: *Beentje (1994) p. 443*
Tanzania: the roots are decocted to make a remedy for malaria: *Burkill (2000) p. 234; Haerdi (1964) p. 163*
Zambia: the roots are used as a remedy for fever and to ease breathing: *Fowler (2000) pp. 126, 812; Storrs (1979) p. 188*
TOXICITY:
Stem bark extracts tested on rats showed toxic effects on vital organs: *Agunu et al.(2004)*

Stenotaphrum dimidiatum L. Brongn. (Poaceae)
DESCRIPTION:
Perennial mat-forming grass, common at roadsides
VERNACULAR NAMES:
E: Buffalo Grass, Pimento Grass, Pemba Grass
PLANT PARTS USED:

whole plant
TRADITIONAL USES:
<u>Brazil</u>: the grass is used as a sudorific:*Williams (1965) pp. 287-8,* quoted in *Burkill (1994) pp. 365-6*

Sterculia africana (Lour.) Fiori [*S. guerichii* K. Schum.; *S. triphaca* R.Br.] (Malvaceae)
DESCRIPTION:
Tree up to 12m high with short stout bole, widespread at low altitudes, esp. Zambezi and Luangwa Valley, usually in rocky places, sometimes on riverine alluvium
VERNACULAR NAMES:
E: Tick-tree; **Afr**: Bosluisboom; **B/ I/ Ku/ Ny/ Tk/ Tu**: mulele (1, 8, 19, 20) **I**: moza (23) **I/To**: mulyozya, mozya (4, 19) **K**: mulende (23), mwemwe-tuseko (19) **Lo**: mukuku-buyu (22) **Lo/ To**: mukosa (3d, 19, 14) **Nm**: mhoja, mhozia, msavala (22) **Ny**: mgoza (3, 3e, 4), mlele (3e, 14) **Tk**: munzumina (8) **To**: mupoposya (22) **Tu**: mugozya (20)
PLANT PARTS USED:
roots, bark, leaves
TRADITIONAL USES:
<u>E. Africa</u>: the roots, bark and leaves are boiled with other plants, and used as an inhalant to relieve fever: *Kokwaro (1976) p. 209*

Sterculia quinqueloba (Garcke) K. Schum. (Malvaceae)
DESCRIPTION:
Small/ medium tree up to 12m high, widespread at lower altitudes, us. on rocky slopes; on anthills in rainy areas
VERNACULAR NAMES:
E: Large-leaved Star Chestnut; **Afr**: Grootblaarsterkastaling; **B**: kapombo (3, 9a, 14) **B/ I/ Le**: mweemwe (3, 3a, 3c, 4, 9a, 10, 14, 19) **I**: mukose, ŋunza (4) **I/ K/ Ln/ To**: mulende (3a, 14, 23) **I/ To**: mutuba-tuba (19) **Le/ To/ Tu**: mulele (10, 17) **Lo**: mukuku-biya (3a) **Lo/ To**:mukosa (3d, 19, 14) **Ny**: mlele-zombo (3e, 14), msamba-mfu (14), msamba-mfumu (3), mseta-nyani (10), mulele-nzombo (1) **Ny/ To**: mulele (3a, 10) **Ny/ Tu**: mgoza (3, 10, 14) **Tk/ To**: mubungu-bungu (8, 17) **To**: mulezya, muzuma-ngoma (17)
PLANT PARTS USED:
roots, leaves
TRADITIONAL USES:
<u>Tanzania</u>: a decoction of roots and leaves is drunk against malaria: *Haerdi (1964) p. 86*

Stereospermum kunthianum Cham. (Bignoneaceae)
DESCRIPTION:
Small tree up to 13m high, found in *cipiya* woodlands
VERNACULAR NAMES:
E: Pink Jacaranda; **B**: kayubule (3, 3a, 3c, 10, 14), mukulu-kayubule (3c, 14) **I/ To**: mutese (3, 4, 10, 14), mutesi (8), mutesyu (10, 23) **Ku**: mutelela-nzovu (1) **Lb**: mbute (23) **Ln**: mutoka-toka (19) **Lo**: mupafu (3d, 14) **Lv**: mulinde (19) **Nm**: msangwa (22), mtelela (7c, 22) **Ny**: kafupa, mponda-njovu, mtutumuko, muramo, mwana-bere, nkokona-simba (9), kavunguti, mlaka-njovu (3e, 9, 10, 14), msunga-wantu, mtelezi, mwinguti (3e, 14), mtala-njovu (3e), mulaka-njovu (3, 3a, 9), mzenda-nzovu (3e) **To**: mtse (3a)
PLANT PARTS USED:
bark

TRADITIONAL USES:

Tanzania: a bark decoction, combined with the bark of *Dalberghia boehmii*, is used to treat children in malarial rigor. The liquid is drunk, and used as a bath: *Burkill (1985) p.266; Haerdi (1964) p. 149*

TOXICITY:

Stem bark extract when tested on rats & mice produced no mortalities at max. oral dose of 8gm/kg: therefore well tolerated & relatively safe by oral route: *Ching et al.(2008)*

Strophanthus eminii Asch. & Pax [*S. wittei* Staner] (Apocynaceae)

DESCRIPTION:

Shrub or small tree up to 3m high, found in open woodland on dry rocky hills in Lake Mweru valley

VERNACULAR NAMES:

E: Emin's Strophanthus; **Nm**: mbele-bele, msungu-lula, muveri-veri, mwele-wele

PLANT PARTS USED:

roots

TRADITIONAL USES:

E. Africa: feverish children are bathed in a root infusion. The pounded roots are soaked in water and patients inhale the vapour while bathing in the liquid: *Kokwaro (1976) p. 28*

TESTS:

In vivo tests on mice in 2009 showed antimalarial activity. The ID_{50} value was 1560 mg/kg body weight: *Innocent et al. (2009)*

TOXICITY:

The seeds contain toxic glycosides & are therefore used as an arrow poison: *PROTA (2008) p. 545*

Strychnos henningsii Gilg. [*S. holstii* Gilg. var. *procera* (Gilg. & Busse) P. A. Duvign.] (Strychnaceae)

DESCRIPTION:

Small, shrubby tree 3m high, to large tree up to 21m high, found in low-altitude dry riverine fringes, in scrub and on anthills

VERNACULAR NAMES:

E: Red Bitterberry, Coffee Bean Strychnos, Natal Teak, Walking Stick; **Afr**: Rooibitterbessie

PLANT PARTS USED:

rootbark

TRADITIONAL USES:

Madagascar: the rootbark is decocted to treat malaria: *Mulhovo*

TOXICITY:

The plant is rich in alkaloids, & in Madagascar the stem bark & roots are used to poison rats & mice, as well as for criminal purposes: *PROTA (2008) p. 570*

Strychnos innocua Delile [*S. huillensis* Gilg. & Busse] (Strychnaceae)

DESCRIPTION: Shrub or tree up to 13m high, with many stems when growing in thickets; widespread in woodland and rocky hills

VERNACULAR NAMES:

E: Powder-bark, Monkey-Orange; **B/ Lb**: mulungi (2, 3, 3c, 14, 20, 23), mulungi-kome (3c, 14) **I/ Le/ To**: kawi (4, 17) **I/ Le/ To**: muteme (2, 3, 4, 17, 19) **K/ Ln**: mukuka-mpombo (14) **Ku**: ntemya (1) **Ku/ Ny**: kambele (1, 3e, 14) **Ln**: muhwila (6), mukunka-mpombo (3)

Lo: kateme (23), muzimbikolo (3d, 14) **Lo/ Tk**: mungolo (19, 23) **Lv**: mukolo (6, 22) **Nm**: mhundu (7c, 22), mpundu, muhonsia, muhundu, mumundu (22) **Ny**: kabulu-kulu (2, 3, 3e, 14), kamwela-lumba, mteme, mtulu-tulu (3e, 14), mgulu-gulu, mkaye, mkhuyu-khuyu, msongoro (9), m'kwakwa (22), nteme (1) **To**: kalungi (14), mutu (14, 19), mwabo (2, 3)

PLANT PARTS USED:

leaves, roots

TRADITIONAL USES:

Tanzania: the leaf sap is drunk against malaria: *Haerdi (1964) p. 128*

Sudan: the roots are pounded in water, used to bathe the patient: *Bissett, pp. 214-6, Broun et al.(1929) p. 242,* quoted in *Burkill (1995) p. 538*

Strychnos spinosa Lam. [*S. lokua* A. Rich.] (Strychnaceae)

DESCRIPTION:

Shrub or small tree, us. 4-5m, up to 10m high. Widespread in open woodland

VERNACULAR NAMES:

E: Spiny Monkey-orange; **Afr**: Doringklapper; **B**: kamino (3, 3c, 14), mapuma-ngulube (3, 10), musaye (3, 3c, 14, 20) **B/ Ku**: sansa (1, 3c) **B/ M**: sansa (3, 10, 14) **I**: kawi, mazimbili (4), muzimbilili (4, 5, 10, 17) **I/ N/ Tk/ To**: muntamba (3, 4, 5, 8, 10, 14, 23) **K**: kapwi (3, 10, 14), munkulwa-iba (10) **K/ Ln**: katonga, munkulu-nkulu (14) **Lb**: musimbilili (10) **Ln**: cikoli, kana-njamba, muhuma, muhwila (21), kampobela (10, 23), mubila, mwijimbe (14), mubilo (3), mutungi (6, 22) **Lo**: mukolo, mwimbili (3, 3d, 14), muholu-holu (10, 16, 23), muhulu-hulu (19, 22), muyimbili (10, 23, 22) **Lv**: mukole (10), mutunga, mwitwi (6) **M**: isanza (10) **Nm**: mkome, mwage (22), mubale (1a, 22) **Ny**: dzayi (10), maye, mzayi (3, 3e, 14), mkhuyu-khuyu, mtembe, mtonga, mwaye (9), mzimbili, temya (3e, 14) **Tk**: muntanoba, nimbwa (10), muthamba (8, 10) **To**: muwi (23), mwaabo (14, 17), nsala (22) **Tu**: muzimbili (8), temya (10)

PLANT PARTS USED:

leaves, bark, roots, fruit

TRADITIONAL USES:

Côte d'Ivoire: the bark of the twigs is decocted and used as a febrifuge: *Kerharo et al. (1950) p. 182*

Ghana: a root decoction is used to treat fever: *Abbiw (1990) pass.,* quoted in *Burkill (1995) p 543*

Nigeria: the root, with the root of *Afromosia laxiflora,* is used as a febrifuge: *De Wildeman (1946) p. 5,* quoted in *Watt et al. (1962) p. 730*

S. Africa: the Zulu use the root or green fruit, mixed with the root of *Leonotis leonurus* and other plants, as a febrifuge: *Burkill (1995) p. 543; Palgrave (2002) p. 931; Watt et al. (1962) p. 730*

Tanzania: the leaf sap is drunk against malaria: *Haerdi (1964) p. 128*

W. Africa: the roots are used a remedy: *Burkill (1995) p. 543*

TESTS:

In vitro tests in Benin on the leaves in 2009 revealed an ID$_{50}$ 15.6 mg/ml efficacy, with low toxicity: *Bero et al., (2009)*

TOXICITY:

The fruit pulp is edible, but the seeds are highly toxic.

Stylosanthes fruticosa (Retz.) Alston [*S. mucronata* Willd.] (Fabaceae, Papilionoideae)

DESCRIPTION:

Small shrubby perennial herb, erect or prostrate, with stems up to 1m long. In *Mopane*, *Acacia* or *Miombo* woodland and wooded grassland, often on sandy soils
VERNACULAR NAMES:
E: Wild Lucerne
PLANT PARTS USED:
whole plant
TRADITIONAL USES:
<u>Senegal</u>: an infusion of the whole plant is used as a febrifuge: *Berhaut (1976) p. 531,* quoted in *Burkill (1995) p. 449*
<u>E. Africa</u>: the whole plant is administered for general fever and debility: *Burkill (1995) p. 449*
TOXICITY:

Swartzia madagascariensis Desv. [*Bobgunnia madagascariensis* (Desv.) J. H. Kirkbride & Wiersema] (Fabaceae, Papilionoideae)
DESCRIPTION:
Small/medium deciduous spreading tree up to 8m high. Widespread in open woodland, especially on sandy soil
VERNACULAR NAMES:
E: Snake-bean; **Afr:** Slangboom; **B/ I/ K/ Lb/ Le/ Ny:** ndale (1, 2, 3, 3a, 3c, 3e, 4, 10, 14, 17, 23) **B/ M:** cilonde (10, 22) **I/ Lb/ To/ Tu:** mulundu (1c, 2, 3, 3a, 4, 10, 14, 17) **I/ To:** kasokolowe (19) **Lo:** masyaga-syela (22) **Lo/ N:** musyaka-syela (2, 3a, 3d, 10, 16) **Lo/ N/ Tk:** musyaka-syala (3, 10, 23) **Ln:** kapwipu (6, 21, 22), kapwipwi (2, 3, 10, 14) **Lv:** mutete (6, 10, 22) N: caronde, mundale (10) Nm: kasanda (22) Ny: cinye-nye, cotsa-soka, dzungu, mdya-kamba (9), kampanga (3a, 9), kasokoso, mdundo, ngolo (3e), mcelekete (3, 3e, 14) **Tk/ To:** m'onjolo (10) **Tk/Tu:** mongolo (8) To: muya-ngula, m'ongola (10), muyongolwa (17), muyungula (23)
PLANT PARTS USED:
bark
TRADITIONAL USES:
<u>W. Africa</u>: the bark is used as a febrifuge: *Burkill (1995) p. 639*

Synsepalum brevipes (Baker) T. D. Penn. [*Pachystela brevipes* (Baker) Engl.] (Sapotaceae)
DESCRIPTION:
Bush to much-branched evergreen tree, 4 – 20m high, in riverine forest and low-altitude, dry-layered forest
VERNACULAR NAMES:
E: Stemfruit, Miraculous Berry; **B:** munombe-nombe (23), munumbi-numbi (3, 3c), munselemba, musambya (3c)
PLANT PARTS USED:
roots
TRADITIONAL USES:
<u>Tanzania</u>: a root decoction is drunk as a remedy for malaria: *Burkill (2000) p.56; Haerdi (1964) p. 116*

Synsepalum passargei (Engl.) Pennington [*Vincentella passargei* (Engl.) Aubrev.] (Sapotaceae)
DESCRIPTION:

Shrub or much-branched evergreen tree 2 – 8m high, with drooping branches, in fringing forest
VERNACULAR NAMES:
E: Yellow Miraculous-berry; **B**: muncimbili, mupalapati (3, 3c, 20)
PLANT PARTS USED:
roots, leaves
TRADITIONAL USES:
Tanzania: a decoction of the roots and leaves is drunk, and the body bathed with the liquid, as a remedy for malaria: *Burkill (2000) p. 560; Haerdi (1964) p. 117*

Syzigium cordatum Hochst. ex C.Krause (Myrtaceae)
DESCRIPTION :
An ever green waterloving tree up to 8-15m high, found near streams, in swampy areas and riverine forest.
VERNACULAR NAMES ;
E : Waterberry; **Afr** : Waterbessie; **B**: mukute, musyingu (3c, 14), musafwa, namunsi (22)
B/ Lb/ Le: mufinsa (3b, 3c, 10, 14, 19, 20) **I/ K/ Ln/ Lv**: musombo (3, 3b, 4, 14, 22) **K**: musambo (10) **Ln**: musomba (22) **Lo**: gaanumukela (10, 22), mutoya (3, 3b, 3d), mutoye (14, 22) **Lv**: muhlu, mundukundu (22) **M**: nyengele (10) **Ny**: mcisu (3, 3e, 9, 10, 14), msinyika (3, 3b, 10), mnyomwe, nanyole, nyenyewe (9), msombo, mwenye (3e, 14), muhlu (22), nyowe (3e, 9, 10, 14) **Ny/ To**: katope (3, 3b, 3e, 4, 9, 14) **To**: munonya-nansi (22)
Tu: musompe (8)
PLANT PARTS USED:
roots
TRADITIONAL USES:
Zambia: the roots are pounded and mixed with porridge, and applied to tattoos on every joint: *Vongo*

Tamarindus indica L. (Fabaceae, Caesalpinioideae)
DESCRIPTION:
Widespread in tropical Africa, a semi-evergreen tree up to 25m. tall, growing in riparian or mopane woodland, alluvial flats, termite mounds and scarp slopes.
VERNACULAR NAMES:
E: Tamarind; **B**: musyisyi; **B/ Ku**: bwembe, mwembe; **I/ To**: musyika; **Ku/ Ny/ Tu**: muwembe; **Nm**: mkwaju, msyisyi; **Nm/ Ny**: mkwesu; **Ny**: bemba, mbwemba, mwemba; **Tu**: mukwadju
PLANT PARTS USED:
roots, bark, fruit, leaves, twigs
TRADITIONAL USES:
Brazil: both fruit pulp and leaf are used as diaphoretics: *De Wildeman (1948),* quoted in *Watt et al. (1962)*
Burkina Faso: 4 bunches of leafy twigs are boiled in 10 lit. water for 15 min. Bathe the body twice daily, and drink a little, for 4 days. *de la Pradilla (1988)*
E. Africa: a root decoction is drunk for fevers: *Gelfand et al.(1985); Kokwaro (1976)*
E. Sudan and Senegal: the plant is used as febrifuge: *Gelfand et al. (1985)*
Eritrea: the fruit pulp paste is sold in markets as a malaria remedy: *Cortesi (1936),* quoted in *Watt et al. (1962)*
Madagascar: a leaf decoction is used against fevers: *Watt et al. (1962)*

Malaysia: a leaf decoction is taken for fevers: *Burkill (1935); Burkill (1995)*

Mexico: the fruit and leaves are used to make a bath for the relief of symptoms: *Frei (1998)*

W. Africa: the bark, roots, fruit pulp, and leaves are all used as a febrifuge: *Berhaut (1975b), Irvine (1952a),* quoted in *Burkill (1995); Dalziel (1937)*

Zaire: a leaf decoction is used as a febrifuge: *Leonard (1952),* quoted in *Burkill (1995); Gelfand et al. (1985)*

Zambia: the fruit pulp is used to make a refreshing drink to reduce fever: *Palgrave (2002); Storrs (1979) (1979)*

TOXICITY:

The leaves were found to contain *luteoline, apigenine, orientine, isorientine, vitexine* and *pinitol* : *de la Pradilla (1988)*

Tests on albino rats indicated no toxic effects: *Abukakar et al. (2011).* Showed as cytotoxic when tested on snails at 50–100 ppm: *Elsheikh et al. (1990)*

Teclea nobilis Delile (Rutaceae)

DESCRIPTION:

Evergreen shrub or tree up to 13m high, in riverine forest and fringing evergreen forest

VERNACULAR NAMES:

E: Giant Cherry-orange; **Nm**: mdimu (1a, 22) **Ny**: mkulu-kulu (9)

PLANT PARTS USED:

leaves

TRADITIONAL USES:

E. Africa: the leaves are put in a pot with water and covered tightly. The pot is steamed and the vapour inhaled as a remedy for fever: *Kokwaro (1976) p. 197; Watt et al. (1962) p. 924*

TOXICITY:

Tests of hexane extracts on rodents did not produce any toxic effects in animals at various doses: *Al-Rehaily et al. (2001).* Brine shrimp tests showed 156.6 mg/ml: *Moshi et al. (2010)*

Tecomaria capensis (Thunb.) Spach [*Tecoma capensis* (Thunb.) Lindley; *Tecomaria nyassae* (Oliv.) K. Schum.; *T. shirensis* Baker] (Bignoniaceae)

DESCRIPTION:

Large many-stemmed shrub or small tree, up to 10m high, in margins of evergreen forest, bushveld, scrub, and along stream banks and rocky hills

VERNACULAR NAMES:

E: Cape Honeysuckle; **Afr**: Kaapse Kanferfoelie; **B**: matongo (3, 3c), mulyatongo (3c, 23), mutongwe (23), namukosi, zunda (22) **Ln**: bululu (3) **To**: mubwali-bwali (23) **Tu**: mtwate, mutwate (22)

PLANT PARTS USED:

bark

TRADITIONAL USES:

S. Africa: the Soto of the Northern Transvaal administer the powdered bark in cases of high fever: *Burkill (1985) p. 268; Watt et al. (1962) p. 144*

Tephrosia purpurea (L.) Pers. (Fabaceae, Papilionoideae)

DESCRIPTION:

Common perennial wasteland weed with woody stem and purple flowers

VERNACULAR NAMES:

E: Fish Poison, Wild Indigo; **Nm**: ludumyo (7c) **Tk**: kandongo-ndongo (8) **Tu**: cilala- nkanga

(8)
PLANT PARTS USED:
whole plant
TRADITIONAL USES:
India: the dried herb is used as a remedy for "bilious fevers": *Burkill (1995) p. 459; Watt et al. (1962) p. 656*
Senegal: the plant is used as a febrifuge: *Burkill (1995) p. 459; Kerharo (1964b) p. 581, (1974) pp. 476-7*
TOXICITY:
The *N*–butanol extract of the plant when tested on hamsters & monkeys showed antileishmanial activity with no toxic side effects: *Sharma et al.(2003)*

Teramnus labialis (Lin. f.) Spreng. subsp. arabicus Verdc. (Fabaceae, Papilionoideae)
DESCRIPTION:
Slender, twining herb up to 3m high; found in Mwinilungu district
VERNACULAR NAMES:
E: Blue Wiss, Horse Vine, Rabbit Vine; **Ku**: kalunde-zonde (1)
PLANT PARTS USED:
fruit
TRADITIONAL USES:
India: the fruit is used as a febrifuge: *Chadha (1976) p. 157,* quoted in *Burkill (1995) pp. 463, 639*

Tetracera alnifolia Willd. (Dilleniaceae)
DESCRIPTION:
Large woody climber up to 15m high; in evergreen fringing forest in Zambezi Valley
VERNACULAR NAMES:
Ln: cilengwa (3)
PLANT PARTS USED:
leafy stems
TRADITIONAL USES:
Côte d'Ivoire: the leafy stems are macerated in palm wine and drunk to ameliorate fever pains: *Burkill (1985) p. 651; Kerharo et al. (1950) pp. 37-8*

Tetradenia riparia (Hochst.) Codd [*Iboza riparia* (Hochst.) N. E. Br.] (Lamiaceae)
DESCRIPTION:
Tall aromatic shrub up to 3m high with white/pink/lilac flowers, found along riverbanks, forest margins, dry wooded valleys and hillsides
VERNACULAR NAMES:
E: Ginger Bush; **Ny**: cingambilo, cinkambira, cizama-moto, mfutsa-waukulu, piri-piri (9) **Tk**: mulekwa-mbweni (8)
PLANT PARTS USED:
leaves
TRADITIONAL USES:
S. Africa: the Zulu people use one dose of a leaf-infusion against malaria: *Bryant (1909) 2:1; Watt et al. (1962) p. 516*
TOXICITY:
Extremely toxic to humans when taken in excessive doses: *Bodenstein (1977)*

Tragia benthamii Bak. (Euphorbiaceae)
DESCRIPTION:
Herbaceous climber or trailer, wth stinging hairs
VERNACULAR NAMES:
Ny: cinyanya, lyundu-mula, mzaza (9)
PLANT PARTS USED:
roots
TRADITIONAL USES:
Malawi: a root decoction is drunk against fever: *Morris (1996) p. 309*

Treculia africana Decne subsp. africana (Moraceae)
DESCRIPTION:
Evergreen tree (rarely a shrub) 5 – 20m high, bearing fruit 40cm in diameter, weighing 8 –
14 kilos; in riverine forest, moist and swampy forest
VERNACULAR NAMES:
E: African Breadfruit, Wild Jackfruit, African Boxwood; **Ny**: lyaja, mjaya, njale (9)
PLANT PARTS USED:
roots, bark
TRADITIONAL USES:
Tanzania: root and bark decoctions are drunk as a remedy for malaria: *Haerdi (1964) p. 69*
TOXICITY:
Safe when anti-nutrients eliminated: *Ugwu et al. (2006)*

Trema orientalis (L.) Blume [*Celtis orientalis* L.; *T. guineensis* (Schumacher & Thonn.)
Ficalho] (Celtidaceae)
DESCRIPTION:
Rapidly-growing, light-demanding evergreen tree up to 15m high; widespread as pioneer in
mutesyi and gallery forests in wetter parts
VERNACULAR NAMES:
E: Pigeonwood; **Afr**: Hophout; **B**: mutumpu (3c, 14) **K**: kambombo (14) **Ny**: mcende(3e,
14), mjaja-juni, mpefu, mpepu, mtesa (9) **Tk/ To**: mululwe (14, 23)
PLANT PARTS USED:
bark, aerial parts, leaves
TRADITIONAL USES:
Madagascar: the bark is used as a fever cure: *Githens (1948) p. 107;* the aerial parts are
decocted for the same purpose: *Rasoanaivo et al. (1992) p. 125*
Nigeria: the leaves are cooked in earthenware pots, and the decoction is collected in vessels
or used as a steam bath: *Bhat (1990) p. 388*

Tricalysia pallens Hiern [*T. myrtifolia* S. Moore] (Rubiaceae)
DESCRIPTION:
Irregularly- branched evergreen shrub up to 9m high, widespread in *mutesyi* forests
VERNACULAR NAMES:
E: Velvet Jackal-coffee; **B**: mukwikwa (3, 3c)
PLANT PARTS USED:
roots
TRADITIONAL USES:

Tanzania: a root decoction is drunk against malaria: *Haerdi (1964) p. 147*

Tricalysia sp. aff. **Tricalysia coricea** subsp. nyassae (Rubiaceae)
VERNACULAR NAMES:
not known
PLANT PARTS USED:
roots, leaves
TRADITIONAL USES:
Tanzania: a root decoction mixed with leaf juice is drunk, and the body bathed with a root decoction: *Haerdi (1964) p. 147*
Madagascar: the aerial parts are decocted as a febrifuge: *Raoanaivo (1992) p. 124*

Trichilia emetica Vahl [*T. roka* Chiov.] (Meliacheae)
DESCRIPTION:
A large handsome evergreen tree, up to 20m. tall, widespread in fringing forest, in damp riparian forest and woodland,often planted as shade tree
VERNACULAR NAMES:
E: Natal/Cape Mahogany, Christmas Bells; **Afr**: Rooiessenhout; **B/K** musikisyi, musyikisyi; **B/ Ku/ Ny**: musikisi; **I/ Lo/ Tk/ To/ Tu**: musikili, musigili; **Lo**: umsikili; **Ny**: misikisi, msicisi, mwabvi, msikisi; **Ny/ Tu**: Msikizi, Musikizi
PLANT PARTS USED:
roots, root bark, leaves, whole plant
TRADITIONAL USES:
Burkina Faso: (a) a handful of the root bark scrapings is macerated for 3 days with a small handful of red peppers and 3 crumbled millet biscuits. The liquid to be drunk 4x daily for 2 days, with a pinch of the dried powdered sediment. (b) a piece of root 10cm x 2cm is boiled for 20 min. to produce a glass of liquid, used as a wash twice daily for 2 days: *de la Pradilla (1988)*
Côte d'Ivoire: the plant is used to treat malaria: *Kerharo et al. (1950)*
Malawi: a leaf infusion is drunk to relieve fevers: *Morris (1996)*
Mozambique: the leaves are used to treat malaria: *Mulhovo (1999)*
S. Africa: the root is used as a remedy for fever: *Githens (1948); Watt et al. (1962)*. The bark is used as a purgative enema to produce sweating: *Watt et al. (1962)*
TESTS:
In vitro tests on Sudanese materials in 1999 revealed a very high level of antiplasmodial activity, IC$_{50}$ 2.5µg/ml: *El Tahir et al., (1999b)*
In vitro South African tests on leaves and twigs in 2004 showed an IC$_{50}$ 3.5 µg/ml: *Clarkson et al. (2004); Pillay et al. (2008)*
TOXICITY:
Highly toxic. *In vitro* bacterial & mammalian cell assays showed that the plant caused both DNA damage & chromosomal aberrations: *Fennell et al.(2004)*
Bryant included the species in his list of potentially dangerous medicines: *Bryant (1909)*. The whole seeds are poisonous, though said to be edible when husked: *Storrs (1979); Watt et al. (1962); Gelfand et al. (1985)*. Deaths have been reported: *Neuwinger (2000)*. Tests of leaves on Brine shrimp showed LC$_{50}$ 0.1389 and on Vero cell line 0.0933: *Brendler et al., (2010)*

Trichodesma zeylanicum (Burm.f.) R. Br. (Boraginaceae)
DESCRIPTION:
Erect herb or shrub up to 2m high, with well-developed taproot and blue/ white flowers
VERNACULAR NAMES:
E: Camel Bush, Cattle Bush; **B**: zungubele (1) **B/ Ku**: cilungu-nduwe (1) **Nm**: igunguru (7c)
Tu: cawaya, koswe (20)
PLANT PARTS USED:
flowers
TRADITIONAL USES:
Philippines: the flower is used as a sudorific: *Watt et al. (1962) p. 150*
TOXICITY:
The seeds contain *Pyrrolizidine* alkaloids, which can have serious long-term effects: *PROTA (2008) p. 616, 617*

Tridax procumbens L (Asteraceae)
DESCRIPTION:
Invasive noxious weed, hairy stem and leaves, yellow flowers
VERNACULAR NAMES:
E: Coat Buttons, Tridax Daisy; **Ny**: camsala, kalimwendo, namasa-kata (9), katendo (3e)
PLANT PARTS USED:
leaves
TRADITIONAL USES:
E. Africa: the Maasai chew the leaves, followed by a drink of water, as a remedy for malaria: *Glover & Samuel 3255,* quoted in *Burkill (1985) p. 499; Kokwaro (1976) p. 71*

Triumfetta rhomboidea Jacq. (Malvaceae)
DESCRIPTION:
Herb or shrub up to 1.3m; hairy and prickly. A weed of cultivation
VERNACULAR NAMES:
Nm: ilenda-lyahasi, sye-wiye (22) **Ny**: cinyambata, cisapira, khatambuzi (?), masace, mpala, mpempwe
PLANT PARTS USED:
leaves, roots
TRADITIONAL USES:
Côte d'Ivoire: a leaf infusion is used as a febrifuge: *Bouquet (1974) p. 170,* quoted in *Burkill (2000) p. 213*
Tanzania: a root decoction is used with other plants against malaria: *Burkill (2000) p. 213; Haerdi (1964) p. 84*
TOXICITY:
Ethanolic extracts of the plant were tested on rats, and found to be safe at dose 1000mg/kg, with no signs of toxicity: *Duganath et al. (2011)*

Triumfetta welwitschii Mast. var. welwitschii [*T. welwitschii* Mast. var. *decampsii* (De Wild & T. Durand) Brenan] (Malvaceae)
DESCRIPTION:
Perennial herb, growing annual stems from a woody rootstock, with bright yellow flowers. Conspicuous on burnt roadsides, grassland and woodland
VERNACULAR NAMES:

I: sikantyo (4, 5) **Tk**: cikolokozya (8)
PLANT PARTS USED:
tuber
TRADITIONAL USES:
Zimbabwe: a decoction of the tuber is mixed in milk and drunk as a fever remedy: *Gelfand et al. (1985) p. 185*
TESTS:
In vitro tests on the leaves revealed a high level of antiplasmodial activity (IC $_{50}$=3.6µg/ ml): *Clarkson et al. (2004).*
TOXICITY:
No reports of toxic effects on humans

Tulbaghia alliacea L.f. (Alliaceae)
DESCRIPTION:
Geophyte with rhizome up to 10cm long, smelling of onion when bruised
VERNACULAR NAMES:
E: Wild Garlic; **Afr**: Wilde Knoffel, Knoflook
PLANT PARTS USED:
roots
TRADITIONAL USES:
Transkei: the bruised roots are used to prepare a bath to reduce the temperature in fevers. The early Cape colonists used the bulb, decocted or made into soup, as a febrifuge: *Watt et al. (1962) p. 717*
TOXICITY:
Aqueous extracts showed cytotoxicity to certain cells, with possible adverse effects on cattle: *Treurnicht, F.T.(1997)*

Tylophoria sylvatica Decne.(Apocynaceae)
DESCRIPTION:
Herbaceous twiner, with dull purple flowers, growing in thickets
VERNACULAR NAMES:
Not known
PLANT PARTS USED:
whole plant
TRADITIONAL USES:
Nigeria: the plant is decocted as a remedy for long-standing fever: *Thomas,* quoted in *Burkill (1985) pp. 240-1*

Uapaca kirkiana Muell. Arg. (Euphorbiaceae)
DESCRIPTION:
Small much-branched evergreen tree up to 10m high, with dense round crown and edible orange yellow fruit. Widespread and abundant, often forming groves, often at edge of dry dambos
VERNACULAR NAMES:
E: Sugarplum, Masuku fruit; **B**: kanono, masukulu-bemba, masuku-nkuma, mukokola (10)
B/ I/ K/ Ku/ Lb/ Lo/ M/ Ny/ To/ Tu: musuku (1, 1c, 3, 3b, 3c, 4, 10, 14, 17, 19, 20, 23) **B/ I/ K/ Lb/ Tk/ To/ Tu**: musyuku (10) **I**: busuku (4), lusukwe (10), masuku (1c) **I/ Lo**: cilundu (3, 3b, 3d, 10, 14) **K**: masungwa (23), mukonkola (10), mukusu (3, 3b, 10, 19)

Ku: kasuku (1) **Ln**: kabofa (3, 3b, 14), mubula (10, 19, 23), musyukusu (19) **Lo**: muhaka (3d, 14), muhuku (16), mukuku (19), musugu (10) **Lv**: mupopolo (6, 10) **N**: muhuka (10) **Ny**: mpoto-poto (3e, 14), msuku (3, 3b, 3e, 9, 10, 14), mtoto (9) **To**: mubula (23)
PLANT PARTS USED:
roots
TRADITIONAL USES:
Tanzania: a root decoction is drunk against malaria: *Haerdi (1964) p. 99*

Uapaca nitida Muell. Arg. (Euphorbiaceae)
DESCRIPTION:
Small tree up to 7m high; widespread, often forming dense groves with other *Uapaca* spp, esp.near edges of dambos
VERNACULAR NAMES:
B/ K/ Ku/ Lb/ Le/ Ny: musokolowe (1, 3, 10, 14, 19) **B/ Ny**: kasokolowe (3, 3e, 9, 10, 14, 20, 23) **B/ Tu**: musokolowe-wafita (10, 20) **I**: cinunda (4, 10) **I/ K/ Lo/ Tk/ To**: mulundu (4, 10, 19, 23) **I/ Tk/ To**: munundu (4, 8, 17) **K**: musokobele (3, 14) **Le**: mulobelobe, musokolobe (17) **Ln**: mudengula (14) **Lo/ N/ Ny**: mulengo (3d, 9, 10) **Lo/ To**: cilundu (3, 3d, 14) **Ny**: khologo, mscera, msokolowe, mtoto (9), msecela (3e) **Ny/ Tu**: mnengo (3e, 8, 14) **To**: cilunda (17), mununda (10, 17), patwe (10)
PLANT PARTS USED:
roots
TRADITIONAL USES:
Tanzania: a root decoction is drunk against malaria: *Haerdi (1964) p. 100*
TESTS:
Kirby et al. investigated various extracts of the root bark for *in vitro* activity against the chloroquine and pyrimethamine-resistant *P.falciparum* K1 strain. Their results indicated good activity in the ethanolic and hexane extracts, but not in the aqueous extract. Some *in vivo* activity was also noted, using *P.berghei* in mice, but the toxicity was quite high. Steele et al. found that the *in vitro* activity could be attributed to betulinic acid, a triterpene, but that the compound did not exhibit any *in vivo* activity: *Schwikkard et al. (2002) p. 682*
TOXICITY:
See above

Uraria picta (Jacq.) DC. (Fabaceae, Papilionoideae)
DESCRIPTION:
Erect perennial herb up to 1.8m high with pink/violet/white flowers, found in floodplains and marshy areas in woodlands and grasslands
VERNACULAR NAMES:
Not known
PARTS USED:
leaves, roots
TRADITIONAL USES:
Nigeria:a leaf infusion is used as a febrifuge: *Burkill (1995) p. 467; Dalziel (1937) p. 266*
India: a root decoction is used against feverish chills: *Chadha (1976) p. 413,* quoted in *Burkill (1995) p. 467*

Urena lobata L. (Malvaceae)
DESCRIPTION:

Shrubby herb with pink flowers and deeply-lobed leaves; found on shores of lakes and rivers
VERNACULAR NAMES:
E: Caesar's Weed, Burr Mallow, Congo Jute, Hibiscus/Pink/Chinese Burr; **B**: citelela (3c, 9a)
Ku: lumanda-sanga (1) **Lo**: lukuku (19, 20)
PLANT PARTS USED:
roots, whole plant
TRADITIONAL USES:
Côte d'Ivoire: the whole plant is reputed as a febrifuge: *Kerharo et al.(1950) p. 66*
India: The root is crushed and given with water in cases of malaria after delivery of a child:
Singh & al. (1993), p. 71

Vangueria infausta Burch. [*V. tomentosa* Hochst.; *V. lasioclados* K. Schum.; *V. rupicola*
Robyns] (Rubiaceae)
DESCRIPTION:
Deciduous shrub 2m high, with light brown edible fruit. Widespread, in thicket and *cipiya*
woodland
VERNACULAR NAMES:
E: Wild Medlar; **Afr**: Velde Mispel; **I**: ngawe (4, 23) **Ku**: mapolo-yakalulu, mcoka, mucoka
(1) **Lb**: leyambaso, yina-nana (22) **Ku/ Le/ Ny/ To**: mungayi (1, 23) **Lo**: mubilo,
mumbwe-ngelenge, mupumu-lenge (23), mutoma-toma (3d) **Nm**: msada (22) **Ny**: cilya-
ndembo, kamwazi, macenda-kalulu, mjukutu, mkanda-ndembo, mpuluku-tutu, mziru (9),
mpfilu, muyogo-yogo (22), mpolo, mvilu (3e) **Tk**: mubu, mububa (23) **To**: mububu (3, 19,
23), umbubo (19, 23)
PLANT PARTS USED:
roots
TRADITIONAL USES:
Malawi & S. Africa: a decoction of the roots is used as a remedy for malaria: *Morris (1996)*
p. 463; Palgrave (2002) p. 1078
Zimbabwe: people of the Cibi district mix the roots with two other ingredients as remedy for
malaria: *Thompson 9, p. 33,* quoted in *Watt et al. (1962) p. 906*
TESTS:
An aqueous leaf extract was tested *in vitro* against a chloroquine-sensitive strain of *P.*
falciparum, and showed activity of IC$_{50}$ 10 – 20mg/ ml: *Nundkumar et al. (2002)*
TESTS
Tests on brine shrimps showed no toxicity at LC$_{50}$ 144.7 mg/ml: *Moshi et al.(2010)*

Verbena officinalis L. subsp. africana R. Fern. & Verdc. (Verbenaceae)
DESCRIPTION:
Perennial erect herb up to 1m high, with lobed leaves and mauve flowers
VERNACULAR NAMES:
E: Common Vervain
PLANT PARTS USED:
whole plant
TRADITIONAL USES:
Tropical countries generally: the plant is widely used as a diaphoretic and fever remedy:
Ambasta (1992) p. 671, quoted in *Burkill (2000) p.270; Quisumbing (1951) pp. 805-6,*
quoted in *Watt et al. (1962) p. 1054*

Vernonia amygdalina Delile [*V. randii* S. Moore] (Asteraceae)
DESCRIPTION:
Shrub or small tree up to 8m high, sometimes flowering as herb 1m high; on riverine
alluvium in drier areas, or on anthills on Plateau
VERNACULAR NAMES:
E: Bitter Tea Vernonia; **Afr**: Bittetee; **B**: cisobo (3, 3c) **Ny**: mfutsa, msangu-sangu, tsoyo
(9)
PLANT PARTS USED:
leaves, bark, roots
Africa: the leaves are widely used for fever as a quinine-substitute. For Ethiopia, see
Getahun (1975); for Guinea, see *Pobéguin (1912) p. 65;* for Nigeria, see *Ainslie sp. no. 225,
Oliver-Bever (1960) pp. 40, 90,* and *Singha (1965);* for Tanzania, see *Tanner 4876A,* all
quoted in *Burkill (1985) p. 502*
Angola: the bark of both root and stem, which is very bitter, is used as a tonic in fevers:
Watt et al. (1962) p. 296
E. Africa: the leaves are pounded to extract the juice, which is drunk to relieve fever:
Kokwaro (1976) p. 72 . Sometimes the paste is mixed with the leaves of *Ambrosia maritima*:
Adjanohoun, J.E., et al. (1993)
Ethiopia: the whole plant is decocted to make a medicine for malaria: *Gedif et al. (2002)*
Rwanda: The leaves are used to treat malaria: *Van Puyvelde et al. (1975); Chagnon (1984).*
The leaves are pounded and macerated in water to treat malaria: *Hakizamungu et al.
(1988).* Fresh leaves are ground, cooked in water with salt, and the liquid drunk to treat
malaria: *Ramathal et al. (2000)*
S. Africa: the bark and roots are taken by people suffering from fevers: *Palgrave (2002) p.
1134*
Uganda: the leaves are infused to treat malaria: *Hamill et al. (2000)*
Zimbabwe: the roots are powdered and eaten as a remedy for fever: *Gelfand et al. (1985)*
TESTS:
In vivo tests on leaves in 1988 found 67% inhibition of plasmodial activity in mice*:
Hakizamungu et al. (1988)*
In vitro tests on the leaves in 1990 found no antimalarial activity up to 499 μg/ml: *Weenen
et al. (1990),* cited in *Ramathal et al. (2000)*
In vitro tests on the leaves in 1996 found significant antimalarial activity, with IC_{50} value of
3.47μg/ml: *Bogale et al. (1996)*
In vitro tests on leaf extracts in 1999 produced an inhibition of between 70% and 100%
when tested at 2 to 0.6 mg/ml. At a concentration of 6 μg/ml, quinine dihydrochloride
showed a higher antiplasmodial acticity: *Tona et al.(1999)*
In vitro tests on the leaves found an IC_{50} of 10mg/ml: *de Madureira et al. (2002)*
TOXICITY:
In vivo tests on mice and rabbits found toxic effects from steroidal saponins at an ID_{50} of
1.12 g/kg: *Ohigashi et al. (1991), Akah et al. (1992), Igile et al. (1995)*
Tests of leaves on Vero cell line showed LC_{50} >1: *Brendler et al. (2010)*

Vernonia brachycalyx O.Hoffm. (Asteraceae)
DESCRIPTION:
Herb up to 4m high, found on dry forest edges or semideciduous bushveld
VERNACULAR NAMES:
Not known

PARTS USED:

leaves

TRADITIONAL USES:

Kenya: a leaf infusion is drunk against malaria by the Kipsigi and Maasai: *Beentje (1994) p. 566; Kokwaro (1976) p. 72*

TESTS:

In vitro tests have revealed antiplasmodial activity in *V. brachycalyx: Oketch-Rabah et al. (1999)*

The roots of *V. brachycalyx* yielded two 5-methylcoumarins, *2'-epicycloisobrachy-coumarinone epoxide*, and *cycloisobrachy-coumarinone epoxide*, both of which showed antiplasmodial activity against chloroquine-sensitive and chloroquine-resistant strains of *P.falciparum in vitro*. The leaves of *V.brachycalyx* yielded an active sesquiterpene dilactone, *16,17-dihydrobrachycalyxolide*: *Schwikkard et al. (2002) p. 681*

TESTS:

Brine shrimp tests on ethanol extracts showed only mild toxicity, at LC_{50} 33.9mg/ml: *Moshi et al.(2010)*

Vernonia colorata (Willd.) Drake subsp. colorata (Asteraceae)

DESCRIPTION:

Shrub or small tree, up to 6m high; in open woodland, riverine fringes, edges of vleis and pans

VERNACULAR NAMES:

E: Lowveld Tree Vernonia, Star-flowered Bitter Tea; **B**: cisobo (3, 3c) **Nm**: kilula-mkunja, msayo (22)

PLANT PARTS USED:

leaves, roots

TRADITIONAL USES:

Côte d'Ivoire: a leaf decoction is mixed with honey and drunk in frequent small doses as a febrifuge: *Kerharo et al.(1950)*

W. Africa: the roots are said to contain alkaloids, and are used to treat fevers: *Palgrave (2002); Watt et al. (1962)*

Zimbabwe: it is used against fever, but no details are given: *Gelfand et al.(1985)*

TOXICITY:

Highly toxic, according to *in vitro* bacterial & mammalian cell assays showing that the plant caused both DNA damage & chromosomal aberrations: *Fennell et al.(2004)*

The LD_{50} tested on mouse is about 10g/kg; tests of leaves on Brine shrimp showed LC_{50} 0.0748 and on Vero cell line 0.0359: *Brendler et al., (2010) pp 249, 287*

Vernonia natalensis Sch. Bip. ex Walp. (Asteraceae)

DESCRIPTION:

Herbaceous species up to 1m high, with velvety silky foliage, in vleis, grassland, and *miombo* woodland.

VERNACULAR NAMES:

E: Silver Vernonia

PLANT PARTS USED:

bark, roots, leaves

TRADITIONAL USES:

S.Africa: the bark is used against fevers and malaria: *Githens (1948)*. A leaf decoction is used to treat fevers: *Neuwinger (2000)*.

Tanzania: a root decoction with leaf juice is drunk for malaria: *Haerdi (1964)*

TESTS:

In 2003, *in vitro* tests on lipophilic extracts revealed strong antiplasmoidal activity with an IC_{50} value of 12.1 μg/ml: *Kraft et al. (2003)*

In vitro tests on the leaves revealed a high level of antiplasmodial activity (IC_{50} 4.7μg/ml): *Clarkson et al. (2004)*

TOXICITY:

No toxicity revealed: *Chagnon (1984)*

Vernonia spp.

VERNACULAR NAMES:

Ku: matongwe (1); **Ln**: isoda(3); **Lo**: mubalusika (3); **Ny**: cinkala-matongwe, cipanda-mwavi, futsa, kalonga, kaluwa-luwa, kanyungwa, kapila-pila, kazota, macula, msyaba, soyo (3e), mathodwa (22); **Tu**: soyo (8)

PLANT PARTS USED:

leaves

TRADITIONAL USES:

E. Africa: leaves are mixed with water & drunk as a malaria remedy: *Kokwaro (1976) p. 76*

Malawi: the Chewa use a leaf infusion, drunk while hot, to treat relapsing fever: *Watt (1992) p. 299*

Vetiveria nigritana (Benth) Stapf. [*V. zizanioides* (L.) Nash var. *nigritana* (Benth.) A. Camus] (Poaceae)

DESCRIPTION:

Robust perennial grass, clumped, with culms up to 2m high; at stream-sides and swampy flood-plains

VERNACULAR NAMES:

E: English Vetiver, Black Vetiver grass; **I**: luhundu (19) **I/ To**: luvundu (4, 17, 19) **Le**: muyobo (17) **Lo**: liyamba (3d, 19), samba (19) maamba (16, 19) **Ny/ Tu**: kaluvela (3e, 8) **Tk**: makelele (8)

PLANT PARTS USED:

roots

TRADITIONAL USES:

W. Africa: the roots are used against fever: *Burkill (1994) p. 488*

TOXICITY:

Tests on molluscs showed cytotoxic activity: *Kela et al.(1989)*

Vitex buchananii Baker ex Gurke (Verbenaceae)

DESCRIPTION:

Creeper, liana, or small tree up to 6m high, in patches of evergreen forest and bush clumps, with bright orange fruit

VERNACULAR NAMES:

E: Orange-fruit Fingerleaf

PLANT PARTS USED:

roots

TRADITIONAL USES:

Tanzania: a root decoction is drunk, and the body bathed with a leaf decoction: *Haerdi (1964) p. 153*

Wahlenbergia denticulata A.DC. (Campanulaceae)
DESCRIPTION:
Slender herb with small blue flowers, found in open sandy soils
VERNACULAR NAMES:
Not known
PLANT PARTS USED:
leaves
TRADITIONAL USES:
Zimbabwe: patients are bathed in a leaf infusion: *Gelfand et al.(1985) p. 231*

Waltheria americana var. americana L. (R.Br. ex Hosaka [*W. indica* L.] (Malvaceae)
DESCRIPTION:
Common hairy savanna weed up to 2m high with clusters of yellow flowers
VERNACULAR NAMES:
E: Sleepy Morning; **B**: kacencete (3c) **Lo**: mulembwe (3d) **Ny**: kabanti (3e)
PLANT PARTS USED:
leaves, roots, whole plant
TRADITIONAL USES:
Cuba, Philippines: the plant has been used as a febrifuge: *Watt et al. (1962) p. 1017*
Senegal: an infusion of the crushed roots is drunk morning and evening as a remedy for malaria and other fevers: *Kerharo p. 102*
Tropical Africa: The leaves are infused against fever: *Burkill (2000) p. 560*
W. Africa: the plant is administered to children to increase their resistance to fevers: *Watt et al. (1962) p. 1017.* Root preparations are used against fever: *Hedberg et al.(1983) p.251; Kerharo (1964)*
TESTS:
Plant extracts exhibited moderate antiplasmodial activity of ID_{50} <50 against *P.falciparum* 3D7 chloroquine-sensitive strain: *Jansen et al. (2010)*

Withania somnifera (L.) Dunal (Solanaceae)
DESCRIPTION:
Herb or woody shrub up to 2m high, with yellow flowers and red fruit
VERNACULAR NAMES:
E: Indian Ginseng, Winter Cherry, **B**: mutuntula (3c) **B/ Lo**: mutuntulwa (3) **Nm**: kuviya (1a, 22), mtemuwa-syimba (7c), mgeda (1a)
PLANT PARTS USED:
roots
TRADITIONAL USES:
S. Africa: the Zulu administer an enema made from the decorticated roots to feverish infants: *Watt et al. (1962) p. 1010*
TESTS:
Methanol and aqueous extracts from the plant were tested against resistant strains of *P. falciparum*: *Kirira et al., (2006)*
TOXICITY:
One report from the Netherlands describes thyrotoxicosis following treatment by the plant:

Van der Hooft et al.(2005), cited in *PROTA (2008) pp. 631, 633*

Xenostegia tridentata (L.) D. F. Austin & Staples subsp. angustifolia (Jacq.) D. F. Austin & Staples [*Merremia angustifolia* Jacq.; *M. tridentata* (L.) Hallier f. subsp. *angustifolia* (Jacq.) van Oostr.] (Convolvulaceae)
DESCRIPTION:
Variable polymorphic perennial, with prostrate or twining stems up to 2m long and lemon-white flowers, in open woodland, grassland, roadsides, disturbed soil
VERNACULAR NAMES:
E: Arrowleaf Morning Glory; **Nm**: migwa-sungu (1a, 22) **To**: nkunzano (5b)
PLANT PARTS USED:
whole plant
TRADITIONAL USES:
Tanzania: the plant is macerated and steeped for a week to make a wash for infants with malaria: *Burkill (1985) p. 550; Haerdi (1964) p. 179*

Xeroderris stuhlmannii (Taub.) Mendonca & E. P. Sousa [*Ostryoderris chevallieri* (Dunn) Roberty; *O. stuhlmannii* (Taub.) Dunn ex Harms] (Fabaceae, Papilionoideae)
DESCRIPTION:
Medium-sized tree up to 30m high, widespread in deciduous woodland and bush in hotter and drier areas
VERNACULAR NAMES:
E: Wind-pod; **Afr**: Vlerkboom; **B/ Ny/ Tu**: mutondo-tondo (3, 3a, 20) **I**: mulombe (4, 17) muvwa-malowa (19) namatuli (17, 23) **I/ Lo/ To**: muzwa-malowa (3, 3a, 3d, 14, 19) **K**: muniya (3a) **Ku**: mbangozi (1) **Ku/ Ny/ Tu**: citondo-tondo (1, 3e, 8, 14, 20) **Le**: muswa-mawa (19) **Nm**: luengi, mnene, mnye- nye, mubale (22) **Ny**: ciyumbu, mlonde, mtutumuko, nya-maroba (9), mlombeya (3e, 14), ntondo-tondo (1) **Tk**: simwami (8) **To**: mufungu, muhunu (23), muntundulu (19), musolo-koto (22), mutundula (3, 14)
PLANT PARTS USED:
leaves
TRADITIONAL USES:
E. Africa: the boiled leaves are plastered all over the patient's body as a febrifuge: *Burkill (1995) p. 482*
TOXICITY:
Reported in Burkill as poisonous at certain times of year

Ximenia americana L. (Olacaceae)
DESCRIPTION:
Shrub or small bushy spiny tree, up to 4m high, in evergreen mountain forest, thickets and open woodland, on anthills on Plateau
VERNACULAR NAMES:
E: Blue Sourplum; **Afr**: Blausuurpruim; **B**: mulebe (3, 14) **K/ Lo/ To**: muŋomba (2, 3, 3d, 14, 23), mutente (2, 3, 14, 23) **Ku/ Ny**: kamulebe (1, 3e), mtengene (1) **Ln**: musoŋa-soŋa (21) muvulama (2, 3, 14) **Lo**: mulutuluha (3d, 23), mulutulwa (14), muŋomba-gomba, mutente, musyongwa-syongwa (3d) **Lo/ Ln**: musongwa-songwa (3, 14) **Lo/ Tk**: muŋomba (8) **Nm**: mdogo (22), mpingi, mtundwa (1a, 22) **Ny**: kamulebe, ntengele (14), mdya-ncembele, ngomba (3e), mpinji, mtengere (9), mtundu (9, 14), mulundu-lukwa (3, 14) **To**: muconfwa (2, 14), musomvwa (22) **Tu**: ntengeni (8)

PLANT PARTS USED:
roots, leaves
TRADITIONAL USES:
Tanzania: the root is used as a febrifuge: *Bally (1937), p. 18; Watt et al. (1962) p. 803*
Central Africa: the leaves, which contain tannin and resins, are used as a fever remedy: *Githens (1948), p. 110; Watt et al. (1962) p. 803*
West Africa: the freshly cut root is applied locally for feverish headaches, and the bark rubbed on the skin to relieve fever: *Watt et al. (1962) p. 803*
TOXICITY:
Aqueous stem bark extracts tested on mice revealed the presence of cardiac glucosides, flavonoids, saponins & tannin. The results suggest that the extracts are not acutely toxic to the mice: *Maikai et al. (2008)* . Tests on molluscs showed cytotoxicity: *Kela et al. (1989)*

Ximenia caffra var. caffra Sond. (Olacaceae)
DESCRIPTION:
Spinous shrub up to 5m high; widespread in open woodland, esp. in hotter and drier areas
VERNACULAR NAMES:
E: Kaffir Plum, Sourplum; **Afr**: Suurpruim; **B**: mulebe (3, 10, 23) **I**: lumina-nzoka (4, 5, 10, 17) **I/ K/ Lb/ Le/ Tk/ To**: muŋomba (3, 4, 5, 8, 10, 17, 23) **K/ Lb/ Tk**: mutente (3, 10, 23) **Ln**: lusongwa-songwa (10), musombolo-mbo (6) **Lo**: mulutuluha (3d, 23), mulutulwa (3), musambya (3d), musyonga-wasyonga (16) **Lv**: mufulombo-mbo (10), musongo-songo (6, 22) **M**: mulebe (10) **N**: muŋomba-ŋomba, musyongwa-syongwa (10) **Nm**: mkubwa, mnembwa (22) Ny: mdya-ncembele (3e), mpinji (9, 10), mtumburuka, pinji (9), mpinji-pinji (23), mungomba (1), ngomba (3e), umtu-nduka (22) **To**: muŋombwa (10, 23)
PLANT PARTS USED:
roots, leaves
TRADITIONAL USES:
Tanzania: a root decoction and leaf sap are drunk against malaria: *Gelbrand, p. 245; Haerdi (1964) p. 108, 109; Watt et al. (1962) p. 803*
Zambia: the leaves are chewed against malaria: *Fowler (2000) p. 827; (2002b) p. 42*
TOXICITY:
Brine shrimp tests gave an LC_{50} figure of 11.25mg/ml, showing high toxicity: *Moshi et al.(2004)*

Xylopia acutiflora (Dunal) A.Rich (Annonaceae)
DESCRIPTION:
Shrub or tree, 2 – 20m high, sometimes scrambling, in fringing and evergreen forest
VERNACULAR NAMES:
E: Mountain Spice, African Pepper
PLANT PARTS USED:
bark
TRADITIONAL USES:
Côte d'Ivoire: the bark is used to treat fever pains: *Bouquet (1974) p. 19,* quoted in *Burkill (1985) p. 129*

Xysmalobium undulatum (L) Aiton f. [*Asclepias undulata* L.] (Apocynaceae)
DESCRIPTION:
Robust geophytic herb up to 2m high, sprouting annually from rootstock, with yellowy

flowers and hairy light-green fruit, in open or moist grassland
VERNACULAR NAMES:
E: Milk Bush, Milkwort, Wild Cotton; **Afr**: Bitterhout, Bitterwortel, Melkbos
PLANT PARTS USED:
roots
TRADITIONAL USES:
S.Africa: The root infusion is used to treat malaria: *Watt et al.(1962) p.139*
Zambia: the root has been widely used as remedy for malaria, typhoid, and other fevers: *Haapala et al.(1994) p. 22*
TOXICITY:
The glucoside *xysmalobin* contained in the root is very toxic to animals: Watt et al.(1962) p. 139. Tests of leaves on Brine shrimp showed LC_{50} 0.0836 and on Vero cell line 0.0261: *Brendler et al., (2010) pp 271, 287*

Zanha africana (Radlk.) Exell [*Dialiopsis africana* Radlk.] (Sapindaceae)
DESCRIPTION:
Small deciduous tree up to 7m (rarely up to 15m); widespread, us. in *miombo* woodland
VERNACULAR NAMES:
E: Velvet-fruit Zanha; **B**: cana-lume (22), cibanga, mutikalamo (3c) **B/ Lb/ Ny**: cibanga-lume (1, 3, 3b, 3c, 7a, 10, 14) **B/ Ku**: muzumba-mulyamba (1) **I/ K**: sanga-lwembe (3, 3b, 4, 5, 14, 17) **Ku**: mataudi (1) **Ku/ Ny**: mulya-mbesyi (1) **Lb**: musanga-lemba (10) **Le**: sanga-lwendo (17) **Ln**: kalayi (3), kayombo-yombo (10), muyombo-yombo (3, 3b, 14) **Lo**: musanga-lwembe (3, 3b, 3d, 7a, 14) sanga-lumbe (10) **Lv**: muyombo-wazimo (10) **M**: cibanga-lumya, musakala (10) **Nm**: mkalya (7c) **Ny**: canga-luce, mtala-banda, mtutumuko (9), canga-lume (3e, 23), cibukuzela (3, 3b, 3e, 7a, 14), mju-ju, mlimbauta, mtalala (9, 10), mkanga-luce, msefu (3e), mzakaka (3) **Tk**: sanga-lwembwe (8) **To**: mumvunyu (17), mutalala (3, 3b, 7a, 14) **Tu**: caangalume (8)
PLANT PARTS USED:
leaves, roots
TRADITIONAL USES:
Tanzania: the ashes of the leaves are rubbed into scarifications on the temples as a relief for headaches and malaria: *Haerdi (1964) p. 124*
Zimbabwe: the powdered root is inhaled as a remedy for fever: *Gelfand et al.(1985) p. 179*
TOXICITY:
Seed reported to be poisonous, containing 10.5% *saponin*. A root infusion caused renal irritation & fatal coma: *Watt et al. (1962) p. 930* . In vitro tests showed trypanocidal activity with IC_{50} value <20mg/ml: *Nibret et al. (2009)*

Zanha golugensis Hiern (Sapindaceae)
DESCRIPTION:
Tree up to 25m high, with round much-branched crown, in open woodland, often on anthills, densely wooded ravines, close-canopy evergreen forest
VERNACULAR NAMES:
E: Smooth-fruit Zanha; **Nm**: mkalia (1a, 22) **Ny**: comba-muluzi, mkwilyo, mtutumuko (9)
PLANT PARTS USED:
bark
TRADITIONAL USES:
Tanzania: a bark decoction is taken for malaria: *Burkill (2000) p.34; Haerdi (1964) p. 125*

Zanthoxylum chalybeum Engl. [*Fagara chalybea* (Engl.) Engl.] (Rutaceae)
DESCRIPTION:
Small tree up to 13m high, strong curved spines on bole and branches; from Solwezo to Lake Bangweule, mostly on anthills
VERNACULAR NAMES:
E: Kundanyoka Knobwood; **B**: cipolo (23), pupwe (3, 3c, 7a, 14, 23), pupwe-culu (3c, 14)
Ln: cipupa (3, 7a, 14), musamba-mayele (11, 11a) **Nm**: mlungu-lungu (22) **Ny**: mcodzi (9), mulungu-culu (3, 3e, 7a, 9, 14)
PLANT PARTS USED:
bark
TRADITIONAL USES:
E. Africa: a decoction of the bark is drunk as a remedy for malaria: *Kokwaro (1976) p. 196*
Zambia: the root bark has been used as a substitute for quinine: *Haapala et al. (1994) p. 23*
TESTS:
Aqueous and other extracts of the roots, leaves and stem bark were tested *in vitro* against *P. falciparum*, and showed strong antiplasmodial activity of $IC_{50}<10mg/ml$: *Gessler et al. (1994)*
Aqueous and methanol extracts were tested *in vitro* against chloroquine sensitive and resistant strains of *P. falciparum*, and showed activity at $IC_{50}3.65$ mg/ml: *Rukunga et al. (2008)*
Root bark extracts of the plant were tested in *vitro* against the chloroquine-resistant *P.falciparum* strain (W2), and showed activity of $IC_{50} <15mg/$ ml: *Muganga et al. (2010)*

Ziziphus abyssinica Hochst. ex A. Rich. (Rhamnaceae)
DESCRIPTION:
Shrub or small tree up to 13m high, in open woodland, wooded grassland, and along river banks, with edible shiny-red berries June-Sept.
VERNACULAR NAMES:
E: large jujube; **B**: kala-nangwa, kangwa (14), kalobwe, kankole-nkole (3c) **B/ Ku/ Ny**: kankande (1, 9, 14) **I**: mukwa-mbanziba, mungwa-malembe, mutantwa-sokwe (4, 5), sikomwa (?) (4, 17) **I/ K**: kankona (4, 5, 10, 14, 19) **I/ Le/ Lo**: mukona (4, 5, 19, 17, 23) **Ln/ Lo**: mukwata (3, 3d, 14) **Ln/ Lv**: mupundu-kaina (6, 22) **Lo**: monganga (22), mukalo (3d, 10, 14, 19), mutjelele (10), mucaluwe (3d, 14), muyambwe, mukanane (19) **Lo/ To**: musawu (10, 19) **Ny**: mlasya-wantu (3e, 14), mpenjere (9) **Tk**: mucecete (8) **To**: mooba-nkonono (17), mucenjete (10), mwicecete (14) muzau, mucejete (10, 17), mujejete (17)
Tu: kankande (10), muceceni (8)
PLANT PARTS USED:
bark, roots
TRADITIONAL USES:
Zambia: An infusion of the bark and roots is drunk to treat malaria: *Vongo (1999)*
TOXICITY:
Tests on molluscs showed cytotoxic activity: *Kela et al.(1989)*

Ziziphus mucronata Willd. subsp. mucronata (Rhamnaceae)
DESCRIPTION:
Small to medium tree, up to 12m high, with spreading canopy, in variety of habitats, often on anthills

VERNACULAR NAMES:
E: Buffalo-thorn; **Afr**: Blinkblaar-wag-'n-bietjie; **B**: kangwa, kalana-ngwa (3b) **I**: mausu (1c, 4) **K**: kankona (3b) **Ln**: cikwata (21), munkona (3b) **Lo**: mukala (22), mukalo (3d, 23), mumbukusyu (3d) **Nm**: ilogero (7c), kagowele, kalembo, mgugunu (22) **Ny**: kankande (3b, 9, 23), msawu-wathengo (9), mukusau (1) **Tk**: muceceti (8) **Tu**: muceceni (8)
PLANT PARTS USED:
whole plant
TRADITIONAL USES:
<u>Côte d'Ivoire</u>: the plant has a reputation as a febrifuge: *Kerharo et al.(1950) p. 143*
TOXICITY:
Highly toxic. *In vitro* bacterial & mammalian cell assays showed that the plant caused both DNA damage & chromosomal aberrations: *Fennell et al. (2004)*
Extracts were evaluated for potential genotoxic effects. The 90% methanol extract was mutagenic in strain TA98: *Elgorashi et al. (2003)*

Zornia glochidiata DC. (Fabaceae, Papilionoidaceae)
DESCRIPTION:
Annual herb, becoming much-branched and straggling, 10-70m high, with small yellow/reddish bristly flowers, in grassland, woodland, riverbanks, roadsides and old cultivation, sandy soils
VERNACULAR NAMES:
Ny: kandulwa, tengwa (9) **To**: mukala-lusaka (5b)
PLANT PARTS USED:
whole plant
TRADITIONAL USES:
<u>Senegal</u>: the plant is pounded in water and the lather mixed with butter. This is rubbed on the body to relieve feverish chills: *Berhaut (1967) p. 17,* quoted in *Burkill (1995) p. 483; Dalziel (1937) p. 271*

SECTION 2: BLACKWATER FEVER REMEDIES

Asparagus sp.
VERNACULAR NAMES:
K: mukuta-nkwaji (3) **Ku**: kalupo-potwa, nakumanga (1) **Ku/ Ny**: kalipo-poto, kalipo-potwa (1, 3e) **Ln**: kaswama-maji (3) **Lo**: ilutwa (3) **Ny**: mkola-nkanga (3e), nakumanga (3, 3e) **To**: mutanda-myoba (3)
PLANT PARTS USED:
roots
TRADITIONAL USES:
<u>Zambia</u>: in the Mufulira district, a root decoction is used as a remedy: *Watt et al. (1962) p. 690*

Bauhinia reticulata DC. [*Piliostigma reticulatum* Hochst.] (Fabaceae, Caesalpinioideae)
DESCRIPTION & VERNACULAR NAMES: See under "malaria remedies"
PLANT PARTS USED:
leaves, bark, root
TRADITIONAL USES:
<u>Zimbabwe</u>: the plant is used as a remedy for malaria & blackwater fever: *Watt et al. (1962) p. 560*

Cannabis sativa L. (Cannabaceae)
VERNACULAR NAMES: See under "malaria remedies"
PLANT PARTS USED:
whole plant
TRADITIONAL USES:
<u>Zimbabwe</u>: the plant is used as a remedy: *Watt et al. (1962) p. 762*

Cassia abbreviata Oliver subsp. abbreviata [*C. beareana* Holmes; *C. granitica* Baker f.; *C. abbreviata* Oliver var. *granitica* (Baker f.) Baker f.] (Fabaceae, Caesalpinioideae)
DESCRIPTION & VERNACULAR NAMES: See under "malaria remedies"
PLANT PARTS USED:
whole plant
TRADITIONAL USES:
<u>E. Africa</u>: the plant is used as a remedy: *Bally (1937) p. 15.*

Citrullus lanatus (Thunb.) Matsum & Nakai (Cucurbitaceae)
DESCRIPTION:
Vine-like scrambler bearing large fruit
VERNACULAR NAMES:
E: Watermelon; **B**: citanga-menzi (1), citatakunda, ntanga (3c, 9a), litanga, mabiliki, makabe (20) **B/ I/ Lo**: namu-nwa (3d, 4, 18, 20) **B/ Ny/ Tu**: ntanga (3c, 3e, 9a, 20) **I/ Lo/ To**: matanga (18, 19) **Lb/ Le/ K**: namu-kondowe (18, 19) **Ln**: nsowa (18, 19) **Lo**: namu-coko (18, 19) **Lv**: katanda, matanda (18, 19), matangala (19), matengala (18) **Ny**: cimwa-nyanza, civwembe (3e) **Ny/ Tu**: majote, mavwembe (20) **Tu**: ntanga-manze, vimwa-manze (20)
PLANT PARTS USED:
leaves

TRADITIONAL USES:
Tanzania: the leaf sap is drunk as a cure: *Haerdi (1964) p. 79*
TOXICITY:
The sprouting seed produces a mildly toxic substance in its embryo

Diplorhynchus condylocarpon (Muell. Arg.) Pichon [*Diplorrhynchus mossambicensis* Benth.] (Apocynaceae)
DESCRIPTION & VERNACULAR NAMES:
See under "malaria remedies"
PLANT PARTS USED:
roots
TRADITIONAL USES:
Angola: a strong decoction of the root is drunk to relieve blackwater fever: *Haapala et al.(1994) p. 16; Palgrave (1957) p. 25; (2002) p. 952; Smith (2004) p. 27; Watt et al. (1962) p. 83*

Lannea edulis (Sonder) Engl. (Anacardiaceae)
DESCRIPTION & VERNACULAR NAMES:
See under "malaria remedies"
PLANT PARTS USED:
roots
TRADITIONAL USES:
root bark
TRADITIONAL USES:
N. Soutspanberg: a root-bark decoction is drunk in frequent large doses as a cure for blackwater fever: *Watt et al.(1962) p.46*

Lippia javanica (Burm. f.) Spreng. [*L. asperifolia* Rich.; *L. whytei* Moldenke] (Verbenaceae)
DESCRIPTION & VERNACULAR NAMES:
See under "malaria remedies"
PLANT PARTS USED:
whole plant.
TRADITIONAL USES:
Zambia/ Zimbabwe: the plant is used as a remedy: *Watt et al.(1962), p. 1051*
TOXICITY:
Avoided by grazing animals (*Codd 1951*) but eaten to some extent *(Dyer 1937), quoted by Watt et al. (1962).* It contains *icterogenin* and is suspected as a cause of geeldikop: *Steyn (1949).* Ingestion by stock causes photosensitisation: *Watt et al. (1962).* No reports of toxic effects on humans.

Microglossa pyrifolia (Lam.) O. Kuntze (Asteraceae)
DESCRIPTION & VERNACULAR NAMES:
See under "malaria remedies"
PLANT PARTS USED:
leaves
TRADITIONAL USES:
W. Africa: a decoction is used as a specific: *Watt et al. (1962) p. 251*

Plumbago zeylanica L. (Plumbaginaceae)
DESCRIPTION & VERNACULAR NAMES:
See under "malaria remedies"
PLANT PARTS USED:
roots
TRADITIONAL USES:
S.Africa: a cold infusion of the root is used by Europeans to treat blackwater fever: *Bhatia (1933), quoted in Watt et al. (1962); Chhabra et al. (1991)*
Tanzania: the roots are used in decoction to treat blackwater fever: *Chhabra et al. (1991) p.144*
TOXICITY:
The long succulent roots contain an irritant poisonous juice, *plumbagin*: *Watt et al. (1962), Fanshawe et al. (1967)*

Pterocarpus angolensis DC. (Fabaceae, Papilionoideae)
DESCRIPTION & VERNACULAR NAMES:
See under "malaria remedies"
PLANT PARTS USED:
roots
TRADITIONAL USES:
Zambia, Zimbabwe:
A root decoction is used against blackwater fever: *Palgrave (1957) p. 332; (2002) p. 390*

Securidaca longepedunculata Fresen. (Polygalaceae)
DESCRIPTION & VERNACULAR NAMES:
See under "malaria remedies"
PLANT PARTS USED:
leaves, roots, seeds
TRADITIONAL USES:
E. Africa: a root decoction is used: (*Hedberg, p. 241*).

Senna occidentalis (L.) Link [*Cassia occidentalis* L.] (Fabaceae, Caesalpinioideae)
DESCRIPTION & VERNACULAR NAMES:
See under "malaria remedies"
PLANT PARTS USED:
whole plant
TRADITIONAL USES:
W. Africa: the plant was used as specific for blackwater fever & yellow fever: *Ainslie sp. no. 79,* quoted in *Burkill (1995) p. 161*

SECTION 3: RHEUMATIC FEVER REMEDIES

Clausena anisata (Willd.) Hook. f. ex Benth. [*C. inequalis* (DC.) Benth.] (Willd.) Hook. f. ex Benth. [*C. inequalis* (DC.) Benth.] (Rutaceae)
DESCRIPTION & VERNACULAR NAMES:
See under "malaria remedies"
PLANT PARTS USED:
leaves

TRADITIONAL USES:

S. Africa: the leaf has been used as a remedy, but Wicht has tested and found it useless: *Watt et al. (1962) p. 917*

Phyllanthus ovalifolius Forssk. [*P. guineensis* Pax] (Euphorbiaceae)
DESCRIPTION:
Much-branched, straggling shrub up to 4m high, widespread in savanna and evergreen forest
VERNACULAR NAMES:
E: Small-fruited Potato-bush; **B**: mupetwalupe (3) mupetwalupe-unono (3c) **Ny**: msukacuma, mtantha-nyele, njiriti, ntanda-nyerere (9) **Tk**: mutoto-toto (8)
PLANT PARTS USED:
roots
TRADITIONAL USES:
Malawi: the roots are pounded together with those of *kangaluce* and *mthunda*, (*Turraea nilotica?*) and with the bark of *Markhamia acuminata*, and rubbed into an incision: *Morris (1996) p. 306; Williamson (1975) p. 190*
Zambia: the roots are used to treat fever: *Storrs (1979) p. 288*

Salix mucronata subsp. mucronata Thunb. [*S. subserrata* Willd.; *S. safsaf* Trauty.] (Salicaceae)
DESCRIPTION:
See under "malaria remedies"
PLANT PARTS USED:
leaves
TRADITIONAL USES:
S. Africa: a leaf decoction is used to treat rheumatic fever: *Watt et al. (1962) p. 925*

SECTION 4: SCARLET FEVER REMEDIES

Artemisia afra Jacq. ex Willd. (Asteraceae)
DESCRIPTION & VERNACULAR NAMES:
See under "malaria remedies"
PLANT PARTS USED:
whole plant
TRADITIONAL USES:
S. Africa: vapour from a hot infusion is used to steam the throat and as a gargle in scarlet fever: *Watt et al. (1962) p. 201*

SECTION 5: TICK FEVER REMEDIES

Rourea orientalis Baill. [*Byrsocarpus orientalis* (Baill.) Baker; *B. tomentosus* Schell.] (Connaraceae)
DESCRIPTION & VERNACULAR NAMES:
See under "malaria remedies"

PLANT PARTS USED:

roots

TRADITIONAL USES:

E. Africa: a root infusion is used as a prophylactic against tick fever: *Kokwaro (1976) p. 77*

Vernonia sp. (Asteraceae)

PLANT PARTS USED:

leaves

TRADITIONAL USES:

Malawi: the Chewa use a leaf infusion, drunk while hot, to treat tick fever: *Kokwaro (1976); Watt et al. (1962) p. 299*

SECTION 6: TYPHOID FEVER REMEDIES

Allium sativum L. (Alliaceae)

See under "malaria remedies"

PLANT PARTS USED:

whole plant

TRADITIONAL USES:

Upper Volta: a regime is prescribe ed as a prophylactic for infectious diseases, including typhoid; it is taken for 20 days. Its bactericidal property is thought to be the active principle: *Burkill (1995) p. 490; de la Pradilla (1982)v.2 p. 53*

S. Africa: the plant is used against typhoid: *Watt et al. (1962) p. 676*

Capparis sepiaria L. [*C. laurifolia* Gilg & Gilg- Ben.; *C. citrifolia* Lam.; *C. subglabra* (Oliv.) Gilg & Gilg- Ben.] (Cappariaceae)

DESCRIPTION & VERNACULAR NAMES:

See under "malaria remedies"

PLANT PARTS USED:

fruit

TRADITIONAL USES:

India: a paste made from the fruits is given to treat typhoid: *Singh, V.K. & al. (1993), p. 72*

Jatropha curcas L. (Euphorbiaceae)

DESCRIPTION & VERNACULAR NAMES:

See under "malaria remedies"

PLANT PARTS USED:

roots

TRADITIONAL USES:

India: the root is crushed and mixed with water to treat typhoid: *Singh, V.K. & al. (1993), p. 73*

Launaea cornuta (Hochst. ex Oliv. & Hiern) C. Jeffrey (Asteraceae)

DESCRIPTION:

Perennial herb up to 1.3m high with yellow flowers, on alluvial soils and disturbed ground along rivers and roadsides

VERNACULAR NAMES:

E: Bitter Lettuce
PLANT PARTS USED:
roots
TRADITIONAL USES:
E. Africa: the roots are pounded and infused or decocted, the liquid being drunk as a remedy for typhoid: *Kokwaro (1976) p. 67*

Leonotis nepetifolia (L.) Ait.f. (Lamiacae)
DESCRIPTION:
See under "malaria remedies"
PLANT PARTS USED:
whole plant
TRADITIONAL USES:
E. Africa: the plant is said to be tonic and antispasmodic, useful in typhoid conditions: *Watt et al. (1962) p. 520*

Nicotiana tabacum (L.) (Solanaceae)
DESCRIPTION:
Widely planted
VERNACULAR NAMES:
E: Tobacco; **B/ K/ Ku/ Lb/ Le/ Ny**: fwaka (1, 3c, 3e, 9a, 17, 19) **I/ To**: tombwe (4, 15, 17) **Lo**: lifwaka, mufofo (3d) **Ny**: fodya (3e, 9), folo (3e)
PLANT PARTS USED:
juice
TRADITIONAL USES:
S. Africa: the juice left in a pipe after smoking is rubbed on the anus of a person with typhoid: *Watt et al. (1962) p. 987*

Oldenlandia capensis L. f. (Rubiaceae)
DESCRIPTION:
Slender spreading or ascending smooth annual herb, up to 50cm long, with white flowers, found in disturbed and waste land
VERNACULAR NAMES:
E: Madder; **Tu**: iteka (7c)
PLANT PARTS USED:
leaves
TRADITIONAL USES:
E. Africa: a leaf decoction is drunk 2x daily as a remedy for typhoid: *Kokwaro (1976) p. 188*

Ormocarpum trichocarpum (Taub.) Engl. (Fabaceae, Papilionoideae)
DESCRIPTION:
Shrub or small tree up to 5m high, with hairy stems and leaves with mauve flowers, in *Brachystegia, Acacia, Colophospermum* woodland and savanna and dry rocky hillsides
VERNACULAR NAMES:
E: Hairy-caterpillar Pod
PLANT PARTS USED:
leaves, stems
TRADITIONAL USES:

E. Africa: the leaves and young stems are pounded with water to extract sap, which is used against typhoid: *Kokwaro (1976) p. 141*

Oryza sativa L. (Poaceae)
VERNACULAR NAMES:
E: Rice; **B/ Lb/ M/ Ny**: mupunga (3c, 3e, 20) **Ln/ Lv**: loso (19) **Ny**: cambunga (3e) **Ny/ Tu**: mpunga (20)
PLANT PARTS USED:
grains
TRADITIONAL USES:
India: The grains are powdered and kept for a month together with cochineal insects, then taken to treat typhoid: *Singh, V.K. & al. (1993), p. 73*

Senna singueana (Delile) Lock [*Cassia singueana* Delile; *C. singueana* Delile var. *glabra* (Baker f.) Brenan] (Fabaceae, Caesalpinioideae)
DESCRIPTION & VERNACULAR NAMES:
See under "malaria remedies"
PLANT PARTS USED:
whole plant
TRADITIONAL USES:
E. Africa: the plant is used as a remedy: *Kokwaro (1976) p. 258*

Solanum tettense Klotzsch, var. renschii (Vatke) A. E. Gonc. [*S. renschii* Vatke]
DESCRIPTION:
Erect, woody, loosely branched spiny shrub, up to 2m high. Blue flowers and red berries; on slopes, in sandy, loamy or stonysoil
VERNACULAR NAMES:
E: Red Berry; **Afr**: Rooibessie, Bitterappel; **B/ Ku**: mutuntula (1)
PLANT PARTS USED:
roots
TRADITIONAL USES:
E. Africa: roots are used to treat typhoid fever: *Kokwaro (1976) p. 207*
TOXICITY:
Noxious to animals; unknown neurotoxin affects cerebellum leading to epileptic seizures: *Botha et al.(2002)*

Solanum sp. aff. S.indicum (Solanaceae)
VERNACULAR NAMES:
not known
PLANT PARTS USED:
roots
TRADITIONAL USES:
Tanzania: a root decoction mixed with leaf juice is drunk as a remedy: *Haerdi (1964) p. 177*

Withania somnifera (L.) Dunal (Solanaceae)
DESCRIPTION & VERNACULAR NAMES:
See under "malaria remedies"
PLANT PARTS USED:

whole plant
TRADITIONAL USES:
S. Africa: the plant is used as a remedy for typhoid: *Watt et al. (1962) p. 1010*

Xysmalobium undulatum (L) Aiton f. [*Asclepias undulata* L.] (Apocynaceae)
DESCRIPTION & VERNACULAR NAMES:
See under "malaria remedies"
PLANT PARTS USED:
roots
TRADITIONAL USES:
Zambia: the root has been widely used in treating typhoid, among other fevers: *Haapala et al.(1994) p. 22*

SECTION 7: YELLOW FEVER REMEDIES

Aloe spp.
VERNACULAR NAMES:
B: itembusya (3), litembusya (3c) **I**: sengwi-sengwi (4, 8) **Lo**: hopani, licenka (3) **Ln**: cilambwe (3) **Nm**: losa (1a) **Ny**: cintebwe, cintembwe, pozo (3e) **Tk**: tati (8) **To**: kanembe (3)
PLANT PARTS USED:
leaves
TRADITIONAL USES:
E. Africa: a leaf decoction is used to treat the spleen: *Kokwaro (1976) pp. 238, 285*

Antidesma venosum E. Meyer ex Tul. (Euphorbiaceae)
DESCRIPTION:
Much-branched shrub or small tree up to 8m high, with greeny-yellow flowers and glossy purple/black fruit; widespread in several types of woodland and on anthills
VERNACULAR NAMES:
E: Tassel-berry; **Afr**: Voelsitboom; **B**: cibenda, mulamba-bwato (3c), itompo, musamba-mfwa (3, 3c, 9a) **B/ Ku**: mwavi (1) **I**: munyo-nyamanzi (23) **K**: mukena-nene (3, 23), mutoma-mena (23), munya-nyamenda (3) **Ku**: mutanta-nyelele (1) **Ku/ Ny/ Tu**: mbwi-mbwi (1, 3e, 8) Lo: musole, simezi (3d), motoya (22), simeye (22, 23) **Nm**: msekelia, msekera (22) **Ny**: kamba-lame, mapira-kukutu, mdundira, mdya-kamba, mdya-pumbwa, mpika-maundu, mpondesa, mpotogola, mpululu, mpungulira, msanganya, msera, napose (9), kasyika-mpombo, mkolo, msinika, msunduzi (3e) **To**: syonga (22)
PLANT PARTS USED:
seeds
TRADITIONAL USES:
E. Africa: a seed infusion is used for liver complaints: *Kokwaro (1976) pp. 87, 285*
TOXICITY:
Highly toxic. In vitro bacterial & mammalian cell assays showed that the plant caused both DNA damage & chromosomal aberrations: *Fennell et al.(2004)*

Capsicum annuum var. frutescens Kuntze. (Solanaceae)
VERNACULAR NAMES:
See under "malaria remedies"

PLANT PARTS USED:
fruit
TRADITIONAL USES:
Zambia: the fruit is used as a remedy for yellow fever: *Nair (1967) p. 97*

Carica papaya L. (Caricacaeae)
VERNACULAR NAMES:
See under "malaria remedies"
PLANT PARTS USED:
leaves
TRADITIONAL USES:
Nigeria: a leaf extract is taken orally with salt for yellow fever: *Bhat (1990), p. 385*

Clutia abyssinica Jaub. & Spach [*C. glabrescens* Knauf; *C. pedicellaris* (Pax) Hutch.] (Euphorbiaceae)
DESCRIPTION & VERNACULAR NAMES:
See under "malaria remedies"
PLANT PARTS USED:
roots
TRADITIONAL USES:
E. Africa: a roots decoction is mixed with milk and drunk as a remedy for liver pains: *Kokwaro (1976) pp. 88, 285*

Craterispermum schweinfurthii Hiern [*C. laurinum* auct. non (Poir.) Benth.; *C. reticulatum* De Wild.] (Rubiaceae)
DESCRIPTION & VERNACULAR NAMES:
See under "malaria remedies"
PLANT PARTS USED:
leaves, bark
TRADITIONAL USES:
W. Africa: the leaves & bark are used as remedies for yellow fever: *Githens (1948) p. 114; Watt et al. (1962) p. 898.*

Crotalaria incana L. (Fabaceae, Papilionoideae)
DESCRIPTION:
Erect annual suffrutescent/ perennial herb with yellow flowers, found at roadsides, pastures, and other disturbed areas
VERNACULAR NAMES:
E: Fuzzy Rattlebox, Fuzzy/ Woolly/ Rattlepod
PLANT PARTS USED:
roots
TRADITIONAL USES:
Trinidad: a root infusion is taken as a remedy for yellow fever: *Wong (1976) p. 126,* quoted in *Burkill (1995) p. 313*
TOXICITY:
Poisonous: *Watt et al. (1962) p. 578*

Cussonia arborea Hochst. ex A.Rich. [*C. kirkii* Seem.] (Araliaceae)

DESCRIPTION & VERNACULAR NAMES:
See under "malaria remedies"
PLANT PARTS USED:
plant
TRADITIONAL USES:
<u>Côte d'Ivoire -Upper Volta</u>: Leafy twigs are used as a remedy: *Burkill (1985) p. 212; Kerharo et al.(1950) p. 172*

Cyperus articulatus L. (Cyperaceae)
DESCRIPTION & VERNACULAR NAMES:
See under "malaria remedies"
PLANT PARTS USED:
roots
TRADITIONAL USES:
<u>E. Africa</u>: the bitter root has been used as an anti-emetic in yellow fever: *Watt et al. (1962) p. 373*

Dissotis rotundifolia (Sm.) Triana [*Heterotis rotundifolia* (Sm.) Jac.-Fel.]
(Melastomataceae)
DESCRIPTION:
Procumbent or scandent creeper, with angular stems up to 90cm long rooting at the nodes and showy purple/crimson flowers 2" wide ; found in damp forest places and along streams
VERNACULAR NAMES:
E: Trailing / Dwarf Tibouchina, Spanish Shawl, Pink Lady
PLANT PARTS USED:
whole plant
TRADITIONAL USES:
<u>Côte d'Ivoire</u>: the crushed plant is infused to make a drink and a lotion as a remedy: *Kerharo et al.(1950) p. 46*

Ehretia obtusifolia Hochst. ex A.DC. [*E. fischeri* Gurke; *E. coerulia* Gurke] (Boraginaceae)
DESCRIPTION:
Deciduous shrub or small tree up to 6m high with straggling branches and orange/red fruit. Found in wooded grassland, woodland, thickets, and rocky areas
VERNACULAR NAMES:
E: Stamperwood; **B**: lunsa (3c) **I**: mufumpe, munkulungu (4, 23), munkuyu-cuulu (4, 17)
Ku: makula-mamena-ncembele, mamena-ncembele (1) **Lo**: muhongo (3d) **Nm**: syene (7c)
Ny: cimyangu (1, 3e) **Tk**: mukweyi-culu (8) **To**: mucuncu-cuulu(17), mugende-gende (23), musunsu-culu (3) **Tu**: cimyangu (8)
PLANT PARTS USED:
roots
TRADITIONAL USES:
<u>E. Africa</u>: a root decoction is drunk for spleen pains: *Kokwaro (1976) pp. 41, 285*

Euclea racemosa Murr. subsp. schimperi (A.DC.) F. White (Ebenaceae)
DESCRIPTION:
Slender evergreen shrub or small tree up to 12m high, with dark green glossy foliage, found in dune scrub

VERNACULAR NAMES:
E: Bush/ Dune Guarri; **Afr**: Duineghwarrie; **B**: mufitu-mbwe (3c) **Nm**: muenya (22) **Ny**: ndule, nyambuzi (9)
PLANT PARTS USED:
roots
TRADITIONAL USES:
E. Africa: root bark (decocted?) is used to treat spleen pains: *Kokwaro (1976) pp. 41, 285*

Gomphocarpus fruticosus (L.) Ait. f. (Asclepiadaceae)
DESCRIPTION:
Perennial slender shrub or shrubby herb up to 2m high, found near Cilanga, on roadsides and open terrain
VERNACULAR NAMES:
E: Milk Bush, Milkweed
PLANT PARTS USED:
roots
TRADITIONAL USES:
E. Africa: root decoction is drunk 3x daily to treat liver troubles: *Kokwaro (1976) pp. 31, 285*

Grewia villosa Willd. (Malvaceae)
DESCRIPTION:
Deciduous shrub up to 4 m. high, covered in pale silky hairs, with yellow clustered flowers and dull red fruit. Found in dry mixed woodland, rocky crevices
VERNACULAR NAMES:
E: Mallow Raisin, Rossberry, Round-leaf grewia; **Afr**: Malvarosyntjie; **Lo**: mutalu, namulomo (?19) **Nm**: msarasi (7c)
PLANT PARTS USED:
leaves
TRADITIONAL USES:
E. Africa: leaves are used for spleen troubles: *Kokwaro (1976) p. 214*

Lippia javanica (Burm. f.) Spreng. [*L. asperifolia* Rich.; *L. whytei* Moldenke] (Verbenaceae)
DESCRIPTION & VERNACULAR NAMES:
See under "malaria remedies"
PLANT PARTS USED:
whole plant
TRADITIONAL USES:
W. Africa: a decoction of the plant is regarded as a specific for yellow fever: *Watt et al. (1962) p. 251*

Maesopsis eminii Engl. (Malvaceae)
DESCRIPTION:
Medium-large quick-growing deciduous tree, up to 35m high, with straight clean bole, planted at Ndola
DESCRIPTION & VERNACULAR NAMES:
E: Umbrella Tree; **Ny**: ndunga (14)
PLANT PARTS USED:

leaves
TRADITIONAL USES:
Côte d'Ivoire: a leaf decoction is used as a diuretic and purgative against yellow fever: *Kerharo et al.(1950) p. 142*

Mangifera indica L. (Anacardiaceae)
DESCRIPTION & VERNACULAR NAMES:
See under "malaria remedies"
PLANT PARTS USED:
leaves
Côte d'Ivoire: the leaves are used as a febrifuge: *Kerharo et al.(1950) p. 168*

Microglossa pyrifolia (Lam.) O. Kuntze (Asteraceae)
DESCRIPTION & VERNACULAR NAMES:
See under "malaria remedies"
PLANT PARTS USED:
whole plant
TRADITIONAL USES:
W. Africa: a decocotion of the plant is used as a specific for yellow fever: *Watt et al. (1962) p. 250*

Momordica charantia L. (Cucurbitaceae)
DESCRIPTION & VERNACULAR NAMES:
See under "malaria remedies"
PLANT PARTS USED:
whole plant
TRADITIONAL USES:
Côte d'Ivoire: the crushed plant is stirred into water and used to bathe the patient. Eyedrops are made from the crushed leaves: *Kerharo et al.(1950) p. 42*

Ocimum gratissimum L. subsp. gratissimum [*O. gratissimum var. suave* (Willd.) Hook f.; *O. urticifolium* Roth] (Lamiaceae)
DESCRIPTION & VERNACULAR NAMES:
See under "malaria remedies"
PLANT PARTS USED:
leaves
TRADITIONAL USES:
Ghana: the leaves are used to treat fever with jaundice: *Ampofo p. 1,* quoted in *Burkill (1995) p. 25*

Paullinia pinnata L. (Sapindaceae)
DESCRIPTION & VERNACULAR NAMES:
See under "malaria remedies"
PLANT PARTS USED:
leafy twigs
TRADITIONAL USES:

Côte d'Ivoire: a decoction of leafy twigs forms part of a complex treatment for jaundice and yellow fever: *Bouquet (1974) p. 160,* quoted in *Burkill (2000) p. 28; Kerharo et al. (1950) pp. 164-5*

Salvadora persica L. (Salvadoraceae)
DESCRIPTION & VERNACULAR NAMES:
See under "malaria remedies"
PLANT PARTS USED:
roots
TRADITIONAL USES:
E. Africa: a root decoction is used to cure spleen trouble: *Kokwaro (1976) p. 198*

Senna occidentalis (L.) Link [*Cassia occidentalis* L.] (Fabaceae, Caesalpinioideae)
DESCRIPTION & VERNACULAR NAMES:
See under "malaria remedies"
PLANT PARTS USED:
leafy twigs
TRADITIONAL USES:
Côte d'Ivoire: a decoction of leafy twigs is drunk, the dregs used as a rub. The leaf juice is used as eye drops: *Kerharo et al.(1950) p. 105*

Tamarindus indica L. (Fabaceae, Caesalpinioideae)
DESCRIPTION & VERNACULAR NAMES:
See under "malaria remedies"
PLANT PARTS USED:
fruit
TRADITIONAL USES:
Côte d'Ivoire: the Ando use the fruits to treat yellow fever: *Visser pp. 45,46,* quoted in *Burkill (1975) p. 174*

Trema orientalis (L.) Blume [*Celtis orientalis* L.; *T. guineensis* (Schumacher & Thonn.) Ficalho] (Celtidaceae)
DESCRIPTION & VERNACULAR NAMES:
See under "malaria remedies"
PLANT PARTS USED:
leaves/ bark
TRADITIONAL USES:
Côte d'Ivoire:a leaf decoction is drunk, and also used in a bath:*; Visser p. 75,* quoted in *Burkill (2000) p. 226;* a decoction of the leaves and bark is used as a drink, gargle, lotion, fumigator, and vapour bath: *Kerharo et al. (1950) p. 129*

Vahlia dichotoma (Murray) Kuntze (Vahliaceae)
DESCRIPTION:
Erect branching annual 12 – 50cm high, with lanceolate leaves and yellow to white/pink flowers, found in damp fields and sandy river banks along Luangwa
VERNACULAR NAMES:
Ku: mwelu (1)
PLANT PARTS USED:

Roots
TRADITIONAL USES:
<u>Kenya</u>: the root is boiled with meat, and the soup drunk as a jaundice remedy: *Burkill (2000) p. 246*

Withania somnifera (L.) Dunal (Solanaceae)
DESCRIPTION & VERNACULAR NAMES:
See under "malaria remedies"
PLANT PARTS USED:
whole plant
TRADITIONAL USES:
<u>E. Africa</u>: the plant is used as a remedy: *Kokwaro (1976) p. 285*
TOXICITY:
In Somaliland, the berry has caused severe gastro-intestinal effects when eaten by children: *Watts et al.(1962) p.1011*; in the Netherlands, thyrotoxicosis resulted from its use: *van der Hooft et al.(2005)*

SECTION 8: INSECT REPELLENTS

Annona senegalensis Pers. [*A. chrysophylla* Bojer] (Annonaceae)
DESCRIPTION & VERNACULAR NAMES:
See under "malaria remedies"
PLANT PARTS USED:
leaves
TRADITIONAL USES:
<u>Nigeria</u>: the leaves are burnt in clay pots in an enclosed room to suffocate mosquitoes and other insects: *Bhat (1990) p. 384*

Cymbopogon citratus (DC.) Stapf. (Poaceae, Graminiaceae)
DESCRIPTION & VERNACULAR NAMES:
See under "malaria remedies"
PLANT PARTS USED:
whole plant
TRADITIONAL USES:
<u>Gabon</u>: the plant is used as an insect repellent: *Walker p.187,* quoted in *Burkill (1994) p. 210*

Datura metel L. (Solanaceae)
DESCRIPTION & VERNACULAR NAMES:
See under "malaria remedies"
PLANT PARTS USED:
whole plant
TRADITIONAL USES:
<u>Sind</u>: the plant is used as an insecticide: *Roark,* quoted in *Burkill (2000) p. 104; Watt p. 958*

Euphorbia tirucalli L. (Euphorbiaceae)
DESCRIPTION:
Dense, succulent bush or tree up to 10m high, yellow-green in appearance; widely planted

and wild in *Acacia* grassland, and forming clumps in *Combretum* woodland
VERNACULAR NAMES:
E: Rubber-hedge euphorbia; **Afr**: Kraalmelkbos; **B**: ipupu (20), lunsonga (2, 3, 3c, 9a) **B/ Ku**: lusonga, nkhazye (1) **I**: mubambala (4, 5, 17) **I/ To**: mulota (4, 19, 17) **K**: kilota (3, 9a) **Ku**: nkalanga (1) **Le**: lilota (17) **Ln**: lunana (3, 9a) **Ny**: mgaci, mkhadze, ngaci (9), kazi (3, 9a), nkadze (3e) **To**: mbala (17)
PLANT PARTS USED:
tree
TRADITIONAL USES:
E. Africa: the tree is planted near houses to deter mosquitoes: *Bally (1937) p. 14; Brenan, Roark,* quoted in *Watt et al. (1962) p. 415*

Euphorbia thymifolia L.(Euphorbiaceae)
DESCRIPTION & VERNACULAR NAMES:
See under "malaria remedies"
PLANT PARTS USED:
whole plant
TRADITIONAL USES:
India: the plant is used as a repellent: *Sastri(1952) p. 227,* quoted in *Burkill (1994) p. 78*

Hyptis spicigera Lam. (Lamiaceae)
DESCRIPTION & VERNACULAR NAMES:
See under "malaria remedies"
PLANT PARTS USED:
whole plant
TRADITIONAL USES:
Sudan: it is burned in houses to get rid of mosquitoes: *Burkill (1995) pp. 11, 638; Dalziel (1937) p. 460; Kerharo et al.(1950) p. 236*

Leucas martinicensis R. Br. (Lamiaceae)
DESCRIPTION & VERNACULAR NAMES:
See under "malaria remedies"
PLANT PARTS USED:
whole plant
TRADITIONAL USES:
W. Africa: this strongly-scented plant is burned to drive mosquitoes from a room: *Brotherton, p. 136,* quoted in *Burkill (1995) p.16; Dalziel (1937) p. 461; Irvine (1955) 5: 34,* quoted in *Watt et al. (1962) p. 521*

Lippia oatesii Rolfe (Verbenaceae)
DESCRIPTION:
Perennial herb or small shrub up to 1.3m high, from woody rootstock in grassland and mixed deciduous woodland , and stony hillsides
VERNACULAR NAMES:
B: nsakafwa (9a, 22)
PLANT PARTS USED:
branches
TRADITIONAL USES:

Zambia: the Bemba people strew branches on the floor of the house, or whisk them about so that the leaf is damaged, to repel mosquitoes. The branches are left at the doorway all night: *Watt et al. (1962) p. 1051-2.* See also entry for *Lippia javanica,* & *Govere et al.(2000)*

Muvumbani
PLANT PARTS USED:
whole plant
TRADITIONAL USES:
Zambia: J.W. Price, a missionary at Kasenga in 1920, recorded the use of this aromatic plant by the Ila people to repel mosquitoes: *Fowler (2000) p. 492*

Nuxia floribunda Benth. (Buddlejaceae)
DESCRIPTION:
Small to medium tree, up to 10m high, in evergreen forest or at forest margins
VERNACULAR NAMES:
E: Forest Nuxia; **Afr**: Bosvlier
PLANT PARTS USED:
whole plant
TRADITIONAL USES:
Kenya: the plant is used as a repellent: *Kokwaro (1976)*

Ocimum americanum L. [*O. canum* Sims] (Lamiaceae)
DESCRIPTION:
Annual erect herb up to 60cm high with white or lavender flowers
VERNACULAR NAMES:
E: Hairy/ Hoary Basil, Basil Lime; **B/ Ku**: lwena (1, 3c, 20), mununka (3c) **Ku**: mukupaimbu (1) **Nm**: msumba-mpungu (1a, 22) **Ny**: canzi, kafavumba, leyani, manunka, mpungabwe, msinyani, nunkamani (9)
PLANT PARTS USED:
leaves, whole plant
TRADITIONAL USES:
Malawi: a leaf infusion is used to repel mosquitoes: *Morris (1996) p. 330*
Tanzania: a leaf is placed under the bed as a mosquito repellent: *Watt et al. (1962) p. 524*
W. Africa: the plant is burned in a room as a mosquito repellent: *Irvine (1955) 5:34,* quoted in *Watt et al. (1962) p. 524*

Ocimum gratissimum L. subsp. gratissimum [*O. gratissimum var. suave* (Willd.) Hook f.; *O. urticifolium* Roth] (Lamiaceae)
DESCRIPTION & VERNACULAR NAMES:
See under "malaria remedies"
PLANT PARTS USED:
oil, whole plant
TRADITIONAL USES:
Kenya: the Maasai people burn the plant as a repellent: *Brown (1929) p. 11,* quoted in *Watt et al. (1962) p. 254*
W. Africa: the plant is traditionally known as the "fever leaf", and grown around houses as a mosquito repellent. A British Resident Officer in N. Nigeria reported in 1903 that three or four pots of the plant around his bed enabled him to sleep without a net. Ocimum oil is

added to insect-repellant preparations: *Ainslie sp. no. 249, Shipley pp. 205-6, Sofowora (1980) pp. 110-2, Walker (1953) p. 279, Wong p. 137,* quoted in *Burkill (1995) p. 25; Dalziel (1937) pp. 462-3*

Ocimum spp. [ex *Becium* spp.]
VERNACULAR NAMES:
Ku: tilwena (1) **Lo**: mukubamwe, saminani (3) **Ny**: kapafumba, lwenye, mkubani (3e), katanda-imbu (1)
PLANT PARTS USED:
whole plant
TRADITIONAL USES:
Zimbabwe: the plant is used as a repellent: *Gelfand et al.(1985) p. 276*

Plectranthus sp.
VERNACULAR NAMES:
not known
PLANT PARTS USED:
whole plant
TRADITIONAL USES:
S. Africa: the plant is used to repel mosquitoes: *Marloth,* quoted in *Watt et al. (1962) p. 524*

Pycnostachys urticifolia Hook.f. (Lamiaceae)
DESCRIPTION:
Evergreen aromatic perennial shrub 1 – 2.5m high, with blue/ white flowers, found in rocky ground or in grassland in transitional zone between vlei and woodland
VERNACULAR NAMES:
E: Large Hedgehog flower, Prickly Salvia; **Ny**: mdya-tsongwe, msikwipi (9)
PLANT PARTS USED:
whole plant
TRADITIONAL USES:
Malawi: the plant is used as a repellent: *Morris (1996)*

Ricinus communis L. (Euphorbiaceae)
DESCRIPTION & VERNACULAR NAMES:
See under "malaria remedies"
PLANT PARTS USED:
oil
TRADITIONAL USES:
The oil is added to paraffin-based insecticidal sprays as an anti-malaria agent: *Burkill (1935) pp. 1907-12; Burkill (1994) p 135; Dalziel (1937) pp. 160-3*

CITATION REFERENCES:

Abalaka, M.E., O.S.Olonitola, J.A.Onaolapo & H.I.Inabo (2009) Evaluation of acte toxicity of *Momordica charantia* extract using wistar rats to detect safety levels etc. in *Int. J. Pure & App. Sc. 2009 iss.1597*

Abbiw, D.K. (1990) *Useful Plants of Ghana* (Royal Botanic Gardens, Kew)

Abubakar, S.M. and M.S.Sule (2010) Effect of oral administration of aqueous extract of *Cassia occidentalis* seeds on serum electrolytes concentration in rats, in *Byero J. of Pure and Applied Sciences, 3(1) pp 183-187*

Abukakar, M.G., A.N.Ukwuani & R.A.Shhu (2008) Evaluation of the toxic effect of *Tamarindus indica* pulp extract in albino rats, in *Science Alert (2011)*

Acamovic,T., C.S.Stewart & T.W.Pennycott (2004) *Poisonous Plants and related toxins* (CAB Publishing)

Adam, J.G. (1966) Les pâturages naturels et postculturaux du Sénegal in *Bull. Inst. Fond. Afr. Noire* 28 A, pp. 450-537

Adjanohoun, E. & L.Ake Assi (1979) *Contribution au recensement des plantes médicinales de Côte d'Ivoire* (Abidjan: Centre Nationale de Floristique)

Adjanohoun, E., M.R.A.Ahyi et al. (1981) *Contribution aux études ethnobotaniques et floristique au Mali* (Paris: A.C.C.T.) cited in *Prelude med. plants database (2011)*

Adjanohoun, E., V.Adjakidje et al. (1986) *Contribution aux études ethnobotaniques et floristique au Togo* (Paris: A.C.C.T.)

Adjanohoun, E., M.R.A.Ahyi et al. (1988) *Contribution aux études ethnobotaniques et floristique en Rép. Pop. du Congo* (Paris: A.C.C.T.)

Adjanohoun, E., V.Adjakidje et al. (1989) *Contribution aux études ethnobotaniques et floristique en Rép. Pop. du Bénin* (Paris: A.C.C.T.)

Adjanohoun, E., M.R.A.Ahyi et al. (1991) *Contribution to ethnobotanical & floristic studies in W. Nigeria* (CSTR-OUA: PHARMEL2 ref.HP10)

Adoum, O.A. (2009) Determination of toxicity levels of some savannah plants using Brine Shrimp test (BST), in *Bayero J. of pure & app. Sc. 2009 2:1, pp 135-8*

Aebi A. (1950) in *Helvetica, Chimica Acta* 33

Afolayan, A.J., M.T.Yakubu, J.R.Appidi & M.Mostafa (2009) in Toxological implications of aqueous extract of *Clematis brachiata* Thunb.leaves in male Wistar rats, in *Af. J. of Pharm. and Pharm. 2009 3:11 pp. 531-538*

Agunu,A.,E.M.Abdurahman, G.O.Andrew & Z.Muhammed (2004) Diuretic activity of the stem-bark extracts of *Steganotaenia araliacea*, in *J. Ethnoph. 2005 96:3 pp.471-5*

Ahmed, O.M. & S.E.I.Adam (1980) The toxicity of *Capparis tomentosa* in goats, in *J. of Comp. Path. 1980 90:2 pp. 187-195*

Ahmed, O.M., S.E.Adam, G.T.Edds (1981) The toxicity of *Capparis tomentosa* in sheep and calves, in *Vet. Hum. Tox. 1981 23:6 pp. 403-9*

Ainslie, J.R. (1937) *A list of plants used in native medicine in Nigeria* (Oxford, Imp. Forest. Inst., Paper 7)

Ajaiyeoba,E., M.Falade, O.Ogbole, L.Okpako, D.Akinboye (2005) *In vivo* antimalarial and cytotoxic properties of *Annona senegalensis* extract, in *Afr.J.Trad., Com. and Alt. Medicines (2006) 3(1), pp. 137-41*

Ajibesin, K.K., B.E.Ekpo, D.N.Bala, E.E.Essien & S.A.Adesanya (2008) Ethnobotanical survey of Akwa Ibom State of Nigeria, in *J. Ethnoph. 2008 vol.115 pp.387-408*, cited in *Prélude med. plants database (2011)*

Akah, P.A., & C.L.Okafor (1992) Blood sugar lowering effect of *Vernonia Amygdalina* Del. in an experimental rabbit model, in *Phytotherapy Research 6, 171-173*

Akinniyi, J.A., & M.U.S. Sultanbawa (1983) A glossary of Kanuwa names of plants with botanical names, distributions, and uses, in *Ann. Borno 1* (Univ. Maiduguri)

Al-Rehaily, A.J., K.E.H.El-Tahir, J.S.Mossa & S.Rafatullah (2001) Pharmacological studies of various extracts & the major constituent, Lupeol, obtained from hexane extracts of *Teclea nobilis* in rodents, in *Nat. Prod. Sciences 2001 7:3, pp.76-82*

Ambasta, S.P. (1992) *The Useful Plants of India* (Delhi: National Institute of Science Communication)

Amico, A. (1977) Medicinal Plants of Southern Zambesia in *Fitoterapia 48, pp. 101-135*

Ampofo, O (1983) *First Aid in Plant Medicine* (Mampong-Akwapim: Ghana Rural Reconstruction Movement)

Amresh, G., G.D.Reddy, C.V.Rao and P.N.Singh (2007) Evaluation of anti-inflammatory activity of *Cissampelos pareira* root in rats, in *J. of Ethnopharm.2007 110:3, pp.526-31*

Andrade-Neto, V.F., M.G.L.Brandão, F.Q.Oliveira, V.W.Casali, B.Njaine, M.G.Zalis, M.A.Oliveira and A.U. Krettli (2004), Antimalarial activity of *Bidens pilosa* L. (*Asteraceae*) ethanolic extract in different localities or plants cultivated in humus soil, in *Phytotherapy Research (2004)*

Anis, M. & Iqbal, M. (1986) Antipyretic utility of some Indian Plants in Traditional Medicine, in *Fitoterapia vol 57 no. 1*

Anon. (1906) *Bulletin of the Imperial Institute, London*: 4

Anon. (1911-14) *Amani* Deutsch-Ostafrika (1911-14) B: 1.8 (2)

Anon. (1939) In *Not. Farm.,* Porto 5: 193

Anon. (no date) *Cassia beareana* (Thomas Christy & Co)

Aplin,T.E.H.(1968) Cyogenetic plants of W.Australia, in *J. Ag.W. Aust.1968 9:7 pp.323-331*

Asase, A., A.A.Oteng-Yeboa, G.T.Odamtten & M.S.J.Simmonds (2005) Ethnobotanical study of some Ghanaian anti-malarial plants, in *J.Ethnoph. 2005 v.99 pp.273-299*

Asase, A. & G.Oppong-Mensah (2090 Traditional antimalarial phytotherapy remedies in herbal markets in S.Ghana, in *J. Ethnoph. 2009: 126, pp.492-499*

Asenjo, C.F. et al. (1945) *Jour. of the Am. Chem. Soc.* 67: 1936
                 (1948) *Puerto Rico Journal of Public Health* 24: 44

Assam, J.P., J.P.Dzoyem, C.A.Pieme & V.B.Penlap (2010) *In vitro* antibiotic & acute toxicity studies of aqua-methanol extract of *Sida rhombifolia*, in *BMC Compl. & Alt. Med. 2010 10:40*

Aubréville, A. (1950) *Flore forestière Soudano-guinéenne, A.O.F, Cameroun, A.E.F.* (Paris: Société d'Éditions Géographiques, Maritîmes et Coloniales)

Avasthi, B.K.& al.(1955) *Jour. of the Am. Pharm. Ass., Sci. edn.*, 44
                 (1955b) *Arch. Pharm., Berl.*, 288

Ayensu, E.S. (1981) *Medicinal Plants of the West Indies* (Algonac, Michigan: Reference Publications)

Ayoola, G.A., H.A.B.Coker, S.A.Adesegun, A.A.Adepoju-Bello, K.Obaweya, E.C.Ezennia, T.O.Atangbayila (2008), in Phytochemical screening and antioxidant activities of some selected medicinal plants used for malaria therapy in Southwestern Nigeria, in *Tropical Journal of Pharmaceutical Research, September 2008; 7(3) pp.1019 – 1024.*

Babayi, H.M., J.J.I. Udeme et al. (2007) Effect of oral administration of aqueous whole extract of *Cassytha filiformis* on haematograms and plasma biochemical parameters in rats, in *J. of Med. Toxicology 2007 3:4 pp. 146-151*

Baerts, M. & J.Lehmann (1989) Guérisseurs & plantes médicinales de la région des crêtes Zaire-Nil au Burundi, in *Mus. Roy. de l'Afr.Cen., Tervuren, Belg, Ann. Sc. Eco. v.18*; cited in

Prelude med. plants database (2011)

Bah, S., A.K.Jäger, A.Adsersen et al. (2007) Antiplasmodial and GABA(A)-benzodiazepine receptor binding activities of five plants used in traditional medicine in Mali, West Africa, in *J. Ethnopharm. 2007 Apr 4; 110 (3): 451 - 7*

Bally, P.R.O. (1937) Native Poisonous & Medicinal Plants of East Africa *Kew Bulletin*: 10 (1938) Heil-und Giftpflanzen der Eingeborenen von Tanganyika, in *F.Fedde, Repertorium speciorum novarum regni vegetabilis*

Bandeira, S.O., F.Gaspar and F.P.Pagula (2001) African ethnobotany and healthcare : emphasis on Mozambique, in *Pharmaceutal Biology 2001, vol.39, Supp., pp. 70-73*

Banzouzi, J.T., R.Prado, H.Menan et al., (2004) Studies on medicinal plants of Ivory Coast: Investigation of *Sida acuta* for *in vitro* antiplasmodial activities and identification of an active constituent, in *Phytomedicine 2004 11 (4): 338 - 41*

Barnish, G. & S.K.Samai (1992) *Some medicinal plant recipes of the Mende, Sierra Leone* (Med. Research Council lab., P.O.Box 81, Bo, Sierra Leone) cited in *Prelude med. plants database (2011)*

Barth, A.T., G.D.Kommers, M.S.Salles, F.Wouters and C.S.de Barros (1994) Coffee Senna (*Senna occidentalis*) poisoning in cattle in Brazil, in *Vet. Hum.Toxicol. vol. 36 (6) pp. 541-545*

Batalha, M.M. & D.S.Van-Dunen (1983) Etat d'avancement de la recherche sur la méd. et la pharmacopée trad.dans la R.P.d'Angola, in *Report of 4th interAfrican Symposium on trad. Pharm. & Med. Plants of Africa, 1979* (Cairo: Univ. Press 1983)

Baumer, M.C. (1975) Catalogue des Plantes utiles du Kordofan (R.e du Soudan) particulièrement du point de vue pastorale in *J. Agr. trop. Bot. appl.* 22: 81-119

Beare, D.R.O'S. & al. (1902) *The Lancet* 2: 282

Bebawi, F.F. & L.Neugebohrn (1991) Review of plants of N.Sudan with special reference to their uses *(Eschborn: D.T.Z.)* cited in *Prelude med. plants database (2011)*

Beentje, H.J. (1994) *Kenya Trees, Shrubs, and Lianas* (Nairobi: National Museums of Kenya)

Behl, P.N., A.Luthra (2002) Bullous eruption with *Calotropis procera* – a medical plant used in India, in *Indian Journal of Dermatology, vol.68, Iss.3, pp.150-1*

Bellakhdar,J. (1997) *La Pharmacopée marocaine traditionelle* (Ibis Press); cited in *Prélude med. plants database (2011)*

Beloin,N., M.Gbeassor, K.Akpagana, J.Hudson, K.deSoussa, K.Koumaglo & J.T.Arnason (2005) Ethnomedicinal uses of *Momordica charantia* in Togo, and (in?) relation to its phytochemistry & biological activity, in *J. Ethnoph. 2005: 96 pp. 49-55* cited in *Prélude med. plants database (2011)*

Berhaut,J.(1967) *Flore du Sénégal,* ed. 2 (Dakar: Clairafrique) (1975) *Flore illustrée du Sénégal, Dicotyledones,* vol. 4, *Ficoidees à Légumineuses* (Dakar) (1976) *Flore illustrée du Sénégal, Dicotyledones,* vol. 5, *Légumineuses Papilionacées* (Dakar)

Berk, L.H. van (1930) *Bijdrage tot de kennis der West-Indische Volksgeneeskruiden* (Utrecht: Proefschrift)

Bero, J., H.Ganfon, M.C.Jonville et al. (2009) *In vitro* antimalarial activity of plants used in Benin in traditional medicine to treat malaria, in *J. of Ethnopharm. 2009 April 21; 122(3) pp.439 - 444*

Beyer, G. (1927) *Festschrift Meinhof* (Hamburg: J.J.Augustin)

Bhat, G.P. & N.Surolia (2001) *In vitro* antimalarial activity of extracts of three plants used in the traditional medicine of India, in *Am.J. Trop. Med. & Hyg. 2001 65:4, pp. 304-308*

Bhat, R.B., E.O.Etejere, V.T.Oladipo (1990) Ethnobotanical Studies from Central Nigeria, in *Economic Botany 44:3 (1990), pp. 382-390*

Bhatia, B.B. et al. (1933), in *Indian J. of med. Res.,* 20

Bissett, N.G. (1970) The African species of *Strychnos*, 1. The Ethnobotany, in *Lloydia 33: 201-43*

Bizimenyera, E.S., G.E.Swan, F.B.Samdumu, L.J.McGaw & J.N.Eloff (2007) Safety profiles of *Peltophorum africanum* extracts, in *S. A.J. Sc. 2007 cap.7*

Blair,S., A.Correa, B.Madrigal, C.Zuluaga, & H.D.Franco (1991) *Plantas antimalaricas. Una revision bibliografica* (Universidad de Antioquia: Editorial)

Blood, D.C.(2007) (ed.) *Saunders Comprehensive Veterinary Dictionary, 3rd edn revised 2007* (Edinburgh : Saunders Elsevir)

Bodenstein, J.W. (1977) *Toxicology of traditional herbal remedies*, in S.A. Medical Journal v.52 p.790

Bogale, M. and B. Petros (1996) Evaluation of antimalarial activity of some Ethiopian traditional medicinal plants againsy P.falciparum *in vitro*, in *Eth.J.Sci., 1996, 19(2): 233-243*

Boiteau, P. (1986) *Médecine traditionelle et pharmacopée. Précis de matière médicale malgache* (Paris: A.C.C.T) cited in Prelude med. plants database (2011)

Bonnefoux, B.M. (n.d.) *Plantes medicinales du Hufla.* (Congregation do Espirito Santo, 1885-1937) cited in *Prelude med. plants database (2011)*

Bossard, E. (1996) *La médicine traditionelle au centre et à l'ouest de l'Angola* (Lisboa: Instituto de investigaçâo ciêntifia tropical) cited in *Prelude med. plants database (2011)*

Bostock, L. (1907) *Transvaal Medical Journal* 2: 273

Botha, C.J. & E.Venter (2002) Plants poisonous to livestock, Southern Africa, CD-ROM *(Univ. Pretoria, Faculty of Veterinary Sc.)*

Botha, E.W., C.P.Kahler, W.J.du Plooy (2005) The effects of *Boophane disticha* on human neutrophils, in *J. of Ethnopharmacology v. 96 iss.3, 2005*

Boulesteix,M., & S.Guinko (1979) Plantes medicinales utilisées par les Gbayas dans la région de Bouar (Empire Centrafricain), in C.A.M.E.S., Libreville, pp. 23-52) cited in *Prelude med. plants database (2011)*

Boulos, L. (1983) *Medicinal Plants of North Africa* (Reference Publns. Inc.); cited in *Prelude med. plants database (2011)*

Bouquet, A. (1969) *Fêticheurs et Médicines traditionelles du Congo-Brazzaville* (Mem. O.R.S.T.O.M.,13)

Bouquet, A. & M.Debray (1974) *Plantes médicinales de la Côte d'Ivoire* (Trav. Doc. O.R.S.T.O.M.,32)

Boury, N.J. (1962) Végetaux utilizés dans la médicine africaine dans la region de Richard-Toll (Senegal), in Adam, J. *Les Plantes utiles en Afrique occidentale,* Notes Afr.

Brandão, M.G.L., M.G.A.Botelho, A.U.Krettli (1985) Quimoterapia experimental antimalarica com produtos naturais: uma abordagem mais racional? in *Ciencia e Cultura: 37 (7) pp. 1152-63*

Brandão, M.G.L., T.S.M.Randi, E.M.M.Rocha, D.R.Sawyer, A.U.Krettli (1992) Survey of medicinal plants in the Amazon, in *J. of Ethnopharmacology 36 (1992) 175-182*

Brandão, M.G.L., A.U. Krettli, L.S.R. Soares, C.G.C. Nery, H.C. Marinuzzi (1997) Antimalarial activity of extracts and fractions from Bidens pilosa and Bidens species (Asteraceae), correlated with the presence of acetelene and flavonoid compounds, in *Journal of Ethnopharmacology 57, pp. 131-8*

Braun, K. (1927) *Arch. Pharm.* (Berlin) 265, 45
        (1930) *Faserforschung* 8: 2, 8, 90

Brenan, J.P.M. & P.J.Greenway (1949) *Checklists of the forest trees & shrubs of the British Empire, part 5: Tanganyika Territory* (Oxford: Imperial Forestry Institute)

Brendler,T., J.N.Eloff, A.Gurib-Fakim & L.D.Phillips (2110) *African Herbal Pharmacopeia* (Mauritius: AAMPS)

Broun, A.F., & R.E. Massey (1929) *Flora of the Sudan* (London: Sudan Govt. Office)

Bryant, A.T. (1909) Zulu Medicine & Medicine Men. In *Annals of the Natal Museum,* (1916) 2:1, ed. by Warren, E. (London: Adlard & Son and West Newman)

Bugmann, N. (2000) *Le concept du paludisme, l'usage et l'efficacité in vivo de trois traitements traditionnels antipalustres dans la region dee Dori, Burkina Faso* (Doctorial thesis, Faculty of Medicine, Univ. of Basle)

Burkill, H.R.M. (1985) *The Useful Plants of West Tropical Africa, vol. 1* (London: Royal Botanic Gardens, Kew)

　　　　　　(1994) *The Useful Plants of West Tropical Africa, vol. 2* (London: Royal Botanic Gardens, Kew)

　　　　　　(1995) *The Useful Plants of West Tropical Africa, vol. 3* (London: Royal Botanic Gardens, Kew)

　　　　　　(2000) *The Useful Plants of West Tropical Africa, vol.5* (London: Royal Botanic Gardens, Kew)

Burkill, I.H. (1935) *A dictionary of the economic products of the Malay Peninsula (London: Crown Agents for the Colonies)*

Burtt Davy,J., & A.C.Hoyle (ed. 1937) *Check-lists of the Forest Trees and Shrubs of the British Empire, No.3*, Draft of first descriptive check-list of the Gold Coast (Oxford: Imp. For. Inst.)

Cai, Y, M.Sun, H.Corke (2003) Antioxidant activity of betalains from plants of the Amaranthaceae, in *J. Agric. Food Chem., 2003:51 (8), pp. 2288-94*

Calore, E. E., M.J.Cavaliere, M.Haraguchi, S.L.Gorniak, M.L.Z.Dagli, P.C.Raspantini and N.M.P.Calore (1997) Experimental mitochondrial myopathy induced by chronic intoxication by *Senna occidentalis* seeds, in *Journal of Neurological sciences, 1997 146 I, pp 1-6*

Calore, E. E., M.J.Cavaliere, M.Haraguchi, S.L.Gorniak, M.L.Z.Dagli, P.C.Raspantini, N.M.P.Calore and R.Weg (1998) Toxic peripheral neuropathy of chicks fed *Senna occidentalis* seeds, in *Ecotoxic. Environ. Saf. 1998 39:1, pp.27-30*

Calore, E.E., N.M.Calore, R.Wegg, M.J.Cavaliere, A.Ruckert da Rosa and S.De Sousa Dias (1999) The Lysosoma enzymes Acid Phosphatase and Cathepsin D in rats intoxicated with *Senna occidentalis* seeds, in *J. Submicro. Cytol.Path. 1999 31:2 pp. 259-64*

Calore, E.E., R.Weg, M.Haraguchi, N.M.Calore, M.J.Cavaliere and A.Sesso (2000) Mitochondrial muscle impairment in muscle fibres of rats chronically intoxicated with *Senna occidentalis* seeds, in *Exp. Toxicol. 2000 52:4, pp.357-63*

Calore, E.E., A.Sesso, H.Correia, M.C.Marcondes and L.Vilela de Almeida (2002) Distribution of cox-negative mitochondria in myofibres of rats intoxicated with *Senna occidentalis* seeds, in *J.Submicros.Cytol.Pathol. 2002 34:2 pp. 227-31*

Cardosa, J., jnr. (1939) *Not. Farm., Porto* 5: 387

Cavaco, A. (1963a) *Chenopodiacées*, in A. Aubréville, *Flore du Gabon* 7:16-20, Paris: Museum National d'Histoire Naturelle

Cavaliere, M.J., E.E.Calore, M.Haraguchi, S.L.Gorniak, M.L.Z.Dagli, P.C.Raspantini, N.M.P.Calore and R.Weg (1997) Mitochondrial Myopathy in *Senna occidentalis*-seed-fed chickens, in *Ecotox.& Envir. Saf. 1937 37:2 pp.181-5*

Chadha, Y.R. ed.(1976) *The Wealth of India, Raw Materials, vol. 10* (New Delhi: C.S.I.R.)

Chagnon, M. (1984) Inventaire pharmacologique general des Plantes médicinales Rwandaises, in *J. of Ethnopharmacology 12 (1984) 239-251*

Chhabra, S.C., R.L.A.Mahunnah, E.N.Mshiu (1987) Plants used in traditional medicine in Eastern Tanzania. I. Pteridophytes & Angiosperms (Acanthaceae to Canellaceae), in *J. of Ethnopharmacology 21 (1987), pp. 253 - 277*

Chhabra, S.C., R.L.A.Mahunnah, E.N.Mshiu (1990) Plants used in traditional medicine in Eastern Tanzania. iv. Angiosperms (Mimosaceae to Papilionaceae), in *J. of Ethnopharm. 29 (1990) pp. 295-323*

Chhabra, S.C., R.L.A.Mahunnah, E.N.Mshiu (1991) Plants used in traditional medicine in Eastern Tanzania. v: Angiosperms (Passifloraceae to Sapindaceae), in *J. of Ethnopharmacology 33 (1991), pp. 143-157*

Chinemana, F., R.B.Drummond, S.Mavi & I.De Zoysa (1985) Indigenous plant remedies in Zimbabwe, in *J. Ethnoph. 1985 14, pp. 159-172*

Ching,F.P., E.K.I.Omogbai, R.I.Ozolua & S.O.Okpo (2008) Antidiarrhoeal activities of aqueous extract of *Stereospermum kunthianum* stem bark in rodents, in *A.J. Biotech. 2008 v.79 pp. 1220-5*

Chopra, R.N. (1933) *Indigenous drugs of India: Their Medical and Ecomomic Aspects* (Calcutta: The Art Press)

Clarkson,C., V.J.Maharaj, N.R.Crouch, O.M.Grace, P.Pillay, M.G.Matsabisa, N.Bhagwandin, P.J.Smith, & P.I.Folb (2004) *In vitro* activity of medicinal plants native to or naturalised in South Africa, in *J. of Ethnopharmacology 92 , pp. 177-191*

Codd, L.E.W. (1951) *Trees & Shrubs of the Kruger National Park, 26:14* (Pretoria: Department of Agriculture, Botanical Survey Memoir)

Coe, F.G. & G.J.Anderson (1996) Ethnobotany of the Garifuna of Eastern Nicaragua, in *Economic Botany 50(1) pp. 71-107*

Colodel, E.M., S.D.Traverso, A.L.Seitz, A.Correa, F.N.Oliveira, D.Driemeier & A.Gava, (2003) Spontaneous poisoning by *Dodonea Viscosa* in cattle, in *Vet. Hum.Toxic.2003 45:3 pp147-8*

Colvin, B.M., L.R.Harrison, L.T.Sangster and H.S.Gosser (1986) *Cassia occidentalis* toxicosis in growing pigs, in *J. Am.Vet. Med. Assoc 1986: 189(4) pp. 423-6*

Comerford, S.C. (1996) Medicinal Plants of two Mayan Healers from San Andres, Peten, Guatemala, in *Economic Botany 50: 3, (1996), pp. 327-336*

Cooper, M.R. & A.W.Johnson (1984) *Poisonous plants in Britain & their effects on animals & men* (London : HMSO)

Cornell University website: www ansci.cornell.edu/plants/medicinal/epazote.ht

Cortesi, I.F. (1936) In *Rassegna Economica Colon. Ital.* 14:71

Creach, P. (1940) Le Balanites aegyptiaca, ses multiples applications au Tchad, in *Revue de Botanique appliqué & d'Agriculture Tropicale,* 20: 578- 593
        (1943) *Chem. Zbl.* 1: 651
        (1944) *Chem. Abstr.* 38: 377

Dagar, H.S., & J.C. Dagar (1991), Plant Folk Medicines among the Nicobarese of Katchal Island, India, in *Economic Botany, 45 (1), 1991, pp. 114-119*

Daisy, P., R.Jasmine, S.Ignacimuthu & E.Murugan (2009) A novel steroid from *Elephantopus scaber* etc., in *Phytomed. Inter. J.of Phytother. & Phytopharm. Pp. 1-7*

Dalziell, J.M. (1937) *The Useful Plants of West Tropical Africa* (London: Crown Agents for the Colonies)

Darmani,H., A.S.Al-Hiyasat, A.M.Elbetieha & A.Alkofahi (2003) Effect of an extract of *Salvadora persica*..on fertility of male & female mice, in *Phytomedicine 2003 10 1 pp. 63-5*

Dawodu 53, *Herbarium,* RBG Kew

Debray, M.H., H. Jaquemin, & R. Razafindrambo (1971) Contribution a l'inventaire des plantes medicinales de Madagascar, in *Trav. Doc. O.R.S.T.O.M.,* 8

Deighton, F.C. *(1954) Herbarium,* RBG Kew

de la Pradilla, C.F. (1982) *Des Plantes qui nous ont guéris, vol. 2* (Ouagadougou, B.P. 1471, Burkina Faso) (1988) *Plantes medicinales contre le paludisme* (Barcelona: ined. ms.)

Delaude,C.,& J.& H.Breyne (1971) Plantes médicinales et ingrédients magiques du grand marché de Kinshasa, in *Africa-Tervuren 1971 17 93-103* cited in *Prelude med. plants database (2011)*

De Madureira, M.D.C., A.P.Martins, M.Gomes, J.Paiva, A.P.da Cunha & V.do Rosario (2002) Antimalarial activity of medicinal plants used in traditional medicine in S. Tomé and Principe islands, in *Journal of Ethnopharmacology 81 (2002)*

De Wildeman, E. (1946) *Mémoires de l'Institut Royal Colonial Belge,* 13
    (1948) *Mémoires de l'Institut Royal Colonial Belge,* 17
    (1949) *Mémoires de l'Institut Royal Colonial Belge,* 18

Dhetchuvi, M.-M. & J.Lejoly (1990) Contribution à la connaissance des plantes médicinales du Nord-Est du Zaire, in *Mitt. Inst. Allg. Bot. Hamburg 1990 23b pp.991-106,* cited in *Prélude med. plants database (2011)*

Diafouka, A.J.P. (1997) *Analyse des usages des plantes médicinales dans 4 régions de Congo-Brazzaville* (Doct. Thesis, Univ. Libre de Bruxelles) ; cited in *Prelude med. plants database (2011)*

Diarra, N. (1977) Quelques plantes vendues sur les marchés de Bamako, in *J. Agr.trad. Bot. Appl.* 24

Di Stasi, L.C., C.A.Hiruma, E.M. Guimaraes, C.M. Santos (1994), Medicinal plants popularly used in Brazilian Amazon. in *Fitoterapia LXV, 6, 1994 pp. 529-540*

D'Orey no. 177, (Herbarium, RBG Kew)

Dornan, S.S. (1916) In *South African Journal of Science* 13: 356
    (1924/5) In *South African Quarterly* 6, 7 p.3

Dragendorff, G. (1898) *Die Heilplanzen der verschiedenen Volker und Zeiten* (Stuttgart: Ferdinand Enker)

Duganath,N., D.Rama-Krishna, G.D.Reddy, B.Sudheera, M.Mallikarjun & P.Beesetty (2011) Evaluation of antidiabetic activity of *Triumfetta rhomboidea* in alloxan induced Wistar rats, in *R.J. Pharm. Biol. & Chem. Sci., 2011 v.2:1 pp. 721-7*

Duke, J.A., & R.Vasquez R. (1994) Amazonian ethnobot. dictionary (Boca Raton: CRC Press)

Durand, J.M. (1959) Les plantes bienfaisantes du Rwanda et du Burundi (Groupe scolaire) cited in *Prélude med. plants database (2011)*

Dykman, E.J. (1908) *De Suid Afrikaanse Kook-Koek-en- Resept Boek, 14th imp.,* (Cape Colony: Paarl Printers)

Editors (1945) *East African Medical Journal*

El Gamal, A.A., O.S.A.Mohamed, S.A.Khalid (2008) *Toxicity of Argemone mexicana seed, seed oil and their extracts on albino rats* (unpub. article, University of Khartoum)

Elgorashi, E., J.L.S.Taylor, A.Maes & J.van Staden (2003) Screening of medicinal plants used in S.A.traditional medicine for genotoxic effects, in *Toxicology Letters 2003 143:2, pp.195-207*

El-Hamidi (1970) Drug-plants of the Sudan Republic in native medicine, in *Pl. Med. 18: 278-80*

El-Kamali, H.H., & K.F.El-Khalifa (1997) Treatment of malaria through herbal drugs in the Central Sudan, in *Fitoterapia LXVIII, no. 6, 1997, pp. 527-530*

El-Kamali, H.H. (2009) *Ethnopharmacology of medicinal plants used in North Kordofan (Western Sudan)*, Ethnobotanical Leaflets 13: pp.89-97 (www. ethnoleaflets. com/ leaflets/sudan.htm)

El-Tahir, A., G.M.Satti and S.A. Khalid (1999a) Antiplasmodial activity of selected Sudanese medicinal plants with emphasis on *Acacia nilotica,* in *Phytotherapy Research 13: 474-8*

El-Tahir, A., G.M.Satti and S.A. Khalid (1999b) Antiplasmodial activity of selected Sudanese medicinal plants with emphasis on *Maytenus senegalensis,* in *J.of Ethnopharm. 1999 Ma; 64 (3): 227-33*

Elufioye,T.O., and J.M.Agbedahunsi (2004) Antimalarial activities of *Tithonia diversifolia* (Asteraceae) and *Crossopteryx febrifuga* (Rubiaceae) on mice *in vivo*, in *Journal of Ethnopharmacology 2004 Aug, 93(2-3) pp.167-171*

EMEA (2000) *European Agency for evaluation of Medicinal products* (http/www.eudra. org/emea. html)

Etkin, N.L.,& P.J. Ross (1991) Recasting malaria, medicine and meals: A perspective on disease adaptation. In Romanucci-Ross L., Moerman D.E., Tancredi L.R. (eds.) *The anthropology of medicine* (2e) (New York: Bergin & Garvey)

Fanshawe, D.B., & C.D.Clough (1967) *Poisonous plants of Zambia* (Lusaka:Government Printer)

Fennell, C.W., K.L.Lindsay, L.J.McGaw, S.G.Sparg, G.I.Stafford, E.E.Elgorashi, O.M.Grace & J.van Staden (2004) Assessing African medicinal plants for efficacy & safety: pharmacological screening & toxicology, in *J. Ethnopharm. 2004 94, pp.205-217*

Ferreira, F.H. (1952) *The trees and shrubs of South Africa 2* (Pretoria)

Ferry, M.P., M.Gessain, R.Gessain, (1974) *Ethnobotanique Tenda* Docum. Centre Rech. anthrop. Mus. Homme, No. 1

Flory, W., C.B.Spainhour jnr., B.Colvin and C.D.Herbert (1992), The toxicological investigation of a feed grain contaminated with seeds of the plant species *Cassia*, in *J.Vet.Diagn.Invest. 1992 4:1, pp. 65-9*

Fowler, D.G. (2000) *A Dictionary of Ila Usage 1860 – 1960.* Monographs from the International African Institute, 5. (Hamburg: Lit-Verlag)

(2002) *The Ila Speaking.* Monographs from the International African Institute, 7. (Hamburg: Lit-Verlag)

(2002b) Traditional Ila Plant Remedies from Zambia. In *Kirkia* 18 (19): 35-48 (Harare: Government Printer)

Francesconi, K., P.Visoottiviseth, W.Sridokchan and W.Goessler (2002) Arsenic species in an arsenic hyperaccumulating fern, *Pityrogramma calomelanos*: a potential phytoremediator of arsenic-contaminated soils, in *Science of the Total Environment Vol.284, Issues 1-3, 4 Feb 2002, pp 27 - 35*

Frei, B., M.Baltisberger, O.Sticher, M.Heinrich (1998) Medical ethnobotany of the Zatopecs of the Isthmus-Sierra (Oaxaca, Mexico): Documentation and assessment of indigenous uses, in *Journal of ethnopharmacology 62 (1998) pp. 149-165*

Ganapaty, S, G.K.Dash, T.Subburaju, P.Suresh (2002) Diuretic, laxative and toxicity studies of *Cocculus hirsutus* aerial parts, in *Fitoterapia vol 73, issue 1, Feb 2002, pp. 28-31*

Gansané A., S.Sanon, L.P.Ouattara et al., (2010) Antiplasmodial activity and toxicity of crude extracts from alternative parts of plants widely used for the treatment of malaria in Burkina Faso, in *Parasit. Res. 2010 Jan; 106 (2) 335 - 40*

Garcia-Barriga, H. (1992) *Flora Médicinal de Colombia,* 2nd edn.(Bogota: Terca Mundo)

Garon,D., E.Chosson, J.-P.Rioult, P.E.de Pecoulas, P.Brasseur & P.Vérité (2007) Poisoning by *Cnestis ferruginea* in Casamance (Senegal), in *Toxicon 200 7 50 pp.189-195*

Gathirwa,J.W., G.M.Rukunga, E.N.Njagi et al.(2007), The *in vitro* antiplasmodial & *in vivo* antimalarial efficacy of combinations of some medicinal plants used traditionally for treatment of malaria by th Meru community in Kenya, in *J. Ethnopharm. 2007*

Gbeassor, M., Y.Kossov, K.Amegbo, C.De Souza, K.Koumaglo & A.Denke (1989) Antimalarial effects of 8 african medical plants, in *J. Ethnopharm. 1989 vol.25 pp. 15-18*

Gedif,T. & H.-J.Hahn (2002) Treatment of malaria in Ethiopian folk medicine, in *Tropical Doctor, October 2002, 32*

Gelfand,M., S.Mavi, R.B. Drummond, & B. Ndemera (1985) *The Traditional Medical Practitioner in Zimbabwe* (Gweru: Mambo Press)

Gessler,M.C., M.H.H.Nkunya, L.B.Mwasumbi, M.Heinrich, M.Tanner (1994) Screening Tanzanian medical plants for antimalarial activity, in *Acta Tropica 56 (1994) pp. 65-77*

Gessler, M.C., D.E.Msuya, M.H.H.Nkunya, L.B.Mwasumbi, A.Schar, M.Heinrich, M.Tanner (1995) Traditional healers in Tanzania: the treatment of malaria with plant remedies, in *J. of Ethnopharmacology 48 (1995) pp. 131 - 144*

Getahun, A. (1975) *Some common medicinal and poisonous plants used in Ethiopian folk medicine* (ined. ms., Herbarium, RBG Kew)

Geyid, A., D.Abebe et al. (2005) Screening of some medical plants of Ethiopia for their anti-microbial properties & chemical profiles, in *J.Ethnoph. 2005: 97, pp. 421-7*; cited in *Prelude med. plants database (2011)*

Giday, M. (2001) An ethnobot. study of med. plants used by the Zay people in Ethiopia, in *CBM: Skriftserie 3, pp. 81-99, Uppsala,* cited in *Prelude med. plants database (2011)*

Giday, M., T.Teklehaymanot, A.Animut & Y.Mekonnen (2007) Medicinal plants of the Shinasha, Agew-Awi & Amhara peoples in NW Ethiopia, in *J. Eth. 2007 110, pp.516-525*

Gill, L.S. & C.Akinwumi (1986) Nigerian folk medicine practices and beliefs of the Ondo people, in *J. Ethnoph. 1986 vol. 18 pp. 257-66*

Githens, T.S. (1948) *Drug Plants of Africa* Africa Handbooks no. 8 (Philadelphia: University of Pennsylvania Press)

Glover & Samuel *Herbarium*, RBG Kew

Glover, P.E., J.Stewart & M.D.Gwynne (1966) Masai & Kipsigis notes on E.African plants, pt.III - Medical uses of plants, in *E.African Ag. & For. J. 1966 32:2 pp.200-207*

Goodson, J.A. et al. (1919) In *Journal of the Chemistry Society* 115: 923

Goodson, J.A. (1922) In *Biochemical Journal* 16: 489

Goossens, V.(1924) *Catalogue des plantes du jardin botanique d'Eala (Congo Belge* (Bruxelles : Min.des colonies) cited in *Prelude med. plants database (2011)*

Gossweiller, J. (1953) Nomes indigenas de plantas de Angola, in *Agronomia Angolana No. 7* (Luanda)

Govere, J., D.N.Durrheim, L.Baker, R.Hunt & M.Coetzee (2008) Efficacy of three insect repellents against the malaria vector *Anopheles arabensis*, in *Med. & Vet. Entomology, 2000, v.14:4 pp. 441-4*

Grace, O.M., H.D.V.Prendergast, A.K.Jager & J.van Staden (2003) Bark medicines used in traditional healthcare in KwaZulu Natal, S.Africa, in *S.A.J.Botany 2003 69:3, pp.301-363*

Graz, B., M.L.Willcox, C.Diakite, J.Falquet, F.Dackuo, O.Sidibe, S.Giane, D.Diallo (2009) Argemone mexicana decoction versus Artisunate-amodiaquine for the management of malaria in Mali: policy and public health implications, in *Transactions of the Royal Society of Tropical Medicine and Hygiene (2007) 101, 1190-1198*

Graziano,M.J., W.Flory, C.L.Seger and C.D.Hebert (1983) Effects of a *Cassia occidentalis* extract in the domestic chicken (Gallus domesticus), in *Am. Journal of Vet. Research 1883 44:7 pp.1238-44*

Greenway, P.J. (1941) In *E. Africa Agr. Journal* 6: 127, 132, 199, 241; 7: 96

(1947) In *E. Africa Agr. Journal* 13: 8, 98, 228

Greshoff, M. (1900) *Buitenzorg: Mededeelingen uit's Lands Plantentuin* 29: 104-106 (Batavia: Kolff.)

(1913) *Buitenzorg: Mededeelingen uit's Departement van Landbouw* 17 (Batavia: Kolff) cited in *Watt et al., (1962)*

Grøenhaug, T.E. et al. (2008) Ethnopharmacological survey of six medicinal plants from Mali, in *J. Ethnobiol. & Ethnomed. 4:26,* cited in *Prelude med. plants database (2011)*

Guerrero, L.M. (1921) p. 149 in *Bul. Bur. For. Phil. Is.*, 22 cited in *Watt et al., (1962)*

Guenoukpati, K.(1983) Ces plantes de chez nous. Multiples recetes de l'economie domestique (RBG Kew: Centre for Econ. Bot.)

Guillarmod, A.J. (1971) *Flora of Lesotho* (J. Kramer) cited in *Watt et al., (1962)*

Gurib-Fakim, A. & J.Guého (1994) *Plantes medicinales de l'île Rodriguez Stanley, Rose-Hill* (Univ.de Maurice: Editions de l'Ocean indien)

Haapala, T., G.P.Mwila, K.Kabamba, & E.Chishimba. (1994) *Some introduction to some Medical Plants found in Zambia.* (Mufulira: Finnish Volunteer Service, Copperbelt Regional Research Station).

Haerdi, F. (1964) Die Eingeborenen-Heilpflanzen des Ukanga-Distriktes Tanganjikas (Ostafrica), in *Acta Tropica Suppl. 8,* pp. 1-278

Haerdi, F., J. Kerharo, & J.G. Adam (1964) *Afrikanische Heilpflanzen* (Basel)

Hakizamungu, E. and M.Weri (1988) L'usage de plantes médicinales dans le traitement du paludisme en médecine traditionelle Rwandaise, in *Bull. Méd. Trad. Pharm., 1988, vol. 2, no. 1, 11-17*

Hamill,F.A., S.Apio, N.K.Mubiru, M.Mosango, R.Bukenya-Ziraba, O.W.Maganyi, & W.D.D. Soejarto (2000) Traditional herbal drugs of S. Uganda, in *J. of Ethnoph. 2000 70 281-300*

Hammiche, H., & K.Maiza (2006) Traditional medicine in Central Sahara : Pharmocopeia of Tassili N'ajjer, in *J. of Ethn.v.105, pp. 358-67;* cited in *Prelude med. plants database (2011)*

Haraguchi,M., E.E.Calore, M.L.Z.Dagli, M.J.Cavaliere, N.M.P.Calore, R.Weg, P.C.Raspantini and S.L.Gomiak (1998) Muscle atrophy induced in broiler chicks by parts of *Senna occidentalis* seeds, in *Veterinary Research Communications vol.22 no.4, pp 265-271*

Harborne, J.B. & G.P.Moss (1999) *Phytochemical Dictionary: a handbook of bioactive compounds from plants (2nd ed.)* (Taylor & Francis)

Harjula, R.(1980) Mirau & his practice. A study of the ethnomedical repertoire of a Tanzanian herbalist (London:Tri-Med Bks Ltd), cited in *Prelude med. plants database (2011)*

Haxaire, C. (1979) *Phytothérapie et Médecine Familiale chez les Gbaya-Kara* (These de doctorat, Univ. de Paris); cited in *Prelude med. plants database (2011)*

Hebert, A. (1896) In *Bulletin de la Société chimique de France* 13: 927

Hedberg, I., O.Hedberg, P.J.Madati, K.E.Mshigeni, E.N.Mshiu, & G. Samuelsson (1983) Inventory of Plants used in traditional medicine in Tanzania. Pt.III. Plants of the families Papilionaceae-Vitaceae, in *Journal of Ethnopharmacology 9*, pp. 237-60

Heine, B. & I.Heine (1988) *Plant concepts & plant use. An ethnobot. survey of the semi-arid & arid lands of East Africa Pt.3: Rendille plants (Kenya)* (Cologne: Verlag Breitenbach) cited in *Prelude med. plants database (2011)*

Heine, B. & B.Brenzinger (1988b) *Plant concepts & plant use. An ethnobot. survey of the semi-arid & arid lands of E. Africa Pt.4: Plants of the Borana (Ethiopia & Kenya)* (Cologne: Verlag Breitenbach) cited in *Prelude med. plants database (2011)*

Hewat, M.L. (1906) *Bantu Folklore* (Cape Town: Maskew Miller)

Hilou, A., O.G.Nakoulma, T.R.Guiguemde (2006) *In vivo* antimalarial activity of extracts from *Amaranthus spinosus* L. and *Boerhaavia erecta* L. in mice, in *Journal of Ethnopharmacology vol.103 issue 2, 16/1/06, pp.236-240*

Holford-Walker, A.F. (1994) *Herbal medicines & drugs used by the Masai*, in Indigenous knowledge & Development Monitor (www.nuffic.nl/ciran/ikdm/)

Holland, J.H. (1922) Useful plants of Nigeria, in *Kew Bull.,* additional series, 9: 1

Hollis, A.C. (1905) *The Masai: their language and folklore* (Oxford: OUP)

Holmes, E. M. (1877-8) Notes on the medicinal plants of Liberia, in *Pharm. J. ser. 3,8*

Ibrahim, H.A. & H.Ibrahim (2009) Phytochemical screening and Toxicity Evaluation on the leaves of *Argemone mexicana*, in *Int.J. of Pure & Applied Sciences 2009 3:2*

Idu, M. & B.C.Ndukwu (2006) Studies of plants used in ethnomedicine in Ethiope Concil area of Delta State, Nigeria, in *Research J. of Botany 2006 1:1, pp.30-43*

Igile, G.O., Oleszeck, W., Burda, S., and Jurzysta, M. (1995) Nutritional assessment of *Vernonia amygdalina* leaves in growing mice, in *J. of Agric. and Food Chem. 43 , 2162-2166*

Innocent, E., M.J.Moshi, P.J.Masimba, Z.H.Mbwambo, M.C.Kapingu and A.Kamuhabwa (2009), Screening of Traditionally Used Plants for *in vivo* Antimalarial Activity in Mice, in *Afr. J. Traditional Complementary Alternative Medicine, Vol.6, (2), pp163-167*

Irvine, F.R. (1930) *Plants of the Gold Coast* (Oxford University Press)

(1956) Cultivated and semi-cultivated leafy vegetables of West Africa , in *Materiae Veg. 2*

(1961) *Woody Plants of Ghana* (Oxford University Press)

Iwalewa, EO, L.Lege-Okuntoye, P.P. Rai, T.T. Iyaniwura and N.L. Etkin (1990) *In vitro* antimalarial activity of leaf extracts of *Cassia occidentalis* and *Guiera senegalensis* in *Plasmodium yoelii nigeriensis*, in *W.African J. of Pharmac. Drug Res. 9 (supp): 19-21*

Iwu, M.M. (1986) Empirical Investigations of Dietary Plants used in Igbo Ethno-medicine, pp. 131 -56, in Etkin, N.L., *Plants in indigenous Medicine & Diet. Behavioural approaches* (New York: Redgrave Publishing Co.)

Iwu, M.M. (1986b) *Handbook of African medicinal plants* (Nigeria: CRC Press) cited in *Prelude med. plants database (2011)*

Jansen O., L.Angenot, M.Tits et al. (2010) Evaluation of 13 plants used against malaria in traditional medicine in Burkina Faso, in *J. Ethnopharm 2010 Apr.29*

Jeruto, P., C.Lukhoba, G.Ouma, D.Otieno & C.Mutai (2008) An ethnobot. study of med. plants used by the Nandi people in Kenya, in *J. Ethn. 2008 16: pp.370-6* ; cited in *Prelude med. plants database (2011)*

Jiofack, T., L.Ayissi, C.Fokunang, N.Guedje & V.Kemeuze (2009) Ethnobotany & Phytomedicine of the upper Nyong Valley forest in Cameroon, in *A.J. Pharmacy & Pharmacology 2009 3 :4, pp. 144-150* cited in *Prelude med. plants database (2011)*

Johns,T., D.E.B.Mhoro, P.Sanaya & E.K.Kimanani (1994) Herbal remedies of the Batemi of Ngorongoro District, Tanzania; a quantitative appraisal, in *Econ. Bot. 1994 48:1 pp. 90-95, cited in Prelude med. plants database (2011)*

Johns,T., R.L.Mahunnah, P.Sanaya, L. Chapman and T. Ticktin (1999) Saponins and phenolica content in plant dietary additives of a traditional subsistence community, the Batemi of Ngorongo District, Tanzania, in *J. of Ethnopharmacology 66, pp. 1-10*

Jonville, M.C., H.Kodja, L.Humeau et al., (2008) Screening of medicinal plants from Réunion Island for antimalarial and cytotoxic activity, in *J. of Ethnopharm. 2008 Sep. 18.*

Joseph, C.C., M.M.Ndoile, R.C.Malima, M.H.H.Nkunya (2004) Larvicidal and mosquitocidal extracts & coumarin, isoflavonoids and pterocarpans from *Neorautanenia mitis*, in *Tr. of R.Soc.of Tropical Medicine and Hygiene, Vol.98, Issue 8, pp 451-455*

Kaendi, J.M. (1994) Coping with malaria and visceral leishmaniasis in Baringo District, Kenya: Implications for disease control, from *Indigenous knowledge in the management of malaria and visceral leishmaniasis among the Tugen of Kenya* from Ind. Know. & Dev. Mon., 1997: www.nuffic.nl/ciran/ikdm/5-/articles/-mungutiart.htm

Kamba, A.S. & L.G./Hassan (2010) Antibacterial screening & brine shrimp toxicity of *Securidaca longepedunculata* root bark, in *Afr.J. Pharm. Sc. & Pharm. 2010 1:1 pp. 85-95*

Karel, L. & Roach, E.S. (1951) *Dictionary of Antibiosis* (New York: Columbia University Press)

Kareru, P.G., G.M.Kenji, A.N.Gachanja, J.M.Keriko & G.Mungai (2007) Tr. Med. among the Embu & Mbeere peoples of Kenya, in *Prelude med. plants database (2011)*

Kareru, P.G., A.N.Gachanja, J.M.Keriko & G.M.Kenji (2007) Antimicrobial activity of some medicinal plants used by herbalists in eastern province, Kenya, in Afr.J. of trad., comp., & alt. med., 2007: vol.5:1 pp.51-55, cited in *Prelude med. plants database (2011)*

Karunamoorthi, K., D.Bishaw, T.Mulat (2008) Laboratory evaluation of Ethiopian local plant *Phytolacca dodecandra* extract for its toxicity effectiveness against aquatic macroinvertebrates in *Eur. Rev. Med. Pharmacological Sciences 2008 ; 12 (60 pp 381-386*

Kassa, M., R.Mshana, A.Regassa and G.Assefa (1998) *In vitro* test of five Ethiopian medicinal plants for anti-malarial activity against *Plasmodium falciparum*, in *Ethiopian J. of Sc. (1998) Vol. 21(1) pp. 81-89*

Kasuku, W., F.Lula, J.Paulus, N.Ngiefu & D.R.Kaluila (1999) Contribution à l'inventaire des plantes utilisées pour le traitement du paludisme à Kinshasa, (RDC) in *Rev. Med. Pharm.1999 13 : 95-102,* cited in *Prelude med. plants database (2011)*

Katuura, E., P.Wako, J.Ogwal-Okeng, R.Bukenya-Ziraba (2007) Traditional Treatment of malaria in Mbarara district, W.Uganda, in *Af. J. Ecol. 2007:45 Suppl.1 pp.48-51* cited in *Prelude med. plants database (2011)*

Kayode, J., O.E.Ige, T.A.Adetogo & A.P.Igbakin (2009) Conservation & biodiversity erosion in Ondo State, Nigeria: Survey of plant barks used in native pharmaceutical extraction in Akoko region, in *Ethnobotanical Leaflets 13 : 665-7*

Kela, S.L., R.A.Ogunsusi, V.C.Ogbogu & N.Nwude (1989) Screening of some Nigerian plants for molluscicidal activity, in *Rev. Elev. Med. Vet. Pays Trop.1989, 42:2, pp.195-202*

Kerharo, J. (1973) Pharmacognosie de quelques graminées senegaleses, in *Bull. Soc. Med. Afr. noire, Lang. Franc, 18: 1-13*

Kerharo, J., & A.Bouquet (1950) *Plantes médicinales & Toxiques de la Côte-d'Ivoire-Haute-Volta* (Paris: Vigot Freres)

Kerharo, J., & J.G. Adam (1963) Deuxième inventaire des plantes médicinales et toxiques de la Casamance (Senegal), in *Ann. Pharm. Franc. 21: pp. 853-70*

Kerharo, J., & J.G. Adam (1964) Plantes médicinales & Toxiques des Peul et des Toucouleur du Sénegal, in *J. Agr. trop. Bot. appl. 11: 384-444, 543-99*

Kerharo, J., & J.G. Adam (1974) *La Pharm. Sénegalaise Traditionelle. Plantes méd. et toxiques* (Paris:Vigot Freres)

Kirira, P.G., G.M.Rukunga, A.W.Wanyonyi et al. (2006) Antiplasmodial activity and toxicity of extracts of plants used in traditional malaria therapy in Meru and Kilifi Districts of Kenya, in *J. of Ethn. 2006 July 19; 106 (3) 403 - 407*

Kling, H. (1923) *Die Seiketrooster* (Cape Town: de Villiers) cited in *Prelude med. plants database (2011)*

Koch,A., P.Tamez, J. Pezzuto & D.Soejarto (2005) Evaluation of plants used for antimalarial treatment by the Maasai of Kenya in *J. of Ethnopharm. (2005) 101: 95-99*

Koita, N. (1991) Comparative study of the traditional remedy "Suma-kala" and chloroquine as treatments for malaria in the rural areas, in *Mshgeni KE, Nkunya MHH, Fupi V, Mahunnah RLA, Mshiu E (Eds) Proceedings of an international conference of experts from developing countries on traditional medicinal plants, Arusha, Tanzania, 18-23 Feb 1990 (Dar-es-Salaam University Press, 1991)*

Kokwaro, J.O. (1976) *The Medicinal Plants of East Africa* (Kampala: East Africa Literature Bureau)

Kone, M.W., K.K.Atindehou, H.Terre & D.Tyraore (2002) Quelques plantes médicinales utilisées en pediatrie traditionelle dans la région de Ferkessedougou (Côte-d'Ivoire), in *Rev.Int.Sci.de la Vie et de la Terre* cited in *Prelude med. plants database (2011)*

Kori, M.L., K.Gaur, V.K.Dixit (2009) Investigation of immunomodulatory potential of *Cleome gynandra* Linn., in *Asian J. of Pharm. & Clin. Res. 2009 2:1*

Kraft, C., K.Jenett-Siems, K.Siems, J.Jakupovic, S.Mavi, U.Bienzle, E.Eich (2003) *In Vitro* Antiplasmodial Evaluation of Medicinal Plants from Zimbabwe, in *Phytotherapy Research 17, pp. 123-128*

Krettli, A.U., V.F. Andrade-Neto and M.G.L.Brandão *Bidens pilosa (Asteraceae) crude extracts are active against P. falciparium in vitro, and in mice infected with P. berghei* (Minas Gerais: Centro de Pesquisas René Rachou, Fiocruz,and Faculdade de Farmácia, Federal University of Minas Gerais, Belo Horizonte, Brasil)

Krüger, N. & M. Krüger (1985) Beobacht. zur traditionellen Medizin der Mende in Sierra Leone, in *Curare, Sonderbund 3 pp. 325-336* cited in *Prelude med. plants database (2011)*

Kudi, A.C., J.U.Umoh, L.O.Eduvie & J.Gefu (1999) Screening of some Nigerian medicinal plants for antibacterial activity, in *J.Ethnoph.1999 67, pp.225-228*

Kumar, I., S.Sridhava, B.P.Singh & S.V.Gangul (1998) Characteristics of Cogon Grass (*Imperata cylindrica*) pollen extracts, etc.,in *Int.Arch.Allergy & Immunol. 117, pp.174-9*

Lachman-White, D.A., C.D.Adama, U.O'D.Trotz (1987) *A Guide to the Medicinal Plants of Coastal Guyana* (London: Commonwealth Science Council)

Laidler,P.W.(1928) The magic medicine of the Hottentots,in *S.Afr.J.Sci. 1928 25, pp.433-47*

Lambert,J.D., J.H.Temmink, J.Marquis, R.M.Parkhurst, C.B.Lugt, E.Lemmich, L.Wolde-Yohannes, D.de Savigny (1991) Endod: safety evaluation of a plant molluscicide, in *Regul Tox. Phamacol. 1991 14 2 pp. 189-201*

Lamidi, M, E.Ollivier et al. (2007) Enquête ethnobotanique auprès des tradipraticiens dans 3 régions du Gabon etc.,1ième partie, in *Ethnopharmacologia 2007 39 pp. 44-51, cited in Prelude med. plants database (2011)*

Leclerc, H. (1930) In *Presses Médicales* 38: 948

Lemordant, D. (1971) Contribution à l'ethnobotanique éthiopienne , in *J. d'Agr.Trop.et de Bot. Appl. 1971 18 1-35, 142-179, cited in Prelude med. plants database (2011)*

Leonard, J. (1952) Cynometreae et Amherstieae, in Boutique, R., *Flore du Congo Belge et du Ruanda-Urundi*, 5: 176-359 (Brussels)

Levin, Y, Y.Sherer, H.Bibi, M.Schlesinger& E.Hay (2000) Rare *Jatropha multifida* intoxication in two children, in *J. Emergency Med. 2000 19:2, pp. 173-5*

Lindsay, R.S. & F.N.Hepper (1978) *Medicinal plants of Marakwet, Kenya* (RBG Kew)

Lorenzi H. (1991) *Plantas Daninhas do Brasil: Terrestres, Aquaticas, Parasitas, Toxicas e Medicinais*, 2nd ed. (Nova Odessa: Editora Plantarum)

Loustalot, A.J. & C. Pagan (1949) Some Cuban Medicinal Plants *El Crisol* 35: 3-5 (San Juan, Puerto Rico)   (1950) Local 'fever' plants tested for the presence of alkaloids in *Chemical Abstracts:* 44 (5): 2179

Lubini, A.(1990) Les plantes utilisées en médicine traditionelle par les Yansi de l'entre Kwilu-Kamtsha (Zaire), in *Mitt. Inst. Allg. Bot.Hamburg 23b 1007-1020* cited in *Prelude* *med. plants database (2011)*

Lusakibanza, M., G. Mesia, G.Tona et al. (2010) *In vitro* and *in vivo* antimalarial and cytotoxic activity of five plants used in Congolese traditional medicine, in *J. Ethnopharm. 2010 June 16; 129(3) pp.398 - 402*

Maberley, J. (1899) In *The Lancet* 157: 874

MacDonald, I., & D.I.Olurunfemi (2009) Research projects: Plants used for medicinal purposes by Koma people of Adamawa State, Nigeria, in *Indigenous Knowledge & Development Monitor 2000, cited in Prelude med. plants database (2011)*

MacFoy, C.A. & A.M.Sama (1983) medicinal plants in Pujehun district of Sierra Leone, in *J. Ethnoph. 1983 v.8 pp. 215-223*

Magogo & Glover, *905, Herbarium, RBG Kew*

Maikai,V.A., P.I.Kobo & A.O.Adaudi (2008) Acute toxicity studies of aqueous stem bark extract of *Ximenia americana*, in *Af. J. Biotech. 2008 v.7 10.pp.1600-03*

Malgras, D. (1992) *Arbres et arbustes guérisseurs des savannas maliennes* (Paris: Editions Karthala) cited in *Prelude med. plants database (2011)*

Mapi, J. (1988) Contribution à l'étude ethnobotanique et analyses chimiques de quelques plantes utilisées en médecine traditionelle dans la region de Nkongsamba (Moungo) (Doc. Thesis, Yaonde Univ.) cited in *Prelude med. plants database (2011)*

Maregesi, S., S.Van Miert, C.Pannecouque at al. (2010) Screening of Tanzanian medicinal plants against *Plasmodium falciparum* and human immunodeficiency virus, in *Planta Med. 2010 Feb; 76 (2) 195 - 201*

Mariano-Sousa, D.P. (2009) *Toxic effects of Senna occidentalis on lymphohematopoetic system: evaluation of its exposure in rats during the growth and pre-natal period,* (Doctoral thesis, University of Säo Paulo, Brazil)

Martin, B.W., M.K.Terry, C.H. Bridges and E.M. Bailey jnr (1981) Toxicity of *Cassia occidentalis* in the horse, in *Vet. Hum. Toxicol. 1981 23:6, pp.416-7*

Masinde, P.S.(1996) *Medicinal plants of the Marachi people of Kenya* (Wageningen:1994 Aetfat Congress) cited in *Prelude med. plants database (2011)*

Maundu, P.M. et al. (1999) *Traditional Food Plants of Kenya* (National Museum of Kenya)

Mesfin, F., S.Demisse & T.Teklehaymanot (2009) An ethnobotanical study of medicinal plants in Wonago Woreda, in *Ethiopia J. of Ethnobiology & Ethnomedicine 2009 5:28*; cited in *Prelude med. plants database (2011)*

Mesia, G.K., G.L.Tona, T.H.Nanga et al., (2007) Antiprotozoal and cytotoxic screening of 45 plant extracts from Democratic Republic of Congo, in *J. of Ethnopharm.2007 Dec. 6*

Milliken, W., (1997) Malaria & antimalarial plants in Roraima, Brazil, in *Tropical Doctor, 1997, 27 (suppl. 1), 20-25*

Milliken, W. (1997b) Traditional Anti-malarial Medicine in Roraima, Brazil, in *Economic Botany 51 (3): 212-237*

Mlambo, V., T.Smith, E.Owen, F.L.Mould, J.L.L.Sikopsana & I. Mueller-Harvey (2004) Tanniniferous *Dichrostachys cinerea* fruits do not require detoxification for goat nutrition, *in sacco* & *in vivo* evaluation, in *Livestock Prod. Scie. (2004) 90:2-3 pp. 135-44*

Molgaard, P., A.Chihaka, E.Lemmich, P.Furu, C.Windberg, F.Ingerslev and B.Halling-Sorensen (2000) Biodegradability of the Molluscicidal Saponins of *Phytolacca dodecandra* in *Regulatory Toxology and Pharmacology Volume 32, Issue 3, December 2000, pp 248-255*

Monzote, L., A.M.Montalvo, R.Scull, M.Miranda & J.Abreu (2007) Activity, toxicity and analysis of resistance of essential oil from *C. ambrosioides* etc., in *Biomedicine & Pharmacotherapy 2007 61:2-3, pp.148-153*

Morris, B. (1996) *Chewa Medical Botany. Monographs from the International African*

Institute, 2. (Hamburg: Lit-Verlag)

Moshi, M.J., J.C.Cosam, Z.H.Mbwambo, M.Kapimgu & M.H.H.Nkunya (2004) Testing beyond ethnomedical claims: brine shrimp lethality of some Tanzanian plants, in *Pharmaceutical Biol. 2004, 42:7, p.547*

Moshi, M.J., Z.H.Mbwambo, R.S.O.Nondo, P.J.Masimba, A.Kamuhabwa, M.C.Kapingu, P.Thomas & M.Richard (2006) Evaluation of ethnomedical claims & Brine shrimp toxicity of some plants used in Tanzania as traditional medicines, in *Af. J. Trad. Compl. & Alt. medicines, 2006:3/3, pp. 48-58*

Moshi, M.J.,E. Innocent, J.J.Magadula, D.F.Otieno, P.K.Mbabazi, A.Weisheit & R.S.O.Nondo (2009) Ethnomed.of the Haya people of Bugabo ward, Kagera region, N.W.Tanzania, in *J. Ethnobiol & Ethnomed. 2009 5:24,* cited in *Prelude med. plants database (2011)*

Moshi, M.J., D.F.Otieno, P.K.Mbabazi & A.Weisheit (2010) Brine shrimp toxicity of some plants used in traditional medicine in Kagera region, NWTanzania, in *Tanz. J. Health Research 2010 12 :1*

Muganga, R., L.Angenot, M.Tits et al. (2010) Antiplasmodial and cytotoxic activities of Rwandan medicinal plants used in the treatment of malaria, in *J. of Ethnopharm. 2010 Mar.2; 128(1) 52 - 57*

Mukinda,J.T.& J.A.Syce (2007) Acute and chronic toxicity of the aqueous extract of *Artemisia afra* in rodents, in *J. of Ethnoph. 2007 112:1 pp. 138-44*

Mulhovo, S.F. (1999) unp. address at 1st RITAM conference at Moshi, Tanzania

Muregi, F.W., S.C.Chhabra, E.N.Njagi et al. (2003) *In vitro* antiplasmodial activity of some plants used in Kisii. Kenya against malaria and their chloroquine potentiation effects, in *J. Ethnopharm. 2003 Feb; 84 (2-3): 235-9*

Musa, K.Y., A.Ahmed, G.Ibrahim, O.E.Ojonugwa, M.Bisalia, H.Musa & U.H.Danmalam (2007) Toxicity studies on the methanolic extract of *Portulaca oleracea*, in *J. Boil.Sc.7 pp. 1293-5*

Muthaura, C.N., G.M.Rukunga, S.C.Chhabra et al. (2007) Antimalarial activity of some plants traditionally used in Meru district of Kenya, in *Phytotherapy Research 2007 May 8*

Nadkarni, K.M. (1927) *The Indian Materia Medica* (Bombay)

Naik, R.M. & al. (1956) *Current Science* 25: 324, 325
          (1958) *Biological Abstracts* 32: 38568

Nanyingi, M.O., J.M.Mbaria, A.L.Lanyasunya, C.G.Wagate, K.B.Koros, H.F.Kaburia R.W.Munenge & W.O.Ogara, (2008) Ethnopharm. Survey of Samburu District, Kenya, in *J. Ethnobiol. & Ethnomedicine 2008 4 :14*

Nair, D.M.N. (1967) *Selected families of Zambian flowering plants.* (University of Zambia)

Natabou Dégbé, F. (1991) *Contribution à l'étude de la médicine et de la Pharmacopée traditionelles au Bénin, etc.* (Doc.Thesis, Univ. Cheikh Anta Diop); cited in *Prelude med. plants database (2011)*

Ndukwu, B.C. & N.B.Ben-Nwadibia (n.d.) Ethnomedicinal aspects of plants used as spices and condiments in the Niger Delta area of Nigeria, www.siu.edu/~ebl/leaflets/niger.htm, cited in *Prelude med. plants database (2011)*

Negm,S., O.El Shabrawy, M.Arbid, and A.S.Radwan (1980), Toxicological study of the different organs of *Corchorus olitorius* L. plant, with special reference to their cardiac glycosides content, in *Z.Ernahrungswiss. 19, 28-32*

Neuwinger, H.D. (2000) *African Traditional Medicine. A dictionary of plant use and application* (Stuttgart: Medpharm GmbH Scientific Publisher)

Nguyen-An-Cu & A.Vialard-Goudou (1953) p. 15 in vol. 4A, *Proc. 8 of the Pac. Sc. Congress*

Nibret, E., M.L.Ashour, C.D.Rubanza & M.Wink (2009) Screening of some Tanzanian medicinal plants for their trypanocidal & cytotoxic activities, in *Phtotherapy Research 2009,v.24:6, pp.945-947*

Nikkon, F., S.Hasan, M.H.Rahman, M.A.Hogue, A.Mosaddik & M.E.Haque (2008) Biochemical, Hematological & Histopathological effects of *Duranta repens* stems on rats, in Asian J. Biochemistry 2008 3: pp. 366-72

Nikkon, F., K.A.Salam, T.Yeasmin, A.Mossadik, P.Khondkar& M.E.Haque (2010) Mosquito-cidal triterpenes from the stem of *Duranta repens*, in *Pharm.Biol.2010 48:3, pp. 264-8*

Nikonorow, M. (1939) In *Acta Polonica Pharmacologica* 3: 23
(1942) In *Biological Abstracts* 16: 633

Njoroge,G.N. & R.W.Bussmann (2006) Diversity & utilisation of ethno-phyto-therapic remedies among the Kikuyus (C.Kenya), in *J.Ethnobiol. & Ethnomed. 10.1186* cited in *Prelude med. plants database (2011)*

Novy, J.W. (1997) Medicinal plants of the eastern region of Madagascar, in *J. of Ethnoph. 1997: 55, pp. 119-26*

Nundkumar, N. and J.A.Ojewole (2002) Studies on the antiplasmodial properties of some South African medicinal plants used as antimalarial remedies in Zulu folk medicine, in *Clin. Pharmacol. 2002 Sep.; 24 (7): 397 - 401*

Obidah, W., M.S.Nadro, G.O.Tifayo, A.U.Wurochekke (2009) Toxicity of crude *Balanites aegyptica*, in *J. of American Science 2009 5:5 pp. 13-16*

Obih, P.O. and J.M.Makinde (1986) An investigation into the antimalarial activity of *Cymbopogon citratus* (Lemon Grass) on Plasmodium berghei berghei in mice, in *Sofowora, A. (Ed) (1986) The state of medicinal plants research in Nigeria* (Ibadan: Nigerian Society of Pharmacology)

Obute, G.C. (n.d.) *Ethnomedicinal plant resources of S.E.Nigeria Leaflets,* (www.siu.edu/~ebl/leaflets/niger.html) ; cited in *Prelude med. plants database (2011)*

Oderinde, R.A., I.A.Ajayi and A.Adewuyi (2009) Preliminary toxicological evaluation and effect of the seed oil of Hura crepitans and Blighia unijugata on the lipid profikle of rats, in *Downloads home>vol. 8, Year 2009*

Odugbemi,T.O., O.R.Akinsulire, I.E.Aibainu & P.O.Fabeku (2007) Medicinal plants useful for malaria therapy in Okeigbo ondo State, SW Nigeria, in *A.J.Trad.Comp. & Alt. Med. vol. 4:2 pp. 191-8*

Oduola,T.,I.Bello, G.Adiosun, A.W.Ademosun, G. Raheem & G.Avwioro (2010) Hepatoxic-ology & nephrotoxicology evaluated in Wistar albino rats exposed to *Morinda lucida* leaf extract, in *N.A.J. Med.Sci. 2010 2:5, p.230*

Ogbonnia, S.O,, S.O.Olayemi, E.N.Anyika, V.N.Enwuru & O.O.Poluyi (2009) Evaluation of acute toxicity in mice & subchronic toxicity of hydroethanolic extract of *Parinari curatellifolia*..seeds in rats, in *Af.J. Biotech. 2009 8:9, pp. 1800-1806*

Ogie-Odia,T.O., O.R.Akinsulire, I.E.Aibainu & P.O.Fabeku (2007) Medicinal plants used for malaria therapy in Okeigbo State, SW Nigeria, in *A.J.Trad.Com.& Alt.Med 2007 4:2 pp. 191-8*

O'Hara, P.J. and K.R.Pierce (1974) A toxic cardiomyopathy cause by *Cassia occidentalis*, in *Veterinary Pathology 1974, 11:2 pp. 110-124*

Ohigashi, H., M.Jisaka, T.Takagaki, H. Nozaki, T.Tada, M.A.Huffman, T.Nishida, M. Kaji and K.Koshimizu (2000) Bitter principle and related steroid glucoside from *Vernonia amygdalina*, a possible medical plant for wild chimpanzees, in *Agric. and Biol. Chemistry 55, 1201-1203*

Okokon, J.E., A.L.Bassey & L.L.Nwidu (2007) Antidiabetic & hypolipidaemic effects of ethanolic root extract of *Setaria megaphylla*, in *Int. J. Pharm. 2007 3 pp.91-5*

Okokon, J.E. & P.A.Nwafor (2009) Antiplasmodial acivity of root extract and fractions of *Croton zambesicus*, in *J. of Ethnoph. 2009 121, pp.74-78*

Okunade, A.L. & J.I.Olaifa (1987) Estragole: An acute toxic principle from the volatile oil of the leaves of Clausena anisata *J. Nat. Prod. 1987 50, 5: 990-1*

Okunade, A.L.(2001) Ageratum conyzoides L. (Asteraceae) in *Fitoterapia 2002 73: pp 1 -16*

Ojewole,J. A. Hypoglycaemic effect of *Clausena anisata* (Willd, Hook) methanolic root extract in rats *J Ethnopharmacol 81:2: pp.231-7*

Olagunju, J.A., A.B.Ogunfeibo, A.O.Ogunbosi & O.A,Taiwo (2004) Biochemical changes elicited by isosaline leaf and stem-bark extracts of *Harungana madagascariensis*, in *Phytother. Res. 2004 18: 7pp. 588-91*

Oliver-Bever, B. (1960) *Medicinal plants in Nigeria* (Nigerian College of Arts, Science, and Technology)

　　　　　　(1983) *Medicinal Plants in Tropical West Africa, II. Plants acting on the nervous system* 7: 1- 93 (Cambridge University)

Omulokoli, E., B.Khan and S.C.Chhabra (1997) Antiplasmodial activity of four Kenyan medicinal plants, in *J.Ethnopharm. 1997 Apr; 56(2): 133 - 7*

O'Neill, M.J, D.H.Bray,P. Boardman, J.D.Philipson, D.C.Warhurst, W. Peters and M.Suffness (1986) Plants as sources of antimalarial drugs: *in vitro* antimalarial activities of some quassinoids, in *Antimicrobial agents and Chemotherapy, 30: 101-104*

Orech, F.O.,T.Akenga, J.Ochora, H.Friis & J.Aagaard-Hansen (2005) Potential toxicity of some traditional leafy vegetables consumed in Nyangoma Div., W.Kenya, in *Af. J. Food Nutr. Sci. 2005: 5, pp.1-13*

Osol, A. & Farrar, G. (1947, 1955) *The Dispensary of the USA, 24th & 25th ed.* (Philadelphia: Lippincott)

Pahwa, R. & V.C.Chatterjee (1989) The toxicity of Mexican Poppy (*Argemone mexicana*) seeds to rats, in *Vet.Hum.Toxicol. 1989 31:6, pp. 555-8*

Palgrave, K.C. (1957) *The Trees of Central Africa* (Salisbury: National Publications Trust, Rhodesia and Nyasaland)

　　　　　　(1983) *Trees of Southern Africa*, 2nd ed.(Cape Town: Struik)

　　　　　　(2002) *Trees of Southern Africa,* 3rd ed.(Cape Town: Struik)

Pammel, L.H. (1911) *Manual of poisonous plants* (Cedar Rapids: Torch Press)

Pappe, L. (1868) *Florae Capensis Medicae Prodromus* 3rd ed. (Cape Town: Brittain) cited in *Prelude med. plants database (2011)*

Pardy, A.A. (1952) In *Bulletin of the Dept. of Agriculture in S. Rhodesia* 1674

Parry, O & C.Matambo (1992) Some pharmacological actions of aloe extracts and *Cassia abbreviata* on rats and mice, in *Cent. Afr. J. Med. 1992 38 ;10 pp. 409-14*

Penugonda,S., S.Girisham & S.M.Reddy (2010) Elaboration of microtoxins by seed-borne fungi of finger millett etc, in *Int.J.Biotech.& Mol.Biol. Research 2010 1:5, pp.62-64*

Pernet, R. (1957) Les plantes médicinales malgaches. Catalogue de nos connaissances chimiques et pharmacologiques, in *Mémoires de l'Insitute Scientifique de Madagascar, Série B*, VIII cited in *Prelude med. plants database (2011)*

Pernet, R. & G. Meyer (1957) *Pharmacopée de Madagascar: publication de l'Institut de Recherche scientifique* (Tananarive-Tsimbazaza); cited in *Prelude med. plants database (2011)*

Perrot, E. et al.(1930) In *Bulletin Scientific & Pharmacologic* 37: 401

　　　　　　(1930) In *Bull. de l' Acad. Royale Médicin Belgique* 103

Phillips, E.P. (1917) *Annals of the South African Museum*: 16, 1

Phillipson, J.D., G.C.Kirby and D.C.Warhurst (1993) Tropical plants as sources of antiprotozoal agents, in *Recent Advances in Phytochemistry 27, pp. 1-40*

Pijper, C. (1919) *Die volkesgeneeskunst in Transvaal* (Leyden: Diss)

Pillay, P., V.J.Maharaj and P.J.Smith (2008) Investigating South African plants as a source of new antimalarial drugs, in *Journal of Ethnopharmacology 119 (2008) 438-454*

Pittier, H. (1978) *Las Plantas Usuales de Venezuela* (Caracas: Fundacion Eugenio Mendoza) cited in *Watts et al.(1962)*

Pobéguin, H. (1912) *Plantes médicinales de la Guinee* (Paris) cited in *Watts et al.(1962)*

Polygenis-Bigendako, M.-J.(1990) *Recherches ethnopharmacognostiques sur les plantes utilisées en médecine traditionelle au Burundi occidental* (Doc.Thesis, Brussels Univ.), cited in *Prelude med. plants database (2011)*

Pomini, L. (1938) *Le piante Officinali e del Sottobosco p.* N.F.- (Federazione dei Fasci Femminili Sezione Prov. Massaie Rurali-Vercelli Anno xvi) cited in *Watts et al.(1962)*

Porteres, R. (s.d.) *Reliquiae*, Lab. Ethnobotanique (Paris)

Potel, A.-M. (2002) *Les plantes médicinales au Sénégal* (Study report at Nguekokh) cited in *Prelude med. plants database (2011)*

Power, F.B. (1914) *The Wellcome Chemical Research Laboratories* (London: Burroughs Wellcome)

PROTA (2008) *Plant Resources of Tropical Africa 11(1) Medicinal Plants 1* (PROTA Foundation/Backhuys Publishers/ CTA Wageningen, Netherlands)

PROTA (2010) *Plant Resources of Tropical Africa :* Vegetables (PROTA Foundation/Backhuys Publishers/ CTA Wageningen, Netherlands)

Quimby, M.W. & G.J.Persinos (1964) Notes on preliminary drug-hunting trip on the Jos Plateau, Nigeria, in *Economic Botany 1964 18:3, pp.266-269*

Quin, P.J. (1954) *Ph.D. Thesis* (University of Witwatersrand)

Quisumbing, E. (1951) *Technical Bulletin of the Philippine Department of Agriculture & Natural Resources*: 16 (Manila)

Rahantamalala, C. (2000) *Contribution à l'inventaire et à l'évaluation en laboratoire de quelques plantes médicinales utilisées dans le traitement du paludisme* (Unpubl. thesis, University of Antananarivo)

Rajesh, P., S. Latha, P. Selvamani, V. Rajesh Kannan, (2010) *Capparis sepiaria* Linn - Pharmacognostical standardization and toxicity profile with chemical compounds identification (GC-MS), in *Int. Jour. of Phytomedicine 2010 2: pp.71-79*

Ramadan, A., F.M.Haraz and S.A.El-Mougy (1994) Anti-inflammatory, analgesic, and antipyretic effects of the fruit-pulp of *Adansonia digitata*, in *Fitoterapia 65, pp. 418-422*

Ramathal, D.C. and O.D.Ngassapa (2000) Medicinal plants used by Rwandese Traditional Healers in Refugee Camps in Tanzania, in *Pharm. Biol. 2001, vol.39, no.2, pp.132-137*

Ramesh, D., T.J.Dennis and M.S.Shingare (1992) Constituents of *Adansonia digitata* root bark, in *Fitoterapia LXIII, no. 3, p. 278*

Rampa, L. (1956) *The Third Eye* (London: Secker & Warburg)

Rao, P.R. & al. (1944) In *Proceedings of the Indian Academy of Science* A, 19, p. 88
(1945) In *Biological Abstracts* 19, p. 162

Raponda-Walker, A. & R. Sillans (1961) *Les Plantes utiles du Gabon. Encyclopédie biologique* (Paris: Editions Paul Lechavlier) *cited in Prelude med. plants database (2011)*

Rasoanaivo, P., A.Petitjean, S.Ratsimamanga-Urverg and A.Rakoto-Ratsimamanga (1992) Medicinal Plants used to treat malaria in Madagascar, in *J. Ethnopharm. 37: 117 - 127*

Ratsimamanga, A.R. (1979) Etat d'avancement de la recherche sur la médecine et la pharmacopée traditionelles à Madagascar, etc., (Cairo: Univ.Press) cited in *Prelude med.*

plants database (2011)

Raymond, W.D. (1939) In *Journal of Tropical Medicine* 42: 29

Rivière, C., J.-P. Nicolas, M.-L.Caradec, O.Désiré & A. Schmitt (2005) Les plantes médicinales de la region nord de Madagascar: une approche ethnopharmacologique, in *Ethnopharmacologia 36 pp.36-50* , cited in *Prelude med. plants database (2011)*

Roark, R.C. (1931) Excerpts from Consular Correspondence relating to insecticide and fish-poison plants, in *U.S.Dept. Agric. Bur.Chem. & Soils*

Rosevear, D.R. (1961) *Gambia trees and shrubs* (ined. ms, Herbarium, RBG Kew)

Rukunga, G.M., J.W.Gathirwa, S.A.Omar et al. (2008) Antiplasmodial activity of the extracts of some Kenyan medicinal plants, in *J. of Ethnopharm. 2008: Nov.8*

Russell, A.B., J.W.Hardin, L.Grand, A.Fraser (1997) *Poisonous plants of North Carolina* (N.Carolina State University)

Saadou, M. (1993) Les plantes médicinales du Niger: premier supplément à l'enquête ethnobotanique 1973, in *Rev. Méd. Pharm. Afr. 1993 Vol.7:1, pp.11-24* cited in *Prelude med. plants database (2011)*

Salawu, O.A., A.Y.Tijani, H.Babayi, A.C.Nwaeze, R.A.Anagbogu & V.A.Agbakwuru (2010) Anti-malarial activity of ethanolic stem bark extract of *Faidherbia albida*...in mice, in *Arch.Appl.Sci.Res.2010 2 (5) pp. 261-268*

Samuelsson,G., M.H.Faraha, P.Claeson, et al. (1992) Inventory of plants used in traditional medicine in Somalia III. Plants of the families Lauraceae-Papilionaceae, in *J.Ethnoph. 1992 v.37 pp 93-112* cited in *Prelude med. plants database (2011)*

Sanon, S., N.Azaz, M.Gasquet et al. (2003) Antiplasmodial activity of alkaloid extracts from *Pavetta crassipes* (K. Schum) and *Acanthospermum hispidum* (DC), two plants used in traditional medicine in Burkina Faso, in *Parasit. Res. 2003 Jul; 90(4): 314 - 7*

Sarkiyayi, S., S.Ibrahim & M.S.Abubakar (2009) Toxicological studies of *Ceiba pentandra* Linn, in *Afr.J.Biochem.Research 2009 3:7 pp. 279-81*

Sastri, B.N.[Ed.] (1952) *Wealth of India. Raw Materials, 3 (D-E)* (New Delhi: C. S. I. R.)

Sathish, K.R., A.A.Rahmann, R.Buvanendran, D.Obeth & U.Panneerselvam (2010) Effect of *Evolvus alsinioides* root extract on acute reserpine-induced orofacial dyskinesia, in *I.J.Pharmacy & Pharmaceutical Sciences 2010 v.2 sup.4 pp.117-20*

Savill, P.S., & J.E.D.Fox (1967, ined.) *Trees of Sierra Leone* (Freetown: Forestry Dept. mimeo.)

Sawyerr, E.S. (1983) *Medicinal Plants of W.Africa v.1 Solomon Is.* cited in *Prelude med. plants database (2011)*

Schippers, R.R. (2004) Launaea cornuta, in *PROTA (2008)*

Schlage, C., C.Mabula, R.L.A.Mahunnah and M.Heinrich (1999) Medicinal Plants of the Washambas (Tanzania): Documentation and Ethnopharmacological Evaluation, in *Plant Biology 2000 2:1 pp. 83 - 92*

Schnell, R. (1960) *Icones Plantarum Africanum, fasc. V, nos. 97-120 (Dakar: I. F. A. N.)*

Schwikkard, S. and F.Van Heerden (2002) Antimalarial activity of plant metabolites, in *Nat.Prod.Rep. 2002, 19, pp.675 - 692*

Sharma, P. and J.D.Sharma (1998) Plants showing antiplasmodial activity, in *Indian Journal of Malariology vol. 35 (June 1998) 57-110*

Sharma,P.,S.Rastogi,S.Bhatnagar,J.K.Srivastava,A.Dube,P.Y.Guru,D.K.Kulshrestha,B.N.Mehrota & B.N.Dhawan (2003) Antileishmanial action of *Tephrosia purpurea* extract, in *Drug Dev.Res.2003 v.60:4 pp.285-293*

Shumba, S. & N.Nyazema (1996) *An in vivo assessment of the toxicity and parasiticidal activity of Croton megalobotrys oil in Plasmodium berghei-infected mice* (Hons. Project Pubs,

Univ. of Zimbabwe)

Silva, T.C., S.L.Gorniak, S.C.S.Oloris, P.C.Raspantini, M.Haraguchi, M.L.Z.Dagli (2003) Effects of *Senna occidentalis* on chick bursa of Fabricius, in *Avian Pathology, vol.32, Issue 6, pp 633 - 637*

Singh, M. and J. Singh (1985) Two Flavonoid glycosides from *Cassia occidentalis* pods, in *Planta Med. 1985 51: 6 pp.525-6*

Singh, P. (1956) *J. Sci. Industr. Res. (India)* 156: 259

Singh, V.K., Z.A.Ali (1993) Folk medicines in primary health care: common plants used for the treatment of fevers in India, in *Fitoterapia LXV, 1, 1994, pp. 68 -73*

Singha, S.C. (1965) *Medicinal Plants of Nigeria* (Apapa: Nigerian National Press)

Smith, A. (1888) *A contribution to South African materia medica* 3rd ed., 1895 (Lovedale Press)

Smith, E.W. & A.M.Dale (1920) *The Ila-speaking peoples of Northern Rhodesia* (London: Macmillan & Co. Ltd.)

Smith, P. & Q. Allen (2004) *Field Guide to the Trees and Shrubs of the Miombo Woodlands* (RBG, Kew)

Smith,Y.R.A. (2009) Determination of chemical composition of *Senna siamea* (Cassia Leaves) in *Pakistan J. of Nutrition 2009 8:2, pp. 119-21*

Spencer, C.F. et al. (1947) Survey of plants for anti-malarial activity, in *Lloydia* 10: 145

Ssegawa, P., J.M.Kasenene (2007) Medicinal plant diversity & uses in Sango Bay area, S.Uganda, in *J. Ethn. 2007 113 pp.521-540*

Staner, P. & R.Boutique (1937) *Matériauxl pour l'étude des plantes médicinales indigènes du Congo Belge* (Mém.Inst. Royal Col. Belge) cited in *Prelude med. plants database (2011)*

Storrs, A.E.G. (1979) *Know your Trees: some of the common trees found in Zambia.* (Ndola: The Forest Department)

Subramanian, S. (1955) *Journal of the American Pharmacological Assoc., Science Ed.*

Sugiyama, Y., & J. Koman (1992) The flora of Bossou : its utilisation by chimpanzees & humans, in *African. Study Monographs 1992 13 :3 pp.127-169*

Suliman, H.B., I.A.Wasfi and S.E.Adam (1982) The toxicity of *Cassia occidentalis* to goats, in *Vet. Hum. Toxicol.1982 24:5, pp. 326-30*

Suliman, H.B. and A.M.Shommein (1986) The toxic effect of roasted and unroasted beans of *Cassia occidentalis* in goats, in *Vet. Hum. Toxicol.1986 28:1, pp. 6-11*

Tabuti, J.R.S., K.A.Lye & S.S.Dhillion (2003) Traditional herbal drugs of Bulamogi, Uganda : plants, use & administration, in *J. Ethnoph. 2003: 88, pp.19-44*

Tabuti, J.R.S. (2008) Herbal medicines used in the treatment of malaria in Budiope country, Uganda, in *J.Ethnoph. 2008 v.116 pp. 33-42* cited in *Prelude med. plants database (2011)*

Tadeg, H., E.Mohammed, K.Asres & T.Gebre-Mariam (2005) Antimicrobial activities of some selected traditional Ethiopian medicinal plants used in the treatment of skin disorders, in *J.Ethnoph. 2005: 100 pp. 168-75* cited in *Prelude med. plants database (2011)*

Tadesse, D. (1947) Collecting medicinal plants at Bahir Dar Awraja, in *Germiplasm Newsletter 1988 18 pp.25-27*

Tadesse, D. (1994) Traditional use of some medicinal plants in Ethiopia.vol.1: 273-93, in *Proceedings of 13th plenary meeting of AETFAT, 1991, Zomba,* cited in *Prelude med. plants database (2011)*

Tasaka, A.C., R.Weg, E.E.Calore, I.L.Sinhorini, M.I.Z.Dagli, M.Haraguchi and S.L. Gorniak (2000)Toxicity Testing of *Senna occidentalis* Seed in Rabbits, in *Veterinary Research Communications, vol. 24, no. 8, pp 573 - 582*

Taton, A. (1971) *Boraginaceae*, in P.Bamps, Flore du Congo, Rwanda, et du Burundi,

Spermatophytes (Brussels: I.N.E.A.C)

Tchoumbougnang,F., P.H.A.Zollo, E.Dagne & Y.Mekonnen (2005) *In vivo* activity of essential oils from *Cymbopogon citratus* and *Ocimum gratissimum* on mice infected with *Plasmodium berghei*, in *Planta Medica, 2005.*

Teklehaymanot,T., M.G.Medhim & Y.Mekonnen (2007) Knowledge & Use of Medicinal Plants by people around Debre Libanos Monastery, in *J.of Eth. 2007:11 pp. 271-83* ; cited in *Prelude med. plants database (2011)*

Teklehaymanot,T. (2009) Ethnobotanical study of knowledge & medicinal plants used by the people in Dek Island in Ethiopia, in *Eth.J. of Ethnopharm. 2009:124, pp.69-78*; cited in *Prelude med. plants database (2011)*

Terrac, M.-L (1947) *Contribution à l'étude des plantes médicinales de Madagascar, de la Réunion et de l'Ile Maurice* (Doct. Thesis, Paris University)

Thaiyah, A.G., P.N.Nyaga, J.M.Maribei, D.Nduati, P.G.Mbuthia & T.A.Ngatia (2010) Experimental *Solanum incanum* poisoning in goats, in *Bull. Animal Health & Prod. in Afr. V.58:1*

Thomas, K.D., A.E.Caxton-Martins, A.A.Elujoba & O.O.Oyelola (1991) Effects of an aqueous extract of cotton seed (Gossypium barbadense L.) on adult male rats, in *PubMed 1991*

Thomas, N.W. (1945) *Field notes - reliquiae, 1690* (Herbarium, RBG Kew: Nigeria Series)

Thompson, J.B.Blake (1931) *Nada* 9: 33

Thring, T.S.A. & F.M.Weitz (2006) Medicinal plant use in the Bredadoro/Elim region of the Southern Overberg in the W.Cape Province of S.Africa, in *J. Ethno. 2006 vol.103 pp.261-75*

Thunberg, C.P. (179?) *Travels in Europe, Africa & Asia made between the years 1770 and 1779* (London: W. Richardson)

Tibiri, A, J.T.Banzouzi, A.Traore, G.O.Nacoulma, I.P.Guissou & B.Mbatchi (2007) Toxicological assessment of Methanolic stem bark and Leaf extracts of *Entada Africana*, in *Int.J. Pharm. 2007 3:5 pp.393-99*

Titanji, V.P.K., D.Zofou & M.N.Ngemenya (2008) The Antimalarial Potential of Medicinal Plants Used for the Treatment of Malaria in Cameroonian Folk Medicine, in *Afr. J. Trad. Compl. Altern. Med. 2008 5:3, pp. 302-321*

Togola, A.D.Diallo, S.Dembele, H.Barsett & B.S.Paulsen (2005) Ethnopharmacological survey of different uses of 7 med. plants from Mali, in the regions Doila, Kolokani & Siby, in *J. of Ethnobiol. & Ethnomed. 2005: 1,7* cited in *Prelude med. plants database (2011)*

Tona, L., N.P.Ngimbi, M.Tsakala, K.Mesia, K.Cimanga, S.Apers, T. De Bruyne, L.Pieters, J. Totte and A.J.Vlietinck (1999) Antimalarial activity of 20 crude extracts from nine African medicinal plants used in Kinshasa, Congo, in *J. Ethnopharm. 68 (1999) pp.193-203*

Tona, L., K.Mesia, N.P.Ngimbi, B.Chrimwami, O.Ahoka, K. Cimange, T.De Bruyne, S.Apers, N. Hermans, J.Totte, L. Pieters and A.J.Vlietinck (2001) *In vivo* antimalarial activity of *Cassia occidentalis, Morinda morindoides* and *Phyllanthus niruri*, in *Annals of Trop. Med. and Parasitology, vol. 95, no 1, 47-57 (2001)*

Tona, L., R.K.Cimanga, K. Mesia, C.T. Musuamba, T.De Bruyne, S.Apers, N.Hernans, S.Van Miert, L.Pieters, J.Totté and A.J.Vlietinck (2004) *In vitro* antiplasmodial activity of extracts and fractions from seven medicinal plants used in the D.R.C., in *Journal of Ethnopharm. 93 (2004) 27-32*

Tra-Bi-Fézan, H. (1979) Utilisations des plantes par l'homme dans les forets classees du Haut-Sassandra et de Scio, en Côte d'Ivoire *(Doc. Thesis, Abidjan Univ.)* cited in *Prelude med. plants database (2011)*

Traore, M. (n.d.) *Le recours à la Pharmacopée traditionnelle Africaine dans le Nouveau Millénaire: Cas des Fammes Herboristes de Bamako* (www.codesria.org/Archives/ga10/Abstracts%20GA%201-5/AIDS_Traore.htm) in "Grey Litterature" ; cited in *Prelude med. plants database (2011)*

Treurnicht,F.T.(1997) *Evaluation of the toxic & potential antiviral effects of some plants used by S.Africans for medical purposes* (Stellenbosch: MSc thesis)

Tshiamuene, N., J.Paulus, M.Kabeya, L.Nlandus & K.Kizika (1995) Plantes médicinales à usage domestique cultivées dans deux quartiers de Kinshasa, in *Rev. Méd. Pharm. Afr. 1995: 9:2, pp. 9-14*

Tuleun, C.D., S.N.Carew & J.A.Patrick (2008) Fruit characteristics & chemical composition of some varieties of velvet beans (*Mucuna spp.*) found in Benue State of Nigeria, in *Livestock Research for rural development, 2008 20:10*

Turner, G.A. (1907) *Transv. Med. J.* 2: 273
                    (1908-9) *Transv. Med. J.* 4: 204

Udoh, P.& A.Kehinde (1999) Studies on antifertility effect of pawpaw seeds (*Carica papaya*) on the gonads of male albino rats. In *Phytotherapy research (1999) 13:3 pp. 226-228*

Ueno,H.M., J.T.Doyama, C.R.Padovani and E.Salata (1996) Efeito de *Momordica charantia* l. em camundongos infectados por *Plasmodium berghei*, in *Revista da Sociedade Brasileira de Medicina Tropical 29 (5) pp. 455-460 (1996)*

Ugwu, F.M. & N.A.Oranye (2006) Effects of some processing methods on the toxic components of African Breadfruit (*Treculia Africana*), in *Af.J. Biotech. 2006 v.5:22 pp.2329-2333*

Van der Hooft, C.S., A.Hoekstra, A.Winter, P.A. de Smet & B.H.Stricker (2005) Thyrotoxicosis following the use of ashwagandha, in *Ned.Tijd. Geneesk. 2005 149:47 pp. 2637-8*

Van der Steur, L. (1994) Plantes medicinales utilisees par les Peurl du Senegal Oriental, in *Rev. Med. Pharm. Afr. 1994 8:2 pp.189-200* cited in *Prelude med. plants database (2011)*

Van der Venter, M., S.Roux, L.C. Bungu, J. Louw, N.R.Crouch, O.M.Grace, V.Maharaj, P. Pillay, P.Sewnarian, N. Bhagwandin and P.Folb (2008) Antidiabetic screening and scoring of 11 plants traditionally used in South Africa, in *J. of Ethnophar. 2008, 119:1, pp. 82, 84*

Vanhee, E. (1996) *Contribution a l'étude de plantes médicinales inhérentes au système de santé traditionelle Bamileke, Cameroun* (Lille Univ. Doctoral thesis)

Van Puyvelde, L., V.Pagezy & A.Kayonga (1975) Plantes médicinale et toxiques au Rwanda I in Afr.Méd., 1975 14 (135) pp. 925-930

Vashishtha, V.M., A.Kumar, T.J.John and N.C.Nayak (2007) *Cassia occidentalis* poisoning as the probable cause of hepatomyo-encephalopathy in children in western Uttar Pradesh, in *Indian J. Med. Res. 2007 125 pp. 756-762*

Vashishtha, V.M., T.J.Turner, A.Kumar (2009) Clinical and pathological features of acute toxicity due to *Cassia occidentalis* in vertebrates, in *Indian J. Med. Res. 2009 130:1, pp.23-30*

Vergiat, A.M. (1970) Plantes magiques et médicinales des Féticheures de l'Oubangui (Région de Bangui), in *J. Agr. Trop. Bot. Appl.* 17

Verma, S.K., G.Dev., A.K.Tyagi, S.Goomber, G.V.Jain (2000) *Argemone mexicana* poisoning : autopsy findings of two cases, in *Forensic Science International 2001 115 :1, pp. 135-41*

Vinayagam,A., N.Senthilkumar and A.Umamaheswari (2008) Larvicidal activity of some medicinal plant extracts against malaria vector *Anopheles steveni* , in *Research Journal Parasitology 3 (2): 50-58*

Visser, L.E. (1975) Plantes médicinales de la Côte d'Ivoire, in *Meded. Landb., Wageningen* 75-115

Volkonsky, M. (1937) In *Archives de l' Institut Pasteur Algérien* 15: 427

Vongo, R. (1999) Traditional medicine on antimalarials (Paper presented at RITAM Conference, Moshi, Tanzania)

Voss, K.A. & L.H.Brennecke (1991) Toxicological & hematological effects of sicklepod (*Cassia obtusifolia*) seeds in Sprague-Dawley rats: a subchronic feeding study, in *Toxicon 1991 29 11 pp 1329-36*

Walker, A.R. (1953) Usages pharmaceutiques des plantes spontanies du Gabon, in *Bull. Inst. Centraf.*, n.s. 6, pp. 275-329.

Walker, A.R., & R. Sillans (1961) *Les Plantes utiles de Gabon* (Paris)

Watt, J. M. & al. (1930) *Bantu Studies* 4: p. 47

Watt, J.M., & M.G. Breyer-Brandwijk (1962) *The Medicinal and Poisonous Plants of South and East Africa* (Edinburgh and London: E. and S. Livingstone)

Webb, L.J. (1948) *Coun. sci. ind. Res. Aust. Bull.*

Wehmer, C. (1929-31) *Die Pflanzenstoffe 2nd ed.* (Jena: Fischer)
          (1935) *Supplement*

Wellens, R.P., ed. E. De Wildeman (1938) *Sur des plantes médicinales ou utiles du Mayumbe, Congo Belge* (Brussels : G.Van Campenhout) cited in *Prelude med. plants database (2011)*

Whistler, W.A. (2000) Tropical Ornamentals, a guide *(Portland: Timber Press)*

WHO (2002) Guidebook to registration of pub. health pesticides, Annex 2 (http://app.nea.gov..sg/cms/htdocs/category_sub.asp?cid=107)

WHO (2003) The Africa Malaria Report 2003 (Geneva: WHO/CDS/ MAL/ 2003.1093)

WHO (2005) World Malaria Report 2005 (Geneva: WHO/HTM/MAL/ 2005.1102)

Wicht, J. (1918) *South African Medical Record*

Wickens, G.E. (1969) *A study of Acacia albida Del.* (London: Kew Bulletin 23)

Willcox, M.L.,B.Graz, J. Falquet, O.Sidibé, M. Forster, D.Diallo, (2007) *Argemone mexicana* decoction for the treatment of uncomplicated falciparum malaria, in *Transactions of the Royal Society of Tropical Medicine and Hygiene (2007) 101, 1190-1198*

Willcox,M.L, G.Burford and G.Bodeker (2004) An overview of Ethnobotanical Studies on plants used for the treatment of malaria, in *Willcox M.L., Bodeker G., Rasoanaivo, P. (Eds.) Traditional Medicinal Plants and Malaria (Boca Raton: CRC Press)*

Williams, L.O. (1965) *Useful Plants of Tropical America* (Chicago: Plantiana)

Williamson, J. (1975) *Useful plants of Malawi, rev.* (U. of Malawi)

Wilson, O., J.L.Godwin, J.Z.Fate & M.A.Madusolumuo (2010) Toxic effects of *Grewia mollis* stem bark in experimental rats, in *J.Am.Sc.2010/ 6:12, pp. 1544-48*

Wome, B., (1985) *Recherches ethnopharmacognostiques utilisée en medicine traditionelle a Kisangani (H.-Zaire)* (Thèse doct., Univ. lib. De Bruxelles) cited in *Prelude med. plants database (2011)*

Wong, W. (1976) Some folk-medicinal plants from Trinidad in *Econ. Bot.* 30: 103-142

Yaro, A.H., B.B.Maiha, M.G.Magaji, A.M.Musa, M.Aliyu & O.S.Ibrahim (2010) Analgesic and anti-inflammatory activities of N-butanol soluble fraction of *Cissus cornifolia* Planch, in *Int.J. of Pure and applied sciences 2010 4:1 pp. 57-63*

Zirihi, G.N., L.Mambu, F.Guédé-Guina et al. (2005) *In vitro* antiplasmodial activity and cytotoxicity of 33 West African plants used for treatment of malaria, in *J. Ethnopharm. 2005 Apr.26; 98(3): 281 - 5*

VERNACULAR NAMES LINKED WITH SPECIES
(See note on spelling & punctuation, p. 2)

abyssinian tea (E) *Catha edulis*
african basil (E) *Ocimum gratissimum*
african boxwood (E) *Treculia africana*
african breadfruit (E) *Treculia africana*
african cabbage (E) *Cleome gynandra*
african cucumber (E) *Momordica balsamina*
african ebony (E) *Diospyros mespiliformis*
african false currant (E) *Allophylus africanus*
african mallow (E) *Hibiscus sabdariffa*
african osage-orange (E) *Maclura africana*
african pear (E) *Momordica balsamina*
african pepper (E) *Xylopia acutiflora*
african redwood (E) *Hagenia abyssinica*
african soapberry (E) *Phytolacca dodecandra*
african spider flower/ wisp (E) *Cleome gynandra*
african teak (E) *Milicia excelsa*
african wattle (E) *Peltophorum africanum*
african wormwood (E) *Artemisia afra*
afrika-valstaaibos (Afr) *Allophylus africanus*
air potato (E) *Dioscorea bulbifera*
akafungu (Lb) *Cassia abbreviata*
akako-bamakanga (B) *Asparagus africanus/ A. falcatus*
akemya-nsyinga (B) *Asparagus plumosus*
alligator weed (E) *Alternanthera sessilis*
amale (B) *Eleusine coracana*
american bushmint (E) *Hyptis spicigera*
american rope *Mikania chenopodifolia*
anaboom (Afr) *Faidherbia albida*
angel's trumpet (E) *Datura metel*
annatto (E) *Bixa orellana*
anthill saffron (E) *Elaeodendron transvaalense*
anyenze (B/ Ku/ Ny) *Allium ascalonicum/ A. cepa*
apple-ring acacia (E) *Faidherbia albida*
arrowleaf morning-glory (E) *Xenostegia tridentata*
arrowleaf sida (E) *Sida rhombifolia*
asiatic pennywort (E) *Centella asiatica*
asthma plant (E) *Euphorbia hirta*
asthma weed (E) *Conyza bonariensis*
ayuka (E) *Manihot esculenta*
babumba (B) *Lannea discolor*
bahama pear (E) *Momordica charantia*
balloon vine (E) *Cardiospermum grandiflorum*
balsam apple (E) *Momordica balsamina*
balsamina (E) *Momordica balsamina*
baluku (Lv) *Eleusine coracana*

baobab (E) *Adansonia digitata*
barbados nut (E) *Jatropha curcas*
barbados pride (E) *Caesalpina pulcherrima*
bastard mustard (E) *Cleome gynandra*
beach morning glory (E) *Ipomoea pes-capra*
beggar sticks (E) *Bidens pilosa*
beggar tick (E) *Bidens pilosa*
beka-veka (Ny) *Anthocleista schweinfurthii*
bell mimosa (E) *Dichrostachys cinerea*
bemba (Ny) *Tamarindus indica*
bengal indigo (E) *Indigofera arrecta*
besemtrosvy (Afr) *Ficus sur*
beukebos (Afr) *Lippia javanica*
bishop's children (E) *Dahlia variabilis*
biswa (B) *Ocimum gratissimum*
bitterappel (Afr) *Momordica charantia / Solanum tettense*
bitter apple (E) *Solanum incanum*
bitterbroom (E) *Scoparia dulcis*
bitter gourd (E) *Momordica charantia*
bitterhout (Afr) *Xysmalobium undulatum*
bitter lettuce (E) *Launaea cornuta*
bitter melon (E) *Momordica charantia*
bitter-tea vernonia (E) *Vernonia amygdalina*
bitter vine (E) *Mikania chenopodifolia*
bitterwortel (Afr) *Xysmalobium undulatum*
bitter yam (E) *Dioscorea bulbifera*
bittetee (Afr) *Vernonia amygdalina*
biyo (Ny) *Steganotaenia araliacea*
black beniseed (E) *Hyptis spicigera*
black fellows (E) *Bidens pilosa*
black jar (E) *Bidens pilosa*
black sesame (E) *Hyptis spicigera*
black vetivergrass (E) *Vetiveria nigritana*
blausuurpruim (Afr) *Ximenia americana*
blinkblaar-wag-'n-bietjie (Afr) *Ziziphus mucronata*
bloodwood (E) *Pterocarpus angolensis*
bloughwarrie (Afr) *Euclea crispa*
blue/ blue-leaved guarri (E) *Euclea crispa*
blue cat's whiskers (E) *Rotheca myricoides*
blue glory-bower (E) *Rotheca myricoides*
blue sourplum (E) *Ximenia americana*
blue wings (E) *Rotheca myricoides*
blue wiss (E) *Teramnus labialis*
bobbin weed (E) *Leucas martinicensis*
bombwe (B) *Pistia stratiotes*
bosbeukenhout (Afr) *Pittosporum viridiflorum*
boskoorsbessie (Afr) *Croton sylvaticus*
bosluisboom (Afr) *Sterculia africana*

bosveldsaffraan (Afr) *Elaeodendron transvaalense*
bosvlier (Afr) *Nuxia floribunda*
botterblom (Afr) *Ranunculus multifidus*
brick-red ochna (E) *Ochna schweinfurthiana*
brimstone tree (E) *Morinda lucida*
bristle-grass (E) *Setaria longiseta*
bristly star bur (E) *Acanthospermum hispidum*
broad-leaved bristle-grass (E) *Setaria megaphylla*
broad-leaved croton (E) *Croton macrostachys*
broom-cluster fig (A) *Ficus sur*
broom creepeer (E) *Cocculus hirsutus*
broom weed (E) *Scoparia dulcis*
buffalo bean (E) *Mucuna pruriens*
buffalo-grass (E) *Setaria megaphylla/ Stenotaphrum dimidiatum*
buffalo thorn (E) *Ziziphus mucronata*
buffelgras (Afr) *Setaria megaphylla*
bukasi (B/ Ku) *Mucuna pruriens*
bule (B) *Eleusine coracana*
bulolo (B) *Eleusine indica*
bulrush (E) *Schoenoplectus senegalensis*
bulugamu (Ny) *Eucalyptus grandis*
bululu (Ln) *Tecomaria capensis*
burgandy (E) *Leea guineensis*
bur mallow (E) *Urena lobata*
bur marigold (E) *Bidens pilosa*
burning-bush combretum (E) *Combretum microphyllum*
bush fig (E) *Ficus sur*
bush guarri (E) *Euclea racemosa*
bush mallow (E) *Abutilon mauritianum*
bushman's tea (E) *Catha edulis*
bush ochra (E) *Corchorus olitorius*
bushveld saffron (E) *Elaeodendron transvaalense*
busosyi (B) *Sesamum indicum & sp.*
busufu (B) *Ceiba pentandra*
busuku (I) *Uapaca kirkiana*
buttercup (E) *Ranunculus multifidus*
butterfly bush (E) *Rotheca myricoides*
buungasiya (I) *Acacia erioloba/ A. tortilis/ Faidherbia albida*
buwa (Ny) *Albizia antunesiana/ A. versicolor/ Tephrosia vogelii*
buyu (N/ Tu) *Adansonia digitata*
bwazi (Ku/ To) *Securidaca longipedunculata*
bwembe (B/ Ku) *Tamarindus indica*
bwengo (K/ Lb/ Le/ To/ Tu) *Sesamum indicum/ S. orientale & spp.*
byelu-byamunyinga (Lv) *Arachis hypogaea*
byelu-byansoke (Lv) *Arachis hypogaea*
caangalume (Tu) *Zanha africana*
cabana (Tk) *Evolvulus alsinoides*
cabbage fern (E) *Platycerium elephantotis*

cabbage tree (E) *Cussonia spicata*
cabweyo (Tk) *Imperata cylindrica*
caesar's weed (E) *Urena lobata*
cafica (Ny) *Ozoroa reticulata/ Kigelia africana*
cakate (Ny) *Parinari curatellifolia*
cala-cankhwale (Ny) *Euphorbia hirta*
calumba (E) *Jateorrhiza palmate*
camakupa (K) *Ozoroa reticulata*
camasala (Ny) *Trichodesma zeylanicum/ Aspilia kotschyi/ Vernonia glabra*
camba (B/ Ny) *Cannabis sativa*
camba-lupili (B) *Myrothamnus flabellifolius*
cambela (Ku/ Tu) *Aerva lanata*
cambulu-luzi (Ku/ Ny/ Tu) *Alternanthera nodiflora*
cambunga (Ny) *Oryza sativa*
cambwe (B/ K) *Acacia polyacantha*
camel bush (E) *Trichodesma zeylanicum*
camel's foot (E) *Bauhinia reticulata/ Piliostigma thonningii*
camsala (Ny) *Tridax procumbens*
cana-lume (B) *Zanha africana*
canary pea (E) *Eriosema psoraleoides*
canasa (Ny) *Myrothamnus flabellifolius*
candelabra flower (E) *Boophane disticha*
candima (Ny) *Elephantorrhiza goetzei*
candimbo (Ny) *Cussonia arborea*
canga (M) *Flacourtia indica*
canga-luce (Ny) *Zanha africana*
canga-lume (Ny) *Zanha africana*
cangongo (Ny) *Combretum molle*
cangu (Ln/ Lo) *Senna obtusifolia*
canje (Ny) *Helichrysum kirkii/ H. panduratum*
canzi (Ny) *Becium obovatum/ Hoslundia opposita/ Leucas martinicensis/ Ocimum americanum*
canzi-campongo (Ny) *Helichrysum panduratum*
cape fig (E) *Ficus sur*
cape gooseberry(E) *Physalis angulata*
cape honeysuckle (E) *Tecomaria capensis*
cape mahogany (E) *Trichilia emetica*
cape poison bulb (E) *Boophane disticha*
carilla plant (E) *Momordica charantia*
caronde (N) *Swarzia madagascarensis*
carrot tree (E) *Steganotaenia araliaceae*
cashew nut (E) *Anacardium occidentale*
cassava (E) *Manihot esculenta*
castor bean (E) *Ricinus communis*
castor-oil plant/ tree (E) *Ricinus communis*
cat's hair (E) *Euphorbia hirta*
cat's whiskers (E) *Cleome gynandra*
cattle bush (E) *Trichodesma zeylanicum*

cawaya (Tu) *Trichodesma zeylanicum*
cengeno (To) *Salvadora persica*
cenja-kulo (Ny) *Bersama abyssinica*
century plant (E) *Boophane disticha*
ceylon leadwort (E) *Plumbago zeylanica*
ch... (see c....)
cheesewood (E) *Pittosporum viridiflorum*
cherry pie (E) *Lantana camara*
chicken weed (E) *Euphorbia thymifolia*
chickweed (E) *Ageratum conyzoides*
china rose (E) *Hibiscus rosa-sinensis*
chinese bur (E) *Urena lobata*
chinese cabbage (E) *Ipomoea aquatica*
chinese creeper (E) *Mikania chenopodifolia*
chinese hibiscus (E) *Hibiscus rosa-sinensis*
chinese lantern (E) *Dichrostachys cinerea*
chinese senna (E) *Senna obtusifolia*
chinese spinach (E) *Ipomoea aquatica*
chirinda redwood (E) *Catha edulis*
christmas bells (E) *Trichilia emetica*
christmas berry (E) *Psorospermum febrifugum*
cia... (see ciya...)
cibanga (B) *Zanha africana*
cibanga-lume (B/ Lb/ Ny) *Zanha africana*
cibanga-lumya (M) *Zanha africana*
cibenda (B) *Antidesma venosum*
cibombo (B) *Acacia polyacantha/ A. hockii*
cibombwe-sala (B) *Albizia amara/ A. anthelmintica/ Entada abyssinica*
cibuku-zela (Ny) *Zanha africana*
cibumbu (I/ To) *Lannea edulis and spp.*
cibumbu (To) *Lannea discolor*
cibumbya (To) *Salix mucronata subsp.mucronata*
cifatukila (Tu) *Cardiospermum helicababum*
cifiko (Ku/ Ny) *Ozoroa reticulata/ Premna senensis*
cifitye (Ny) *Ozoroa reticulata*
cifui (Ln) *Paullinia pinnata*
cifumbe (B/ M) *Piliostigma thonningii*
cifyopola (Ny) *Steganotaenia araliacea*
cigaga (Ny) *Acalypha chirindica/ A. villicaulis/ Maprounea africana/ Ochna atropurpurea/ O. pulchra*
cigolo-golo (Ny) *Albizia anthelmintica*
cigoma (Ny) *Sesbania sesban*
cigombe (Ny) *Eleusine indica*
ciguluka (Ny) *Securidaca longipedunculata*
cigwenembe (Ny) *Albizia antunesiana/ Catunaregam obovata*
cihoyoka (N) *Cyperus articulatus*
cihuma (Ln) *Combretum molle*
cijonga (Lo) *Acacia polyacantha*

cika (Ny) *Paullina pinnata*
cikala-malanga (I) *Gardenia ternifolia/ G. volkensii*
cikala-ngwe (K) *Capparis tomentosa*
cikalolo (Ny) *Albizia anthelmintica*
cikane (Ny) *Cussonea arborea*
cikanga (Ny) *Dalbergia nitidula/*
cikasa (Ln/ Lo) *Cussonia arborea/ C. kirki*
cikasu (Ny) *Syphonochilus kirkii*
cikhaka (Ny) *Momordica foetida*
cikokoti (Tu) *Hypoestes verticillaris/ Isoglossa eylesii*
cikoli (Ln) *Strychnos spinosa*
cikolo-kozya (Tk) *Triumfetta welwitschii/ Tylophoria sylvatica*
cikolola (Ny) *Commiphora mollis/ Heteromorpha arborescens/ H. trifoliata*
cikololo (B) *Gardenia spatulifolia/ G. ternifolia & spp.*
cikololo (Ny) *Albizia anthelmintica*
cikolo-wanga (Ny) *Hymenocardia acida/ Pavetta schumanniana*
cikondi (Ku) *Clematis brachiata*
cikope (Ny) *Holarrhena pubescens/ Macaranga capensis/ Tabernaemontana elegans*
cikowa-nkanga (B) *Smilax anceps*
cikuku (Lv/ Ln/ Lv) *Phyllocosmus lemaireanus*
cikuli (Tu) *Paralepistemon shirensis*
cikulu-mawele (Ny) *Ficus sur*
cikumba (Ny) *Cardiospermum halicacabum/ Cayratia gracilis*
cikumba-mbowo (Ku/ Tu) *Entada abyssinica*
cikumba-mbu (Ny) *Entada abyssinica*
cikumbelo (Ny) *Gymnosporia senegalensis*
cikundu (Ny) *Maesa lanceolata*
cikundu-lima (Ny) *Elephantorrhiza goetzei*
cikwani (Ny) *Albizia adianthifolia/ A. glaberrima*
cikwata (Ln) *Ziziphus mucronata*
cilala-nkanga (Tu) *Tephrosia purpurea/ Indigofera atriceps*
cilama (Lv) *Croton megalobotrys*
cilambwe (Ln) *Aloe spp.*
cilembusya (Ny) *Lannea edulis*
cilengwa (Ln) *Tetracera alnifolia*
cilenje (I/ Tk/ To) *Imperata cylindrica*
cilima (Ny) *Acacia amythethophylla/ Diospyros senensis*
cilime (Ny) *Flueggea virosa*
cilonde (B/ M) *Swarzia madagascariensis*
cilumbwe-lumbwe (B) *Phyllocosmus lemaireanus*
cilunda (To) *Uapaca nitida*
cilundine (I) *Holarrhena pubescens*
cilundu (I/ Lo) *Uapaca kirkiana*
cilundu (Lo/ To) *Uapaca nitida*
cilungu-nduwe (B/ Ku) *Trichodesma zeylanicum*
cilungu-viva (Ln/ Lv) *Combretum collinum*
cilya-ndembo (Ny) *Vangueria infausta*
cilya-tsongwe (Ny) *Leonotis nepetifolia*

cimamba (B) *Sphenostylis marginata*

cimandwe (To) *Diospyros senensis*

cimbowe - (see cimbobwe)

cimboyi (B/ Ln) *Adenia gummifera*

cimono-mono (B/ Ny) *Croton megalobotryx*

cimpempe (To) *Ozoroa reticulata*

cimpulukwa (Lv) *Jatropha curcas*

cimwa-manzi (Ny) *Rhus longipes/ R. anchietae/ Rhus spp.*

cimwande (Ny) *Diospyros senensis*

cimwa-nyanza (Ny) *Citrullus lanatus*

cimwembe-waukulu (Ny) *Erythrocephalum zambezianum*

cimwe-mwe (Ny) *Dicoma sessiliflora* sbsp. *sessiliflora/ Diplolophium zambesianum/ Lepidagathis anderson*

cimyangu (Ny/ Tu) *Ehretia obtusifolia*

cinama (Ny) *Combretum fragrans/ C. molle/ Philenoptera violacea/*

cinangwe - (see cinangwa)

cindele (N) *Sphenostylis marginata sbsp. erecta*

cinga (B) *Azanza garckeana/ Dombeya rotundifolia*

cingambilo (Ny) *Tetradenia riparia*

cingungu (Ny) *Maesa lanceollolata*

cinkala-matongwe (Ny) *Vernonia spp.*

cinkambira (Ny) *Tetradenia riparia*

cinkombwa (Ny) *Croton megalobotrys*

cinkundu (Ny) *Maesa lanceolata*

cinomba (Ny) *Bidens pilosa*

cinsangwi (Ny) *Eleusine indica*

cintebwe (Ny) *Aloe spp.*

cintembwe (Ny) *Aloe spp.*

cinto-mbolozi (Ny) *Cissus cornifolia*

cintukutu (Tu) *Piliostigma thonningii*

cinunda (I) *Uapaca nitida*

cinungu (B) *Commiphora madagascariensis*

cinyaku (Ny) *Argemone mexicana*

cinyala (Lv) *Senna occidentalis*

cinyalala (Lv) *Senna occidentalis*

cinyambata (Ny) *Desmodium velutinum/ Rhynchosia sublobata/ Triumfetta rhomboidea*

cinyanya (Ny) *Acalypha villicaulis/ Tragia benthamii/ T. brevipes/ T. kirkiana*

cinyenye (Ny) *Swarzia madagascariensis*

cinyeu (Ny) *Hoslundia opposita*

cinyewe (Ny) *Dalbergia boehmii*

cipanda-mwavi (Ny) *Vernonia spp.*

cipangala (Ny) *Dichrostachys cinerea*

cipapata (Ny) *Cocculus hirsutus*

cipapaya (B/ Ku/ Ny) *Carica papaya*

cipatikila (Ny) *Cardiospermum halicacabum*

cipaupa (Ny) *Rotheca myricoides*

cipembele (Ku/ Ny) *Catunaregam obovata*

cipembele (Ny) *Canthium glaucum/ Catunaregam obovata*

cipeta (Ny) *Acacia amythethophylla/ Holarrhena pubescens/ Psychotria zombamontana*

cipika-mongu (Ny) *Eleusine indica*

cipisya-wago (Ny) *Catunaregam obovata/ Dalbergia nitidula/ Dichrostachys cinerea*

cipolo (B) *Zanthoxylum chalybeum*

cipolo-polo (I) *Steganotaenia araliacea*

cipombo (Ny) *Cussonia arborea/ Steganotaenia araliacea*

cipopo-lambwe (Ku) *Laportea aestuans*

cipundu (Ny) *Hymenocardia acida*

cipunga-ŋombe (K) *Entada abyssinica*

cipupa (Ln) *Zanthoxylum chalybeum*

ciputa-mazala (Ln) *Erythrina abyssinica*

cipuwe (Ku/ Ny) *Manilkara mochisia*

cipuzi (Ny) *Lagenaria sphaerica*

cipyelo (Ny) *Flueggea virosa*

cisangwa (B/ Lb) *Erythrina abyssinica*

cisanje (Ny) *Securidaca longepedunculata*

cisapira (Ny) *Triumfetta rhomboidea*

cisa-scamavu (Ny) *Clematis brachiata*

ciseke (Lv) *Sesbania sesban*

cisekele (Ny) *Entada abyssinica*

ciseyo (Ny) *Acacia nilotica/ Dichrostachys cinerea*

cisi (Ny) *Acacia sieberiana*

cisilele (Ny) *Momordica foetida*

cisimbiti (Ny) *Combretum molle/ C. psidioides*

cisisi (Ny) *Acacia sieberana/ Imperata cylindrica*

cisitu (Ku/ Ny) *Allophylus africanus*

cisitu (Ku) *Amaranthus spp./ Rhus longipes*

cisiyo (Ny) *Acacia nilotica*

cisobo (B) *Vernonia amygdalina/ V. colorata*

cisole (Lo) *Dichrostachys cinerea*

cisomanga (Ny) *Dicoma sessiliflora ssp sessiliflora*

cisoni (Ny) *Myrothamnus flabellifolius*

cisosoci (Ny) *Bidens pilosa*

cisunga (Ln) *Erythrina abyssinica*

cisungu-mwamba (Ny) *Trichodesma zeylanicum*

cisungwa (B/ Ny) *Erythrina abyssinica & spp.*

cisyalere (Ny) *Lagenaria sphaerica/ Momordica foetida*

citakaci (Ln) *Margaritaria discoidea (?) Phyllanthus spp.*

citale (Ny) *Albizia anthelmintica*

citambala (Ny) *Gloriosa superba*

citanga-menzi (B) *Citrullus lanatus*

citapa-tapa (B) *Pavetta schumanniana/ Psychotria buzica/ P. kirkii/ P. succulenta*

citatakunda (B) *Citrullus lanatus*

citebi-tebi (B) *Cussonia arborea/ C. corbisieri & spp./ Steganotaenia araliacea*

citedze (Ny) *Mucuna aterrima/ M. coriacea/ M. poggei/ M. pruriens/ M. stans/ M. spp./ Rhynchosia luteola*

citelela (B) *Urena lobata*

citemala (Ku) *Entada abyssinica*

citeta (B/ Ku/ Ny) *Elepantorrhiza goetzei*
citeta (B/ Ku) *Albizia adianthifolia/ Peltophorum africanum*
citeta (Ny) *Elephantorrhiza goetzii*
citeya (Ny) *Mucuna poggei*
citimbe (B/ Ku/ Ny) *Piliostigma thonningii*
citondo-mono (B) *Jatropha curcas*
citondo-tondo (Ku/ Ny/ Tu) *Xeroderris stuhlmann*
citongo-lolo (Ny) *Acacia goetzei* ssp *microphylla/ A. macrothyrsa/ A. polyacantha/*
    Albizia harveyii/ Dichrostachys cinerea
citonje (Ny) *Calotropis procera*
citronella grass (E) *Cimbopogon nardus*
citumbula-mala (Ku) *Peltophorum africanum*
citundwe (I) *Holarrhena pubescens*
citungulu (B/ Lb) *Aframomum alboviolaceum & spp.*
cituta (B) *Bridelia duvigneaudi*
cituzi (Ku) *Cissus cornifolia*
civavani (To) *Paralepistemon shirensis*
civumbe (Ku) *Clerodendrum ternatum*
civumulo (Ny) *Mikania chenopodifolia/ Turraea nilotica*
civumulu (Ny) *Jatropha curcas*
civungula (B/ Ku) *Kigelia africana*
civwembe (Ny) *Citrullus lanatus*
ciwalika (Ny) *Croton macrostachys*
ciwanda (Ny) *Ageratum conyzoides/ Blumea spp.*
ciwere (Ny) *Crotalaria recta*
ciwezeze (I/ Lo/ Tk) *Capparis tomentosa*
ciweze-weze (I/ Lo) *Capparis tomentosa*
ciwezyezyi (To) *Capparis tomentosa*
ciwimbi (Ny) *Holarrhena pubescens/ Rauvolfia birrea/ R. caffra/ Tabernaemontana elegans*
ciwindu (Ny) *Bersama abyssinica*
ciwombela (Ku/ Ny) *Aerva lanata*
ciwombola (Ny) *Ochna schweinfurthiana*
ciwombosi (Ny) *Maesa lanceololata*
ciwose (Tu) *Sorghum sp.*
ciwungu (Ny) *Psorospermum febrifugum*
ciyele (Ny/ To) *Faurea delevoi/ F. speciosa & spp./ Protea spp./ Syzygium cordatum*
ciyele (Tu) *Protea angolensis/ P. madiensis*
ciyombo (Ku/ Ny/ Tu) *Brachystegia boehmii*
ciyumbu (Ny) *Dalbergiella nyassae/ Lannea discolor/ L. schimperi/ Xeroderris stuhlmannii*
ciyungu (B) *Gymnosporia senegalensis*
ciyuta (Ny) *Annona senegalensis/ Clutia abyssinica*
cizama-moto (Ny) *Tetradenia riparia*
cizaya (Ny) *Strychnos pungens*
cizeze (Lo) *Phyllanthus reticulatus*
cizutu (Ny) *Kigelia africana*
climbing hempweed (E) *Mikania chenopodifolia*
climbing lily (E) *Gloriosa superba*
coat-buttons (E) *Tridax procumbens*

cobbler's pegs (E) *Bidens pilosa*
coco grass (E) *Cyperus rotundus*
coffee (E) *Coffea arabica*
coffee-bean strychnos (E) *Strychnos henningsii*
coffee weed (E) *Senna occidentalis*
cokolola (Ny) *Commiphora africana*
columboa (E) *Jateorrhiza palmata*
comba-muluzi (Ny) *Zanha golungensis*
comb bushmint (E) *Hyptis pectinata*
combela (Ku/ Ny/ Tu) *Aerva lanata*
comb hyptis (E) *Hyptis pectinata*
combwe (B/ I/ Ku/ Lb/ Le/ Ln/ Ny) *Acacia polyacantha*
combwe (I/ Le) *Acacia sieberiana*
common joyweed (E) *Alternanthera nodiflora*
common morning-glory (E) *Ipomoea purpurea*
common purslane (E) *Portulaca oleracea*
common vervain (E) *Verbena officinalis*
common wire-weed (E) *Sida acuta*
condi (Ny) *Pseudolachnostylis maprouneifolia*
condwe (To) *Albizia anthelmintica*
confetti spike-thorn (E) *Gymnosporia senegalensis*
congo jute (E) *Urena lobata*
congololo (Ny) *Entada abyssinica*
coni (B) *Ochna leptoclada/ O. schweinfurthiana & spp.*
conswe (To) *Capparis tomentosa*
copper-tips (E) *Crocosmia aurea*
coral bush/ plant (E) *Jatropha multifida*
cosi (Ny) *Securidaca longipedunculata*
cosombo (Lo) *Gardenia ternifolia*
cotsa-tsoka (Ny) *Swarzia madagascariensis*
country mallow (E) *Abutilon mauritianum*
cream-of-tartar tree (E) *Adansonia digitata*
creeping lily (E) *Gloriosa superba*
creeping-tick trefoil (E) *Desmodium triflorum*
crowfoot grass (E) *Eleusine indica*
cruel plant (E) *Euphorbia heterophylla*
crystal-bark (E) *Crossopteryx febrifuga*
cuban jute (E) *Sida rhombifolia*
cumba (Tu) *Brachystegia hockii/ B. spiciformis*
cumbe (Ny) *Brachystegia spiciformis*
cunga-linde (M) *Gymnosporia senegalensis*
cunga-nkobwe (B/ Tu) *Capparis erythrocarpos/ C. tomentosa*
cunga-nunsi (B) *Acacia polyacantha*
cungu (B) *Gymnosporia senegalensis*
cutleaf ground-cherry (E) *Physalis angulata*
dangwe (Ny) *Crossopterix febrifuga*
dawi (M) *Flacourtia indica*
dawidjies (Afr) *Cissampelos pareira*

daza (Ny) *Oncoba spinosa*
delele (I/ B/ Ku/ Ny) *Corchorus olitorius*
denganya (Ny) *Acacia nilotica*
desert poinsettia (E) *Euphorbia heterophylla*
devil's gut (E) *Cassytha filiformis*
devil's trumpet (E) *Datura metel*
dikbas (Afr) *Lannea discolor*
dikelenge (Lv) *Hibiscus sabdariffa*
dixie silverback fern (E) *Pityrogramma calomelanos*
dodder (E) *Cassytha filiformis*
doringklapper (Afr) *Strychnos innocua*
dosa (Ny) *Argemone mexicana/ Datura stramonium*
driehoektolletjies (Afr) *Blighia unijugata*
duikerberry (E) *Peudolachnostylis maprouneifolia*
duineghwarrie (Afr) *Euclea racemosa*
dulu (Ku) *Eleusine indica*
dulwelskerwel (Afr) *Bidens pilosa*
dune guarri (E) *Euclea racemosa*
dutchman's pipe (E) *Aristolochia albida*
dwarf copper leaf (E) *Alternanthera sessilis*
dwarf morning-glory (E) *Evolvulus alsinoides*
dwarf ochna (E) *Ochna macrocalyx*
dwarf tibouchina (E) *Dissotis rotundifolia*
dzayi (Ny) *Strychnos spinosa*
dzorba (Ny) *Oncoba spinosa*
dzungu (Ny) *Swarzia madagascariensis*
east african camphor (E) *Ocotea usambarensis*
edlebur (E) *Amaranthus spinosus*
elephant ear (E) *Platycerium elephantotis*
emin's strophanthus (E) *Strophanthus eminii*
english vetiver (E) *Vetiveria nigritana*
enkelgroendoring (Afr) *Balanites aegyptica*
falling stars (E) *Crocosmia aurea*
false assegai (E) *Maesa lanceolata*
false pareira brava (E) *Cissampelos pareira*
false pareira root (E) *Cissampelos pareira*
fever berry (E) *Croton megalobotrys*
fever bush (E) *Dicoma anomala*
fever pod (E) *Holarrhena pubescens*
fever tea (E) *Acacia xanthphloea/ Lippia javanica*
fifi (B// Ku/ Ny/ Tu) *Acacia tortilis*
fifi (Nm) *Artemisia afra*
finger millet *Eleusine coracana*
fire lily (E) *Gloriosa superba*
fire-plant (E) *Euphorbia heterophylla*
fire-sticks (E) *Ficus sur*
fisaka (B) *Salix subserrata*
fish-poison (E) *Tephrosia purpurea*

fiti (Ny) *Combretum adenogonium*
flamboyant (E) *Delonix regia*
flame climbing bush-willow (E) *Combretum microphyllum*
flame-lily (E) *Gloriosa superba*
flame tree (E) *Delonix regia*
flax-leaved fleabane (E) *Conyza bonariensis*
fluweelbosswilg (Afr) *Combretum molle*
fodya (Ny) *Nicotania tabacum*
folo (Ny) *Nicotiana tabacum*
forest croton (E) *Croton sylvaticus*
forest fever-tree (E) *Croton macrostachys*
forest gardenia (E) *Gardenia ternifolia*
forest nuxia (E) *Nuxia floribunda*
forest sword-leaf (E) *Cassiaria battiscombei*
forget-me-not tree (E) *Duranta erecta*
four-leaved bushwillow (E) *Combretum adenogonium*
fried-egg flower (E) *Oncoba spinosa*
fumbwa-musowa (Lo) *Entada abyssinica*
futi (Tu) *Brachystegia spiciformis*
futsa (Ny) *Vernonia spp.*
fuzzy rattlebox/ rattlepod (E) *Crotalaria incana*
fwaka (B/ K/ Lb/ Le/ Ny) *Nicotiana rustica*
fwaka (Ny) *Nicotiana tabacum*
gaanumukela (Lo) *Syzygium cordatum*
gagani (Nm) *Cleome gynandra*
galahungu (Ny) *Kigelia africana*
galanga (Tu) *Capparis tomentosa*
garden purslane (E) *Portulaca oleracea*
garlic (E) *Allium sativum*
geeleendagsblom (Afr) *Commelina africana*
geewortelboom (E) *Steganotaenia araliaceae*
giant cherry-orange (E) *Teclea nobilis*
giant milkweed (E) *Calotropis procera*
gifbol (Afr) *Boophane disticha*
gifkaneedood (Afr) *Commiphora africana*
gingerbush (E)*Tetradenia riparia*
gladdetontelhout (Afr) *Clerodendrum glabrum*
glorybower (E) *Clerodendrum ternatum*
glory lily (E) *Gloriosa superba*
goat's foot (E) *Ipomoea pes-capra*
goat's head (E) *Acanthospermum hispidum*
godogo (Nm) *Entada gigas*
goewerneurspruim (Afr) *Flacourtia indica*
golden ballerina (E) *Crocosmia aurea*
golden chamomile (E) *Chrysanthellum indicum*
gold fern (E) *Pityrogramma calomelanos*
goloka (Ny) *Crossopterix febrifuga*
goveror's plum (E) *Flacourtia indica*

gowo-gowo (Ny) *Holarrhena pubescens*
grey-bark cordia (E) *Cordia goetzei*
grey-leaved saucer- berry (E) *Cordia sinensis*
grootblaarsterkastaling (Afr) *Sterculia quinqueloba*
grootkorsbessie (Afr) *Croton megalobotrys*
grootrooihartboom (Afr) *Hymenocardia acida*
ground nut (E) *Arachis hypogaea*
grysappel (Afr) *Parinari curatellifolia*
grysblaarpieringbessie (Afr) *Cordia sinensis*
gryshout (Afr) *Dicoma anomala*
guaba (Ku/ Ny) *Psidium guajava*
guatemala rhubarb(E) *Jatropha multifida*
guava (E) *Psidium guajava*
gulf leaf-flower(E) *Phyllanthus fraternus*
gulf sandmat (E) *Euphorbia thymifolia*
gwaba (Ku/ Ny) *Psidium guajava*
gwangaka (Nm) *Brachystegia spiciformis*
haakdoring (Afr) *Acacia tortilis*
haak-en-steek (Afr) *Acacia tortilis*
hairy basil (E) *Ocimum americanum*
hairy beggar-weed (E) *Desmodium barbatum*
hairy-caterpillar pod (E) Ormocarpum trichcarpum
hairy guarri (E) *Euclea natalensis*
hairy horseweed (E) *Conyza bonariensis*
hairy spurge (E) *Euphorbia hirta*
hairy tinderwood (E) *Clerodendrum eriophyllum*
hanyansi (Le) *Allium cepa*
hanyisi (To) *Allium cepa*
harige ghwarrie (Afr) *Euclea natalensis*
hashish/hemp (E) *Cannabis sativa*
heart seed (E) *Cardiospermum grandiflorum*
hedge caper bush (E) *Capparis sepiaria*
heiningkapperbos (Afr) *Capparis sepiaria*
hellfire bean (E) *Mucuna pruriens*
hibiscus bur (E) *Urena lobata*
hindwa (Tu) *Nicotiana sp.*
hoko (Nm) *Phytolacca dodecandra*
honyanisi (To) *Allium cepa*
hopani (Lo) *Aloe spp.*
hophout (Afr) *Trema orientalis*
horingpeultjieboom (Afr) *Diplorrhynchus condylocarpon*
horn-pod tree (E) *Diplorrhynchus condylocarpon*
horseradish tree (E) *Moringa oleifera*
horsevine (E) *Teramnus labialis*
horsewood (E) *Clausena anisata*
hound's tongue (E) *Cynoglossum lanceolatum*
huilboom (Afr) *Peltophorum africanum*
humbangayi (Ny) *Heliotropium indicum*

ibamba (B) *Imperata cylindrica*
ibanga (I) *Pterocarpus angolensis*
ibange (B) *Cannabis sativa*
ibbuyu (To) *Adansonia digitata*
ibono-ntelemba (I) *Ricinus communis*
ibu (I) *Phragmites australis*
ibula (I) *Parinari curatellifolia*
ibuzu (I) *Adansonia digitata*
icenje (I) *Diospyros mespiliformis*
ifufuma (Lb) *Strychnos pungens*
ifungu-fungu (Ln) *Kigelia africana*
igunguru (Nm) *Trichodesma zeylanicum*
ikalangwe (I) *Capparis elaeagnoides/ C. tomentosa*
ikombwe (Lb) *Hibiscus sabdariffa*
ikuakua (Ln) *Entada gigas*
ilenda-lyahasi (Nm) *Triumfetta rhomboidea*
ilogero (Nm) *Ziziphus mucronata*
ilumba-lumba (I) *Phyllocosmus lemaireanus*
ilutwa (Lo) *Asparagus sp.*
imono (Ln/ Lv) *Ricinus communis*
imono-mono (U) *Croton megalobotrys*
impili-pili (I) *Capsicum annuum var. frutescens*
impulya (M) *Ricinus communis*
impunga (I) *Sorghum sp.*
imunge (I) *Nicotiana sp.*
incenje (I) *Diospyros mespiliformis*
incincayo (To) *Artabotrys brachypetalus/ A. monteiroae*
indian ginseng (E) *Withania somnifera*
indian goosegrass (E) *Eleusine indica*
indian heliotrope (E) *Heliotropum indicum*
indian pennywort (E) *Centella asiatica*
indian turnsole (E) *Heliotropum indicum*
indongo (I/ Lo/ To) *Arachis hypogoea*
indongwe (To) *Arachis hypogaea*
indululu (I) *Agaricus spp./ Sphenostylis marginata*
ingyansa (B) *Brachystegia boehmii*
ink berry (E) *Cocculus hirsutus*
inkonkoni (To) *Cassia abbreviata*
inkonze (I) *Arachis hypogaea*
inkuzu (I) *Ficus spp./ F. sur*
insanki (Lb) *Pentanisia prunelloides*
intelemba (I) *Ricinus communis*
intuntu (Tu) *Smilax anceps*
inyele (Lo) *Peltophorum africanum*
inyemu (I/ K/ Le/ Ln/ Lo) *Arachis hypogaea.*
inyemu-mafuta (I/ Le) *Arachis hypogaea*
iŋomba (I) *Capsicum annuum var. frutescens*
ipapawo (B) *Carica papaya*

ipogoro (Tu) *Faidherbia albida*
ipoko (B) *Phytolacca dodecandra*
ipopo (I/ Le/ To) *Carica papaya*
ipuŋaŋombe (Lb) *Peltophorum africanum*
ipupu (B) *Euphorbia tirucalli*
iputila-mpanda (Tu) *Pseudolachnostylis maprouneifolia*
isanza (M) *Strychnos spinosa*
isoda (Ln) *Vernonia spp.*
isumbu (B) *Sorghum sp.*
isyungwa (To) *Cleome gynandra*
itawa-tawa (B) *Opilia amentalea*
itchy bean (E) *Mucuna pruriens*
iteka (Tu) *Oldenlandia capensis*
itelembe (Tu) *Diplorrhinchus condylocarpon*
itembusya (B) *Aloe spp.*
itende (To) *Cocculus hirsutus/ Hippocratea indica*
itubetube (I) *Acacia sieberiana*
ituntulwa (I) *Solanum tomentosum*
ivunga (I) *Kigelia africana*
ivuyu (I/ Lo/ To) *Adansonia digitata*
ivy grape (E) *Cissus cornifolia*
iwonge-wonge (Nm) *Datura metel*
jackal berry (E) *Diospyros mespiliformis*
jakkalsbessie (Afr) *Diospyros mespiliformis*
jamaica sorrel (E) *Hibiscus sabdariffa*
jasmine gardenia (E) *Heinsia crinita*
jasmine tree (E) *Holarrhena pubescens*
jasmynkatjiepiering (Afr) *Heinsia crinita*
jeli-jeli (Ku/ Ny) *Hyptis spicigera/ Kalanchoe lanceolata*
jeli-jeli (Ku) *Leonotis nepetifolia*
jelly-leaf (E) *Sida rhombifolia*
jengaluwo-ngako (Nm) *Gloriosa superba*
jerusalem tea (E) *Chenopodium ambrosoides*
jesuit's tea (E) *Chenopodium ambrosoides*
jews' mallow (E) *Corchorus olitorius*
jimpson weed (E) *Datura metel*
jointed flat-sedge (E) *Cyperus articulatus*
jumping-seed tree (E) *Sapium ellipticum*
kaapsekanferfoelie (Afr) *Tecomaria capensis*
kaasur (Afr) *Pittosporum viridiflorum*
kabala-bala (K/ Ln) *Philenoptera bussei/ Phyllanthus reticulatus/ Pseudolachnostylis maprounifolia*
kabala-bala-musalya (Ln) *Pseudolachnostylis maprounifolia*
kabale (Ny) *Hymenocardia acida*
kabalisi (Ny) *Maesa lanceolata*
kabanga-luulu (I) *Cordia goetzei*
kabanti (Ny) *Waltheria americana*
kabaŋa-kacina (Ln) *Gymnosporia senegalensis*

kabeti (Tk) *Eriosema psoraleoides*

kabici (Ny) *Nicotiana spp.*

kabidzu (Tu) Burkea africana

kabofa (Ln) Uapaca kirkiana

kabonya-camba (To) *Evolvulus alsinoides*

kabu (Ny) *Pistia stratiotes*

kabula-sese (B/ Ny) *Dalbergia nitidula*

kabulu-kulu (Ny) *Strychnos innocua*

kabulwe-bulwe (K) *Burkea africana*

kabumbu (B/ Lb/ Lo/ Ny) *Lannea discolor/ L. schweinfurthii*

kabumbu (Lb/ Lo/ Tk) *Lannea edulis* & *spp.*

kabumbu (Lo) *Lannea gossweileri*

kabumbu-mutemwa (Lo) *Boscia angustifolia/ Maerua angolensis*

kabuye (Tk) *Prosospermum febrifugum*

kabwani (Ny) *Conyza sp.*

kacamba (Ny) *Holarrhena pubescens/ Ipomoea batatas/ Phillippia benguelensis/ Solanum tuberosum*

kacayi (Ln) *Eleusine coracana*

kace-nce (Nm) *Abrus precatorius*

kace-ncete (B) *Waltheria americana*

kacitose (Ny) *Momordica foetida*

kadale (Ny) *Combretum adenogonium/ C. hereroense/ C. molle/ C. zeyheri* (see also kalale)

kadhatula (Nm) *Harrisonia abyssinica*

kadongo-ndongo (Tk) *Tephrosia purpurea*

ka-emena (Ny) *Sorghum spp.*

ka-enya (B) *Mucuna pruriens/ M. stans*

kafavumba (Ny) *Ocimum americanum*

kaffertee (Afr) *Helichrysum nudifolium*

kaffirplum (E) *Ximenia caffra*

kaffir-tea (E) *Helichrysum nudifolium*

kafifi (B/ Ku) *Acacia kirkii, A. nilotica/ A. robusta*

kafifi (B/ Ku/ Ny/ Tu) *Acacia tortilis*

kafifi (Ny) *Acacia gerrardi/ A. hockii/ A. tortilis*

kafotoko (Ny) *Diospyros squarrosa*

kafubu (B) *Sorghum sp.*

kafukwakasyi (Lv) *Asparagus racemosus*

kafula-mume (B/ Ln) *Maprounea africana*

kafuluka-lume (B/ Ku) *Diospyros senensis*

kafumbe (B) *Maerua triphylla*

kafumbe (K) *Piliostigma thonningii*

kafundula (K/ Ln/ To) *Dalbergia nitidula*

kafungu (Lb) *Senna singueana*

kafungu-nasya (B) *Senna singueana/ S. petersiana*

kafupa (Ny) *Rotheca myricoides/ Stereospermum kunthianum/ Vernonia adoensis*

kafupaka-cimbwi (Ku/ Ny) *Allophylus africanus*

kafuwa-kanswi (To) *Pollicia campestris*

kafwalume (Lo) *Acacia schweinfurthii*

kafwamba (To) *Eleusine coracana*

kafwaya (Ku/ Ny) *Ageratum conyzoides*
kagowele (Nm) *Ziziphus mucronata*
kajiha-musongo (Lv) *Senna obtusifolia*
kakata (Tk) *Eleusine indica*
kakeke (Lo) *Ficus sur*
kakola-mvula (Lo) *Elephantorrhiza goetzei*
kakolo (B/ Lu/ Ny) *Combretum adenogonium*
kakoma (Lo) *Oncoba spinosa*
kakuhu (Ln) *Rourea orientalis*
kakukulama (I/ Le) *Combretum molle*
kakululema (I) *Combretum molle*
kakumbwe (Lo) *Acacia mellifera*
kakungi (Ny) *Combretum molle*
kakunguni (Ny) *Combretum collinum/ C. molle*
kalahari christmas-tree (E) *Dichrostachys cinerea*
kala-kala (Nm) *Ozoroa reticulata*
kalalitsi (Ny) *Hymenocardia acida*
kalama (B/ Ku/ Ny) *Combretum adenogonium*
kalama (B/ Ku/ Ny/ Tu) *Combretum collinum/ C. molle*
kalamba-tila (B) *Desmodium gangeticum*
kalana-ngwa (B) *Ziziphus abyssinica/ Z. spp.*
kalanga (Ny) *Capparis tomentosa*
kalatongo (To) *Clematis brachiata*
kalayi (Ln) *Zanha africana*
kale (Ny) *Bixa orellana*
kalembo (Nm) *Ziziphus mucronata*
kalenga-ntunzi (Ny) *Clausena anisata*
kali (Ny) *Flueggea virosa*
kalibabwe (Tu) *Senna singueana*
kalilesya (Ny) *Capparis tomentosa*
kalimbwe (Lo) *Ampelocissus africana*
kalimelera (Ny) *Erianthemum dregei*
kalimwendo (Ny) *Tridax procumbens*
kalinguti (To) *Burkea africana*
kalipo-poto (Ku/ Ny) *Asparagus sp.*
kalipo-potwa (Ku/ Ny) *Asparagus sp.*
kaliputi (Ny) *Bidens pilosa*
kalisabwe (Ku/ Tk) *Senna singueana*
kalisace (Ny) *Erianthemum dregei*
kalivute (Ny) *Bidens pilosa*
kaliza-kulu (Ny) *Opilia amentalea*
kalizya-lapami (Tk) *Clerodendrum ternatum*
kalizyala-pami (Tk) *Clerodendrum ternatum*
kalobwe (B) *Zizyphus abyssinica/ Z. spp.*
kalolo (B/ M) *Eleusine indica*
kalonga (Ny) *Vernonia spp.*
kalongo (B/ Ku) *Capparis sepiaria/ C. tomentosa*
kalongwe (B/ Ny) *Dalbergia nitidula*

kalope (To/ Tu) *Syzygium cordatum*

kalukulu (Ku) *Capparis tomentosa*

kalume-kamulama (B) *Combretum molle*

kalume-kamunganunsyi (B) *Acacia polyacantha*

kalume-kandiya (Ny) *Gloriosa superba*

kalunde-zonde (Ku) *Teramnus labialis*

kalundu (B) *Manihot esculenta*

kalundwe (B/Ku) *Manihot esculenta*

kalungawiba (Lb) *Salix mucronata*

kalungi (To) *Strychnos innocua*

kalunguti (Ku) *Abrus precatorius*

kalupo-potwa (Ku) *Asparagus sp.*

kalusapwe (Ny) *Senna singueana*

kalutenta (I/ Tk) *Plumbago zeylanica*

kaluvela (Ny/ Tu) *Vetiveria nigritana*

kaluwa-luwa (Ny) *Vernonia spp.*

kaluya (I) *Dichrostachys cinerea*

kalwerbossie (Afr) *Dicoma anomala*

kambaku-mbaku (Ny) *Combretum adenogonium*

kambala (Tu) *Eleusine coracana*

kambala-cihamba (B/ Ln) *Albizia antunesiana/ A. versicolor*

kambalame (Ny) *Antidesma venosum*

kambale (Ny) *Eleusine coracana*

kambele (Ku/ Ny) *Strychnos innocua*

kambolo (Le) *Gymnosporia senegalensis*

kambombo (K) *Trema orientalis*

kambu-lafita (K/ Ln) *Salix mucronata subsp. mucronata Thunb.*

kambu-lunje (Ny) *Bridelia cathartica*

kambu-mbu (Ku/ Ny) *Pistia stratiotes*

kamcenga (Ku) *Chenopodium ambrosiodes*

kameelspoor (Afr) *Piliostigma thonningii*

kamfule (Lo) *Sesbania sesban*

kaminda (B) *Mucuna pruriens*

kamino (B) *Strychnos spinosa*

kamoto (Ny) *Senna singueana/ Momordica foetida/ Thunbergia kirkiana*

kampanga (Ny) *Swarzia madagascariensis*

kampangala (Ny) *Dichrostachys cinerea*

kampango (Ny) *Swarzia madagascarensis*

kampela (Ny) *Brachystegia spiciformis*

kampobela (Ln) *Strychnos spinosa*

kampokola (To) *Croton megalobotrys*

kampolo-pombwe (B) *Phyllocosmus lemaireanus*

kampombwe (B) *Phyllocosmos lemaireanus*

kampongo (Ny) *Swarzia madagascariensis*

kamponi (Ny) *Commiphora africana/ Brachystegia longifolia/ B. spiciformis*

kamulebe (Ku/ Ny) *Ximenia americana*

kamulebe (Ny) *Ximenia caffra/ X. spp.*

kamutu-luce (Ny) *Bridelia cathartica*

kamwavi (Ku/ Tu) *Crossopteryx febrifuga*
kamwaya (I) *Maprounea africana*
kamwazi (Ny) *Rhus longipes/ Rhynchosia minima/ R. spp./ Turraea nilotica/ Vangueria infausta/ Vernonia petersii*
kamwela-lumba (Ny) *Strychnos innocua*
kamwenge (K/ Ln) *Bersama abyssinica*
kamwengo (Ln) *Bersama abyssinica*
kana-mwalisisi (Ny) *Cissus cornifolia*
kananga (Ny) *Croton gratissimus*
kandambwa (Ku/ Ny) *Ipomoea aquatica*
kandondo (Ku/ Tu) *Sorghum spp.*
kandondo-mutaba (Tu) *Sorghum spp.*
kandula (Ny) *Allophylus africanus*
kandulwa (Ny) *Zornia glochidiata*
kanembe *Aloe spp.*
kanenge (Ny) *Dichrostachys cinerea*
kangodza (Ny) *Eleusine indica*
kangolo (Ny) *Combretum apiculatum/ C. collinum/C. eleagnoides/ C. psidioides*
kangodza (Ny) *Eleusine indica*
kangwa (B) *Ziziphus spp./ Z. abyssinica/ Z. mucronata*
kangwe (Lo/ Ny) *Crossopterix febrifuga*
kanjoma (Tk) *Heteromorpha arborescens*
kankande (B/ Ku/ Ny/ Tu) *Ziziphus abyssinica*
kankande (Ny) *Acacia nigrescens/ Asparagus africanus/ Grewia flavescens/ Ziziphus spp./ Z. mucronata*
kankerblare (Afr) *Ranunculus multifidus*
kankole-nkole (B) *Ziziphus abyssinica*
kankolo (To) *Cardiospermum halicacabum*
kankona (I/ K) *Ziziphus abyssinica/ Z. mucronata*
kankululu (Ku) *Cardiospermum halicacabum/ Melinis repens/ Themeda triandra*
kankulwa (Ny) *Zornia glochidiata*
kankumbwila (To) *Albizia amara*
kanondo (Tu) *Sorghum spp.*
kanono (B) *Uapaca kirkiana*
kansalu-nsalu (B) *Clerodendrum uncinatum/ Dalbergia melanoxylon/ Dichrostachys cinerea*
kansalu-nsalu (Ku) *Albizia anthelmintica/ Dalbergia melanoxylon*
kansalu-salu (K) *Dichrostachys cinerea*
kansonje (To) *Imperata cylindrica*
kanungi (Ln) *Piliostigma thonningii/ Rutidea spp.*
kanunka (Ku) *Bidens pilosa/ Lippia sp./ Melanthera albinerva*
kanunkila (Lo/ To/ Tk) *Croton gratissimus*
kanyange-nyange (Lo) *Hoslundia opposita*
kanyense (B) *Allium cepa*
kanyera (Ny) *Ozoroa reticulata*
kanyole (Ny) *Psychotria zombamontana*
kanyuku (Ny) *Chenopodium ambrosiodes*
kanyungwa (Ny) *Vernonia spp.*
kanzeki (Ln) *Sesbania sesban/ Aeschynomene nyassana*

kanzota (Ku/ Ny) *Bidens pilosa*
kanzyomwa (Tk) *Sesbania sesban*
kaŋa-njamba (Ln) *Strychnos spinosa*
kapaci (Lv) *Bersama abyssinica*
kapaci-kelongolo (Ln) *Bersama abyssinica*
kapafumba (Ny) *Ocimum spp.*
kapakati (B) *Rourea orientalis*
kapalupalu (Lb) *Eriosema affine/ E. psoraleoides*
kapanga (B/ K/ Ku/ M/ Tu) *Burkea africana*
kapansya (Ny) *Arachis hypogaea*
kapapa (B) *Brachystegia boehmii/ B. stipulata*
kapapati (Lo) *Cassia abbreviata*
kapapi (Ku/ Ny/ Tu) *Securidaca longipedunculata*
kapasupasu (To) *Hymenocardia acida*
kapatati (Lo) *Cassia abbreviata*
kapele (B) *Phyllanthus muellerianus/ Rubus spp.*
kapembe (K/ Ku) *Hymenocardia acida*
kapempe (B/ K/ Le/ Lo/ M) *Hymenocardia acida*
kapepe (Ln) *Hymenocardia acida*
kapepi (Ln) *Hymenocardia acida*
kapika-nduzi (Ny) *Phyllanthus muelleranus*
kapila-pila (Ny) *Dioscorea spp./ Vernonia spp.*
kapila-syila (Ny) *Flueggea virosa*
kapira-pira (Ny) *Flueggea virosa*
kapiyo-piyo (Ny) *Eriosema psoraleoides*
kapok (E) *Ceiba pentandra*
kapoloni (Ny) *Heteromorphia arborescens/ trifoliata*
kapolo-polo (B) *Steganotaenia araliacea*
kapolo-pombwe (B/ Ku) *Phyllocosmos lemaireanus*
kapombo (B) *Diospyros virgata/ Sterculia quinqueloba*
kaponi (Tu) *Brachystegia spiciformis*
kapota (Lb) *Psorospermum febrifugum*
kapububa (Tk) *Ranunculus multifidus*
kapuka (Lo) *Croton menyharthii*
kapula (M) *Strychnos spinosa*
kapulu-koso (Lo/ Ny) *Crossopteryx febrifuga*
kapulula (Ny) *Phyllocosmus lemaireanus*
kapulula-mbusyi (B/ Ku/ Ny) *Rourea orientalis/ Phyllanthus reticulatus*
kapululu (I) *Sphenostylis marginata*
kapululu (Ku/ Ny) *Phyllanthus reticulatus*
kapulu-mbuzi (M) *Rourea orientalis*
kaputi-puti (Tu) *Hymenocardia acida*
kapwi (K) *Strychnos pungens/ S. spinosa*
kapwipu (Ln) *Swarzia madagascariensis*
kapwi-pwi (Ln) *Swarzia madagascariensis*
kapyai-pyai (Ny) *Flueggea virosa*
kari (Ny) *Momordica charantia/ Securinega virosa*
kasabeje (Ln) *Gardenia ternifolia*

kasabwe (B) *Sorghum sp.*
kasakolowe (Ny) *Uapaca nitida/ U. sansibarica*
kasakula (Ny) *Pterocarpus angolensis*
kasalala (I) *Senna singueana*
kasalasya (B) *Rhus longipes*
kasana (Ny) *Jateorrhiza palmata*
kasanda (Nm) *Swarzia madagascariensis*
kasandole (Ny) *Senna petersiana*
kasangala (Ln) *Margarita discoidea*
kasansobwanga (B) *Flueggea virosa*
kasekelele (B) *Ricinus communis*
kasela (B/ Ny) *Sorghum sp.*
kasele (K) *Acacia sieberiana*
kaselelele (B) *Ricinus communis*
kasese (Lo) *Helinus integrifolius*
kasiko-kam'umbu (To) *Lannea edulis and spp.*
kasila-butwilo (B/ Ku) *Alternanthera sessilis*
kasingini (To) *Rourea orientalis*
kasisi (K) *Paullinia pinnata*
kasokolowe (B/ Ny/ Tk/ Tu) *Uapaca nitida/ U. sansibarica*
kasokolowe (I/ To) *Swarzia madagascariensis*
kasokopyo (B/ Ku/ Ny) *Bidens pilosa/ B. spp.*
kasokopyo (Ku) *Phyllocosmos lemaireanus*
kasokoso (Ny) *Swartzia madagascariensis*
kasoko-soko (Ny) *Swartzia madagascariensis*
kasonge-kayisi (Lo) *Asparagus africanus*
kasongu (Ln) *Albizia amara*
kasonta (B/ M) *Nicotiana sp.*
kasterolieboom (Afr) *Ricinus communis*
kasuku (Ku) *Uapaca kirkiana*
kaswami-maji (Ln) *Asparagus sp.*
kasyika-mpombo (Ny) *Antidesma venosum*
kasyokwe (B) *Sorghum sp.*
katanda (Lv) *Citrullus lanatus*
katanda-imbo (Tu) *Hoslundia opposita/ Clerodendrum ternatum*
katanda-imbu (Ny) *Ocimum spp.*
katanda-mbo (Ny) *Hoslundia opposita*
katanda-vibanda (Ku) *Croton gratissimus*
katangala-tuzi (Ku) *Cardiospermum halicacabum*
katatula (Nm) *Acacia schweinfurthii/ A. senegal*
kateke (Lo) *Ficus sur*
kateme (Lo) *Strychnos innocua*
katendo (Ny) *Tridax procumbens*
katenga (K) *Dichrostachys cinerea*
katenge (B/ I/ K/ Le/ Ny/ To) *Dichrostachys cinerea*
katenge-nene (Ku) *Lagenaria sphaerica*
katenje (Tk) *Carissa edulis*
katete (Ln) *Acacia hockii/ A. Seyal/ Solanum sp.*

katete (Ln/ Lv) *Ricinus communis*
katika-yengele (B) *Balanites aegyptiaca*
katima (Lo) *Senna occidentalis*
katimbiti (B) *Hoslundia opposita*
katitu (Nm) *Acacia senegal*
katoma (Lo) *Flueggea virosa*
katombwangu (Lv) *Combretum collinum*
katondo (Tu) *Julbernardia paniculata*
katonga (K/ Ln) *Strychnos pungens/ S. spinosa*
katope (Ny/ To) *Syzygium cordatum/ S. guineense* sbsp. *afromontanum/ S. guineense* sbsp.
    huillense/ S. owariense
katuba-tuba (I) *Margaritaria discoidea*
katumbe-tumbe (B) *Harungana madagascariensis*
katumbi (B/ Ku/ Lb) *Psorospermum febrifugum & spp.*
katumbi (B) *Harungana madagascariensis*
katumbwangu (Ln) *Combretum collinum*
katuna (Ln) *Psorospermum spp./ Harungana madagascariensis*
katunya (Ln) *Harungana madagascariensis*
katupe (Ny) *Boophane disticha/ Gnidia chrysantha /Lasiosyphon kraussianus*
kauju (Ny) *Cyperus articulatus*
kauluzi (Ku) *Phyllanthus pentandrus*
kaumbu (B/ Ku/ Ny) *Lannea discolor/ L. schimperi*
kaunda (B) *Combretum molle*
kausenga (Ny) *Bridelia cathartica*
kavulamume (Ln/ Lv) *Maprounea africana*
kavulawuni (Ln) *Maprounea africana*
kavundula (Ku/ Ny/ Tu) *Psorospermum febrifugum/ P. spp.*
kavunguti (Ny) *Stereospermum kunthianum*
kaweyi (K/ Ln/ Lv) *Dichrostachys cinerea*
kawi (I/ Le/ To) *Strychnos innocua*
kawi (I) *Strychnos spinosa*
kawidzu (Tu) *Burkea africana*
kawigi (Ku) *Burkea africana*
kawinga-zimu (Ny) *Rotheca myricoides*
kawizi (Ku/ Ny) *Burkea africana*
kawowo (Ny) *Physostigma mesoponticum/ Sphenostylis marginata*
kawumbu (B/ Ny) *Lannea discolor/ L. katangensis/ L. schweinfurthii*
kayabule (B) *Ochna pulchra/ Stereospermum kunthianum*
kayalika (Ny) *Mikania chenopodifolia*
kayimbe (B) *Lannea discolor*
kayimbi (B/ K) *Burkea africana*
kayimbi (B/ Ny) *Erythrophleum africanum*
kayimbu (Ku) *Clerodendrum capitatum*
kayombo-yombo (Ln) *Zanha africana*
kayubule (B) *Ochna pulchra/ Sterospermum kunthianum*
kayuwe (Lo) *Mucuna pruriens/ Tragia spp.*
kayuye (Lo) *Mucuna pruriens*
kazewe-lezya (Ku/ Ny) *Mukia maderaspatana*

kazi (Ny) *Euphorbia tirucalli*
kazimbili-nkonde (Ku) *Mukia maderaspatana*
kazongwe (Lo) *Rourea orientalis*
kazota (Ny) *Vernonia spp.*
kersuurboom (Afr) *Pittosporum viridiflorum*
khat (E) *Catha edulis*
khata-mbuzi (Ny) *Triumfetta rhomboidea*
khawa (Ny) *Acacia nilotica*
khesya (Ku) *Senna occidentalis*
khobo (Ny) *Commiphora africana*
khologo (Ny) *Uapaca nitida*
khovani (Ny) *Commelina africana/ Persicaria salicifolia*
kiaat (Afr) *Pterocarpus angolensis*
kiawa-ame (Nm) *Euphorbia hirta*
kibale-bale (Ln) *Afzelia quanzensis*
kifumbe (K) *Piliostigma thonningii*
kifungula (K) *Kigelia africana*
kilota (K) *Euphorbia tirucalli*
kilula-mkunja (Nm) *Vernonia colorata*
kinaboom (Afr) *Rauvolfia caffra*
kipolopolo (K) *Steganotaenia araliacea*
kitungulu (Nm) *Allium sativum*
knapsekerel (Afr) *Bidens pilosa*
knoflook (Afr) *Tulbaghia alliacea*
knotweed-leaved milkwort (E) *Polygala persicariifolia*
koedoebessie (Afr) *Peudolachnostylis maprouneifolia*
koejawel (Afr) *Psidium guajava*
kogon grass (E) *Imperata cylindrica*
kola (Nm) *Afzelia quanzensis*
koloko-ndwe (B/ M) *Hibiscus sabdariffa*
kombora-kombora (Tu) *Pterolobium exosum*
kongwe (Ny) *Clematis brachiata/ Crossopterix febrifuga*
kooman (Afr) *Ficus sur*
koorsbossie (Afr) *Dicoma anomala*
koorspeulboom (Afr) *Holarrhena pubescens*
kopepe (Ln) *Hymenocardia acida*
kortpeul (Afr) *Rourea orientalis*
koswe (Ku/ Ny) *Cassia abbreviata*
koswe (Tu) *Trichodesma zeylanicum*
koti (Lo) *Clematis villosa/ Clematis brachiata*
kozi (Ny) *Rourea orientalis*
kraalmelkboss (Afr) *Euphorbia tirucalli*
kuduberry (E) *Peudolachnostylis maprouneifolia*
kukulumbe (Lv) *Cyphostemma buchananii*
kulakasyi (K) *Cassytha filiformis*
kulakazi (K) *Afzelia quanzensis*
kuluvela (Tu) *Veteveria nigritana*
kunai grass (E) *Imperata cylindrica*

kunda (B) *Combretum molle & spp.*
kundanyoka knobwood (E) *Zanthoxylum chalybeum*
kunze (Ku) *Cardiospermum halicababum*
kuŋ-nayi (Ln) *Sesbania sesban*
kurri (Ny) *Syphonochilus kirkii*
kusandore (Ny) *Senna petersiana*
kuviya (Nm) *Withania somnifera*
kyombe (K) *Acacia polyacantha*
kyombwe (K) *Acacia polyacantha*
laeveldmelkbessie (Afr) *Manilkara mochisia*
lafangala (Tu) *Dichrostachys cinerea*
lapeha (Ny) *Dalbergia boehmii*
large-fruited lightning bush (E) *Clutia abyssinica*
large hedgehog flower (E) *Pychnostachys urticifolia*
large-leaved dalbergia (E) *Dalbergia boehmii*
large-leaved rhus (E) *Rhus longipes*
large-leaved star chestnut (E) *Sterculia quinqueloba*
large red-heart (E) *Hymenocardia acida*
lavender cotton (E) *Croton gratissimus*
lavender popcorn (E) *Lantana trifolia*
leadwort (E) *Plumbago zeylanica*
lemba (Le) *Kirkia acuminata*
lemon grass (E) *Cymbopogon citratus*
letaunde (Ny) *Calotropis procera*
leyambaso (Lb) *Vangueria infausta*
leyani (Ny) *Ocimum americanum*
leza (Tu) *Cassia abbreviata*
libangwe (Lo) *Cannabis sativa/ Kotschya spp.*
libanji (Tu) *Cannabis sativa*
liboha (Ny) *Cyphostemma buchananii*
libono-bono (Lo) *Amaranthus spp./ Ricinus communis*
licenka (Lo) *Aloe spp.*
licongwe (Ny) *Acacia karroo/ A. nilotica/ Aloe christiani/ A. cryptopoda/ Argemone mexicana*
licorice weed (E) *Scoparia dulcis*
lifungula (K) *Kigelia africana*
lifwaka (Lo/ N) *Nicotiana tabacum & spp.*
ligoga (Ny) *Gymnosporia senegalensis*
likaka-lyaŋombe (Ny) *Pavetta crassipes*
likalango (Le) *Capparis elaeagnoides/ C. tomentosa*
likamba (Lo) *Manihot esculenta*
likango (Ny) *Helichrysum nudifolium*
likobeza (Lo) *Lantana trifolia/ Lippia javanica*
likobeza-balisana (Lo) *Lantana trifolia/ Lippia javanica*
likongono (Ny) *Gardenia ternifolia*
likulupsya (Ny) *Kalanchoe prolifera/ K. verticillata/ Momordica charantia/ M. foetida/ Pseudolachnostylis maprouneifolia/ Sonchus oleraceus/ Tagetes minuta*
likwaluku (Lo) *Eleusine indica*
likwanya (Ny) *Mucuna poggei/ M. pruriens/ M. stans*

lilamatwa (Lo) *Pupalia lappacea*

lilime-lyaŋombe (Ny) *Argemone mexicana/ Datura stramonium/ Pavetta crassipes/ Senecio latifolius/ Sonchus bipontini*

lilinji (Lv) *Ricinus communis*

lilongo (Lo) *Acacia polyacantha/ Azanza garckeana/ Sesbania sesban*

lilongo (Tk) *Salix mucronata*

lilongwe (Tk/ To) *Salix mucronata*

lilota (Le) *Euphorbia tirucalli*

linjinje (Lv) *Ricinus communis*

lion's ear (E) *Leonotis nepetifolia*

lion's tail (E) *Leonotis leonurus*

lipapau (B) *Carica papaya*

lipunga-ŋombe (Le) *Entada abyssinica*

litaka (Lo) *Phragmites australis*

litandamwe (Lo) *Ocimum gratissimum*

litanga (B) *Citrullus lanatus*

litembusya (B) *Aloe spp.*

livelong (E) *Lannea discolor*

livindwe (Nm) *Acacia polyacantha/ A. robusta*

liyamba (Lo) *Vetiveria nigritana*

lolombwe-lombwe (Lo) *Acalypha chirindica/ A. indica*

long-pod cassia (E) *Cassia abbreviata*

long-tail cassia (E) *Cassia abbreviata*

losa (Nm) *Aloe spp.*

loso (Ln/Lv) *Oryza sativa*

love vine (E) *Cassytha filiformis*

lowveld milkberry (E) *Manilkara mochisia*

lowveld tree vernonia (E) *Vernonia colorata*

luanika (Lo) *Azanza garckeana*

lubaba-ngwe (I) *Mucuna poggei/ Tragia sp/ Caperonia sp.*

lubalala (B) *Arachis hypogaea*

lubamba (I/ Lb) *Imperata cylindrica*

lubanda (Ku/ Ny) *Dicoma sessiflora*

lubanga (B/ K/ Le/ Tu) *Cleome gynandra*

lubanga-luulu (I) *Cordia goetzei*

lubange (I) *Cannabis sativa*

lubangeni (B) *Senna singueana*

lubanji (Tk) *Cannabis sativa*

lubeba (Lb) *Dalbergia nitidula*

lubebya (Tu) *Pseudolachnostylis maprouneifolia*

lubu (I) *Phragmites australis*

lubulu-kutu (B) *Artabotrys monteiroae*

ludumyo (Nm) *Tephrosia purpurea*

luengi (Nm) *Xeroderris stuhlmannii*

lueni (Lv) *Senna occidentalis*

lufangala (Tu) *Dichrostachys cinerea*

lufwiti (Ku) *Boscia angustifolia/ Philenoptera bussei*

luhombo (Nm) *Brachystegia boehmii*

luhundu (I) *Vetiveria nigritana*
lukambamwe (Lo) *Hemizygia bracteosa*
lukata (Tu) *Stereochlaena cameronii/ Eleusine indica*
lukesyu (Lo) *Sorghum spp.*
lukuku (Lo) *Hibiscus sabdariffa*
lukukwa (I/ To) *Hibiscus sabdariffa*
lukuswa-ula ((B/ Ku/ Ny) *Flueggea virosa*
lulya-bazuba (Tk) *Heteromorpha arborescens/ Aeschynomeme trigonocarpa*
lumanda (Ny/ Tu) *Hibiscus sabdariffa*
lumanda-sanga (Ku) *Urena lobata*
lumanyama (To) *Cassia abbreviata*
lumina-nzoka (I) *Ximenia caffra*
lumole (Tk) *Plectranthus cylindraceus*
lumombwe (Lb) *Allophylus africanus*
lumono (B) *Ricinus communis*
lumonokyulu (K) *Steganotaenia araliacea*
lumpampa-lunono (B?) *Sorghum sp.*
lumpangala (B/ Ku/ Ny/ Tu) *Dichrostachys cinerea*
lumpunga (Ny/ Tu) *Flueggea virosa / Securinega virosa*
lumwamwa (Lo) *Maprounea africana*
lunana (Ln) *Euphorbia tirucalli*
lundi (Lv) *Securidaca longipedunculata*
lungamayowa (I) *Evolvulus alsinoides*
lungamunene (Lo) *Faidherbia albida*
lungwe (Lv) *Pistia stratiotes/ Phyllanthus reticulatus*
lungwizi (Ny) *Acacia goetzei* ssp *microphylla/ Phyllanthus muelleranus*
luni (Ny/ Tu) *Cleome gynandra*
lunsa (B) *Ehretia obtusifolia*
lunsonga (B) *Euphorbia tirucalli*
lunsyi-nsamba (B) *Cassytha filiformis*
lunyanya (Ny) *Tragia benthamii*
lupangala (Ny) *Dichrostachys cinerea*
lupape (K/ Lb) *Securidaca longipedunculata*
lupoko (Ny) *Eleusine coracana*
lusili (To) *Sorghum sp.*
lusolo (Ku/ Ny/ Tu) *Dichrostachys cinerea*
lusonga (B/ Ku) *Euphorbia tirucalli*
lusongwa-songwa (Ln) *Ximenia caffra*
lusukwe (I) *Uapaca kirkiana*
lusyalo (Tk) *Holarrhena pubescens*
lutanda (B) *Dicoma anomala & spp.*
lutata (Lv) *Securidaca longipedunculata*
lutembwe (Lo) *Cleome gynandra*
lutete (B) *Phragmites australis (Cav.) Steud. [P.communis Trin.]*
luvundu (I/ To) *Vetiveria nigritana*
luvung'u (Ln) *Anisophyllea boehmii*
luwanika (Lo) *Azanza garckeana*
luwele (Lo) *Pupalia lappacea*

luweni (Lv) *Senna occidentalis*
luwenje (Nm) *Gymnosporia senegalensis*
luwere (Ny) *Crotalaria natalitia*
luwimbwa (B) *Lannea edulis*
luze (Lo) *Cassytha filiformis*
lwando (Ny) *Cassytha filiformis*
lwena (B/ Ku) *Ocimum americanum*
lwenye (Lo) *Ocimum gratissimum & spp.*
lyaja (Ny) *Treculia africana*
lyamba (Lo) *Vetiveria nigritana*
lyandumaka (Lo) *Salix mucronata*
lyandumuka (Lo) *Salix mucronata subsp. subserrata*
lyundumula (Ny) *Clematis brachiata/ Mikania chenopodifolia/ Thunbergia alata/*
    Tragia benthamii
lyuni (Ny/ To) *Cleome gynandra*
maagbitterwortel (Afr) *Dicoma anomala*
maagbossie (Afr) *Dicoma anomala/ Lippia javanica*
maamba (Lo) *Vetiveria nigritana*
maamba (To) *Clematis brachiata*
ma-au (Lb/ Le) *Eleusine coracana*
mabele (B/ Lb) *Ozoroa reticulata*
mabele (I/ Ku/ Ny/ To) *Eleusine coracana*
mabiliki (B) *Citrullus lanatus*
mabono (I/ Lb/ Le/ To) *Ricinus communis*
mabono-bono (Lo) *Ricinus communis*
macende-kalulu (Ny) *Multidentia crassa / Vangueria infausta*
madagascar periwinkle (E) *Catharanthus roseus*
madder (E) *Oldenlandia capensis*
magabu (Nm) *Dalbergia boehmii*
magenge (Ny) *Lannea edulis*
maggot-killer (E) *Clausena anisata*
magic nut (E) *Maprounea africana*
mahangu (Lo) *Eleusine coracana & E. spp.*
majote (Ny/Tu) *Citrullus lanatus*
makabe (B) *Citrullus lanatus*
makamba ((I/ K/ Le/ Ln/ N/ To) *Manihot esculenta*
makaumba (B) *Lannea discolor*
makelele (Tk) *Veteveria nigritana*
makepera (Ny) *Psidium guajava*
makoko-lono (Ny) *Flacourtia indica*
makonga (Lo) *Sorghum sp.*
makuku (Lo) *Hibiscus sabdariffa*
makula (Ny) *Vernonia spp.*
makula-mamena-ncembele (Ku) *Ehretia obtusifolia*
makuwa-kuwa (Lo) *Gloriosa superba*
makwangala (Tk) *Nymphaea lotus*
malaka (Lo) *Lagenaria sphaerica*
mala-oya (Lo) *Allophylus africanus*

malaza (Ny) *Oncoba spinosa*
male (B) *Eleusine coracana*
malgif (Afr) *Boophane disticha*
malindi (Lv) *Ricinus communis*
mallow raisin (E) *Grewia villosa*
malulwe (Tk) *Cassia abbreviata*
malundu (To) *Aristolochia albida*
maluwa (Ku) *Gomphrena celosioides/ Senna occidentalis/ S. siamea*
malvarosyntjie (Afr) *Grewia villosa*
mambahuru (Ny) *Maclura africana*
mamena-ncembele (Ku) *Ehretia obtusifolia*
mampubila (Lo) *Carica papaya*
manama (Ny) *Euclea natalensis*
manceba (B) *Gardenia ternifolia*
maneko (B/ Lo/ To) *Azanza garckeana*
manga (B/ Ku/ Ny) *Mangifera indica*
manga (K/ Ln) *Brachystegia hockeyii/ B. spiciformis*
mangacule (Ny) *Maesa lanceololata*
mangile (Lo) *Pseudolachnostylis maprouneifolia*
mango (E/ Ny) *Mangifera indica*
mangobe (B) *Acacia polyacantha*
mangondo (Ny) *Anisophylla boehmii*
mangonye (M) *Mangifera indica*
maninga (Ny) *Acacia polyacantha*
manioc (E) *Manihot esculenta*
manja-atali (Ny) *Pavetta crassipes*
manjana (Ny) *Syphonochilus kirkii*
mansisye (B/ Ku) *Cocculus hirsutus*
manunka (Ny) *Ocimum americanum*
manyanzya (Tk) *Cyperus articulatus*
maŋomba (I/ Le/ To) *Capsicum annuum*
mao (Lb/ Le) *Eleusine coracana*
mapapaya (B/ M) *Carica papaya*
mapela (B) *Psidium guajava*
mapemba (B/ Tu) *Sorghum sp.*
mapila (Ku) *Sorghum spp.*
mapili-nganya (Ku) *Solanum spp.*
mapira-kukutu (Ny) *Antidesma venosum*
mapirano (Ny) *Paullina pinnata*
mapira-tutile (B/ Tu) *Sorghum spp.*
mapolo-yakalulu (Ku) *Vangueria infausta/ Vangueriopsis lanciflora*
mapolwa (Ku) *Rotheca myricoides*
mapuma-ngulube (B) *Strychnos spinosa*
marijuana (E) *Cannabis sativa*
maroela (Afr) *Sclerocarya birrea*
marula (E) *Sclerocarya birrea*
masace (Ny) *Sida acuta/ Triumfetta rhomboidea/ Securidaca longipedunculata*
masali (B) *Sorghum spp.*

masango-kawereka (Ny) *Albizia anthelmintica*
masawasa (Ny) *Jatropha curcas*
masawo (Ny) *Dolichos trinervatus/ Ziziphus abyssinicus*
masimya (M/ Ny) *Ozoroa reticulata*
masowa-pusi (Ny) *Abrus precatorius*
masuku fruit (E) *Uapaca kirkiana*
masuku-lubemba (B) *Uapaca kirkiana*
masuku-nkuma (B) *Uapaca kirkiana*
masungwa (K) *Uapaca kirkiana*
masyaga-syela (Lo) *Swarzia madagascariensis*
matakala (Ny) *Boophane disticha/ Pterocarpus rotundifolius*
matanda (Lv) *Citrullus lanatus*
matanga (I/Lo/To) *Citrullus lanatus*
matangala (Lv) *Citrullus lanatus*
matanta (Ku) *Oncoba spinosa*
mata-udi (Ku) *Zanha africana*
matengala (Lv) *Citrullus lanatus*
matete (B/ K/ Ku/ Lo/ Ny/ To/ Tu) *Phragmites australis*
matewele (Ny) *Arachis hypogaea*
matholisa (Ny) *Aristolochia petersiana/ Melia azedarach/ Mikania chenopodifolia*
matokwani (Lo) *Cannabis sativa*
matokwe (To) *Azanza garckeana*
matongo (B) *Microglossa pyrifolia/ Tecomaria capensis*
matongwe (Ku) *Nidorella sp./ Vernonia spp.*
matowo (Ku) *Azanza garckeana*
matseka (Ny) *Oncoba spinosa*
matsutula (Ny) *Rhus longipes*
matuwa-ngoma (Ny) *Pavetta crassipes*
mausu (I) *Ziziphus mucronata*
mavwembe (Ny/Tu) *Citrullus lanatus*
mawele (Lb) *Ozoroa reticulata*
mawembe (B) *Mangifera indica*
mawo (Lb/ Le) *Eleusine coracana*
maye (Ny) *Strychnos spinosa/ S. cocculoides*
mayowa (Tu) *Arachis hypogaea*
mazamba (Ny) *Crotalaria natalitia*
mazimbili (I) *Strychnos spinosa*
mbaba-ngoma (Nm) *Balanites aegyptica*
mbala (To) *Euphorbia tirucalli*
mbalala (B/ Ku) *Arachis hypogaea*
mbala-mbala (Ny) *Bridelia cathartica*
mbale (Ny) *Lannea discolor/ L. schweinfurthii*
mbalika (Ny) *Ricinus communis*
mbali-mbali (Lb) *Peltophorum africanum*
mbalitsa (Ny) *Dicoma sessiliflora ssp sessiliflora/ Pterocarpus rotundifolius*
mbamba-sanga (Ku) *Canavalia sp./ Sphenostylis marginata*
mbangonge (Ny) *Carissa edulis*
mbangozi (Ku) *Senna petersiana/ Xeroderris stuhlmannii*

mbangozi (Ny) *Pterocarpus angolensis*
mbanje (Ku/ Ny) *Cannabis sativa*
mbaselwenje (I) *Albizia glaberrima*
mbawa (Ku/ Ny) *Acacia robusta*
mbawa (Ny/ Tu) *Khaya anthotheca*
mbaya-mpondolo (Ku/ Ny) *Balanites aegyptiaca*
mbele-bele (Nm) *Strophanthus eminii*
mbele-mbele (To) *Sesbania sesban*
mbeli-neli (Nm) *Diplorhyncus condylocarpon*
mbewe (Ny) *Indigofera spp./ Ozoroa reticulata*
mbeza-munku (Ny) *Holarrhena pubescens*
mbiricila (Ny) *Syphonochilus kirkii*
mblaka (Ny) *Bersama abyssinica*
mbono (B/ I/ Lb/ Le/ To) *Ricinus communis*
mbonwa (Lo) *Ricinus communis*
mbotyola (Nm) *Steganotaenia araliacea*
mbowa-wacilungu (To) *Amaranthus spinosus*
mbula (Nm/ Ny/ To) *Parinari curatellifolia/ P. excelsa*
mbulambula (Ny) *Clematis brachiata*
mbulu-bunje (Ny) *Lannea edulis*
mbulu-mbunje (Ny) *Cissus cornifolia/ Lannea edulis*
mbuluni (Ny) *Harungana madagascariensis*
mburi (Nm) *Hibiscus micranthus*
mbute (Lb) *Stereospermum kunthianum*
mbuyu (Ny/ Tu) *Adansonia digitata*
mbwa-bwa (Ny) *Cussonia arborea*
mbwani (Ny) *Croton macrostachys*
mbwa-nyanya (Ny) *Pseudolachnostylis maprouneifolia/ Solanum incanum*
mbwasa (Ku) *Acacia robusta*
mbweba-nyani (Ny) *Senna petersiana*
mbwebe (Le) *Phyllanthus pentandrus*
mbwemba (Ny) *Tamarindus indica*
mbwi-mbwi (Ku/Nya/Tu) *Antidesma venosum*
mcala-mira (Ny) *Cassia abbreviata/ Leucas martinicensis*
mcalima (Ny) *Elephantorrhiza goetzei*
mcansira (Ny) *Leucas martinicensis*
mcanzi (Ny) *Lippia javanica*
mcelecete (Ny) *Margaritaria discoidea*
mcelekete (Ny) *Swarzia madagascariensis*
mcende (Ny) *Trema orientalis*
mcenga (Ny) *Brachystegia boehmii/ B. longifolia/ B. spiciformis/ Julbernardia globiflora*
mcenja (B/ Ny) *Diospyros mespiliformis*
mcenja-sumu (Ny) *Diospyros mespiliformis*
mcenje (Ny) *Diospyros kirkii/ D. mespiliformis*
mcenjema (Ny) *Lippia javanica*
mceramila (Ny) *Elephantorrhiza goetzei*
mcezime (Ny) *Acacia nilotica*
mcindula (Ny) *Dalbergia boehmii*

mcinji (Ny) *Bersama abyssinica*

mcisu (Ny) *Syzygium cordatum*

mcoka (Ku) *Pavetta crassipes/ Vangueria infausta/ Vitex doniata*

mcoma-fisi (Nm) *Gymnosporia senegalensis*

mconi (Ny) *Ochna schweinfurthiana*

mdenjele (Ny) *Maesa lanceololata*

mdima (Ny) *Diospyros batocana/ D. kirkii/ D. lycioides/ D. senensis/ D. zombensis/ Ilex mitis/ Mystroxylon aethiopicum/ Phyllocosmus lemaireanus/ Psorospermum febrifugum/ Rhus longipes*

mdima-madzi (Ny) *Catha edulis/ Euclea spp.*

mdime (Ny) *Flueggea virosa*

mdimu (Nm) *Teclea nobilis*

mdogo (Nm) *Ximenia americana*

mdubilu (Nm) *Acacia nilotica*

mdula-msongwa (Nm) *Cussonia arborea*

mdulu (Ny) *Cyperus articulatus*

mdundira (Ny) *Antidesma venosum*

mdundo (Ny) *Swartzia madagascariensis*

mdya-fungu (Ny) *Vitex doniana*

mdya-kamba (Ny) *Antidesma venosum/ Lannea discolor/ L. edulis/ Swarzia madagascariensis/ Tinospora birrea*

mdya-ncembele (Ny) *Ximenia americana/ X. caffra*

mdya-nsefu (Ny) *Pavetta crassipes*

mdya-pumbwa (Ny) *Antidesma venosum/ Senna petersiana/ S. singueana/ Phyllanthus reticulatus/ Rhoicissus revoilii*

mdya-tsongwe (Ny) *Leonotis nepetifolia/ Maesa lanceololata/ Pychnostachys urticifolia*

mdya-tungu (Ny) *Lannea discolor*

melkbos (Afr) *Xysmalobium undulatum*

melon tree (E) *Carica papaya*

mentholatum plant (E) *Plectranthus cylindraceus*

mexican tea (E) *Chenopodium ambrosoides*

mfele (Ny) *Psorospermum febrifugum*

mfifi (M) *Psorospermum febrifugum*

mfika (Ny) *Lantana camara/ Pouzolzia senensis*

mfipa (Ny) *Boscia salicifolia*

mfira (Nm) *Annona senegalensis*

mfuje-anje (Nm) *Carissa edulis*

mfula (Ny) *Sclerocarya birrea*

mfumba-mula (Ny) *Holarrhena pubescens*

mfumbe (Ny) *Piliostigma thonningii*

mfumpu (Ny) *Croton megalobotrys*

mfunda-zizi (Ny) *Brachystegia boehmii*

mfunfwa (Nm) *Combretum collinum*

mfungo (Ny) *Anisophyllea boehmii/ A. pomifera*

mfutamula (Nm) *Entada abyssinica*

mfutsa (Ny) *Cyanthillium cinereum/ Sphenostylis marginata/ Vernonia adoensis/ V. amygdalina/ V. ampla*

mfutsa-wookulu (Ny) *Tetradenia riparia*

mfwete (Ny) *Capparis tomentosa*
mgaci (Ny) *Euphorbia tirucalli*
mgaga (Ny) *Holarrhena pubescens*
mgagati (Nm) *Abrus fruticulosus*
mgalambuti (Ny) *Asparagus africanus/ Galium bussei*
mgamu (Ny) *Sclerocarya birrea*
mgando (Nm) *Burkea africana/ Erythrophleum africanum*
mgaweko (Nm) *Combretum mossambicense*
m'gazu (Nm) *Commiphora africana*
mgeda (Nm) *Withania somnifera*
mgezi-ngono (Ny) *Capparis sepiaria*
mgogote (Nm) *Paullinia pinnata*
mgondo (Ny) *Anisophyllea boehmii/ A. pomifera*
mgongho (Nm) *Sclerocarya birrea*
mgongo (Ny) *Cissus cornifolia*
mgongolo (Ny) *Acacia polyacantha*
mgowo-gowo (Ny) *Crossopterix febrifuga*
mgoza (Ny/ To/ Tu) *Sterculia africana/ S. quinqueloba*
mgugunu (Nm) *Ziziphus mucronata*
mgula (To) *Diospyros mespiliformis*
mgulu-gulu (Ny) *Strychnos innocua*
mgunga (Nm) *Acacia tortilis*
mgupulu (Nm) *Acacia polyacantha*
mguyu-guyu (Nm) *Balanites aegyptica*
mhaka (Ny) *Boscia angustifolia*
mhalalwa-huba (Nm) *Erythrina abyssinica*
mhoja (Nm) *Sterculia africana*
mhora (Nm) *Afzelia quanzensis*
mhozia (Nm) *Sterculia africana*
mhumba (Nm) *Senna singueana*
mhundu (Nm) *Strychnos innocua*
mhungambu (Nm) *Hoslundia opposita*
mhungu (Nm) *Acacia polyacantha*
migunga (Nm) *Acacia tortilis*
miguplu (Nm) *Acacia polyacantha*
migwa-sungu (Nm) *Xenostegia tridentata*
mile-a-minute (E) *Mikania chenopodifolia*
milk-bush (E) *Gomphocarpus fruticosus/ Xysmalobium undulatum*
milkweed (E) *Gomphocarpus fruticosus*
milkwort (E) *Xysmalobium undulatum*
milungulu (Ny) *Kigelia africana*
minuasa-ungu (Nm) *Catunaregam obovata*
miraculous berry (E) *Synsepalum brevipes*
miriva-nhinga (Nm) *Acacia robusta*
mirungulu (Tu) *Kigelia africana*
misbredie (Afr) *Portulaca oleracea*
misekese (B) *Sesbania sesban*
mistletoe (E) *Plicosepalus kalachariensis*

miyembe (B) *Mangifera indica*
miyenu (To) *Clematis brachiata*
miyombo (Nm) *Brachystegia boehmii/ B. spiciformis*
miyombo-botha (Nm) *Brachystegia spiciformis*
mjajajuni (Ny) *Trema orientalis*
mjaya (Ny) *Treculia africana*
mjele-njete (Ny) *Albizia amara/ A. harveyi/ Cassia didymobotrya*
mjenjete (Ny) *Albizia harveyi/ Pseudolachnostylis maprouneifolia*
mjuju (Ny) *Zanha africana*
mjukutu (Ny) *Vangueria infausta*
mkabwa (Ny) *Manilkara mochisia*
mkaiya (Ny) *Acacia pilispina/ A.sieberiana*
mkakama (Ny) *Maesa lanceololata*
mkako (Ny) *Crossopterix febrifuga*
mkala (Nm) *Ozoroa reticulata*
mkala-camba (Ny) *Holarrhena pubescens*
mkala-kala (Nm) *Ozoroa reticulata*
mkala-kate (Ny) *Erythrocephalum zambesianum/ Monotes africanus/ M. glaber/ M. spp./
    Pleiotaxis spp.*
mkala-lwanghuba (Nm) *Erythrina abyssinica*
mkala-ncamba (Ny) *Holarrhena pubescns*
mkala-nga (Ny) *Albizia amara/ A. harveyi*
mkala-nkanga (Ny) *Albizia amara/ A. harveyi*
mkala-ta (I) *Entada abyssinica*
mkala-ti (Nm/ Ny) *Burkea africana*
mkala-wahuba (Nm) *Erythrina abyssinica*
mkalia (Nm) *Cussonia spp./ Zanha golugensis*
mkalula (Nm) *Hoslundia opposita*
mkalya (Nm) *Cussonia arborea/ Zanha africana*
mkanda-mbazo (Ny) *Flacourtia indica*
mkanda-nkuku (Ny) *Paullina pinnata*
mkanda-nyalugwe (Ny) *Abrus precatorius/ Paullina pinnata/ Steganotaenea araliacea*
mkanga (Ny) *Bersama abyssinica*
mkanga-luce (Ny) *Zanha africana*
mkanga-mwazi (Ny) *Carissa edulis*
mkanga-ndembo (Ny) *Multidentia crassa / Vangueria infausta*
mkangano (Ny) *Clausena anisata*
mkansolo (Ny) *Acacia sieberana*
mkarati (Nm) *Burkea africana/ Erythrophleum africanum*
mkaye (Ny) *Strychnos innocua*
mkhadze (Ny) *Euphorbia tirucalli*
mkhala (Ny) *Carissa edulis*
mkhuyu-khuyu (Ny) *Strychnos innocua/ S. spinosa*
mkola (Nm) *Afzelia quanzensis/ Erythrophleum guineense/ E. suaveolens*
mkolando (Ny) *Afzelia quanzensis*
mkola-nkanga (Ny) *Asparagus sp.*
mkola-singa (Ny) *Dalbergia nitidula*
mkole (Ny) *Azanza garckeana/ Dombeya rotundifolia*

mkolo (Ny) *Antidesma venosum*
mkolo-kolo (Ny) *Carissa edulis*
mkolongo (Ny) *Psychotria zombamontana*
mkoma-nyanda (Ny) *Brachystegia boehmii/ B. longifolia*
mkome (Nm) *Strychnos pungens/ S. spinosa*
mkondo-kondo (Ny) *Flacourtia indica*
mkongomwa (Ny) *Afzelia quanzensis*
mkonora (Nm) *Annona senegalensis*
mkora (Nm) *Afzelia quanzensis*
mkoso (Ny) *Burkea Africana*
mkosomola (Ny) *Psorospermum febrifugum*
mkoswe (Ny) *Cassia abbreviata*
mkubani (Ny) *Ocimum spp.*
mkubwa (Nm) *Ximenia caffra*
mkula-ngondo (Ny) *Flueggea virosa/ Securinega virosa/ Hymenocardia acida*
mkula-nsinga (Ny) *Rourea orientalis*
mkulu-kulu (Ny) *Teclea nobilis*
mkulu-kumba (Ny) *Adansonia digitata*
mkuma-jalaga (Ny) *Argemone mexicana*
mkunda-ngulwe (Ny) *Crossopterix febrifuga*
mkundi (Ny) *Cissampelos mucronata/ Parkia filicoidea*
mkungira (Ny) *Alternanthera sessilis*
mkunguni (Ny) *Pittosporum viridiflorum*
mkungu-njila (Ny) *Momordica foetida*
mkunguti (Ny) *Pittosporum viridiflorum*
mkupa-imbu (Ku) *Hemizygia bracteosa*
mkuruya-mnenfuwa (Nm) *Combretum molle*
mkussu (Nm) *Harrisonia abyssinica*
mkuta (Ny) *Adenia gummifera/ Pachycarpus lineolatus*
mkute (Ny) *Combretum collinum/ C. molle/ C. psidioides/ C. zeyheri*
mkuti (Ny) *Brachystegia spiciformis/ B. utilis*
mkuyu (Ny) *Ficus sur*
mkuzya-ndola (Ku) *Chamaecrista absus/ Senna singueana*
mkuzya-ndola (Ny) *Bridelia cathartica/ Phyllanthus muelleranus*
m'kwakwa (Ny) *Strychnos innocua*
mkwale (Ny) *Holarrhena pubescens/ Stephania abyssinica/ Tabernaemontana elegans*
mkwapu-kwapu (Ny) *Cassia abbreviata*
mkwesu (Nm/ Ny) *Tamarindus indica*
mkwidyo (Ny) *Zanha golungensis*
mlago (Nm) *Ozoroa reticulata*
mlaka-njovu (Ny) *Stereospermum kunthianum*
mlama (Nm) *Combretum molle*
mlama-fupa (Ny) *Crossopterix febrifuga*
mlama-nama (Nm) *Combretum molle*
mlamba (Ny) *Lannea edulis*
mlambe (Ny) *Adansonia digitata*
mlandala (Nm) *Combretum collinum*
mlangamia (Nm) *Cassytha filiformis*

mlanga-mpete (Ny) *Ochna macrocalyx/ O. schweinfurthiana*
mlasya-wantu (Ny) *Ziziphus abyssinica/ Z. spp.*
mlaza (Ny) *Borassus aethiopum/ Hyphaene petersiana/ Oncoba spinosa*
mlela (Nm) *Acacia seyal*
mlele (Ny) *Sterculia africana*
mlele-mupili (Ny) *Albizia anthelmintica*
mlele-zombo (Ny) *Sterculia quinqueloba*
mlelwa-huwa (Nm) *Cussonia arborea/ Erythrina abyssinica*
mlembe-lembe (Ny) *Cassia abbreviata*
mleza (Ku/ Ny) *Cassia abbreviata*
mlila-nzeze (Ny) *Albizia versicolor*
mlimbauta (Ny) *Zanha africana*
mlindi-mila (Ny) *Erythrina abyssinica*
mliwa-nhwiga (Nm) *Acacia robusta*
mlombe (Ny) *Pterocarpus angolensis/ Thunbergia lancifolia*
mlombeya (Ny) *Xeroderris stuhlmannii*
mlombwa (Ny) *Pterocarpus angolensis*
mlomo-wambuya (Ny) *Aerva lanata*
mlonde (Ny) *Xeroderris stuhlmannii*
mlonga (Ny) *Acacia polyacantha*
mlonje (Ny) *Adansonia digitata*
mlowe (Ny) *Combretum collinum*
mlozi (Ny) *Adenia digitata/ A. gummifera*
mluja-minzi (Nm) *Combretum adenogonium*
mlukuwa-mhuli (Nm) *Flacourtia indica*
mlulu (Ny) *Cleistachnes sorghoides/ Cyperus articulatus/ Khaya anthotheca*
mluma-nyama (Ny) *Cassia abbreviata*
mlunga (Ny) *Cissus quadrangularis/ Ozoroa reticulata*
mlungamu (Ny) *Ochna macrocalyx*
mlungu-lungu (Nm) *Zanthoxylum chalybeum*
mlunguti (Ny) *Erythrina abyssinica*
mlunguzi (Ny) *Lantana camara*
mlungwe (Ny) *Dalbergia nitidula*
mlusye (Ny) *Gymnosporia senegalensis*
mnama (Nm) *Combretum molle*
mnama (To) *Securidaca longepedunculata*
mnanga (Ny) *Croton megalobotrys*
mnawa (Ny) *Margaritaria discoidea*
mnaziya-polini (Nm) *Parinari curatellifolia*
mndyodyolo (Ny) *Crotalaria recta*
mnembuwa (Nm) *Maclura africana*
mnembwa (Nm) *Ximenia caffra*
mnembwa (Ny) *Ozoroa reticulata*
mnene (Nm) *Xeroderris stuhlmannii*
mnengo (Ny/ Tu) *Uapaca nitida*
mngongho (Nm) *Sclerocarrya birrea*
mnindi (Nm) *Rotheca myricoides*
mnubulu (Nm) *Diospyros mespiliformis*

mnuwake-munda (Ny) *Cyphostemma junceum*
mnya-mbalame (Ny) *Leonotis nepetifolia*
mnya-nyata (Tu) *Diplorhynchus condylocarpon*
mnya-tsongwe (Ny) *Leonotis nepetifolia*
mnyele (Ny) *Peltophorum africanum*
mnye-nye (Nm) *Xeroderris stuhlmannii*
mnye-nyele (Nm) *Acacia hockii*
mnyesani (Ny) *Rauvolfia caffra*
mnyoka (Ny) *Cassia abbreviata*
mnyomwe (Ny) *Syzygium cordatum*
mnyonga-pembe (Nm) *Cussonia arborea*
mobola plum (E) *Parinari curatellifolia*
mobole (B) *Adenia lobata/ Cissus quadrangularis*
moca-ngoko (Nm) *Catunaregam obovata*
mojolo (To) *Gardenia ternifolia*
mokungu (Lo) *Pseudolachnostylis maprounifolia*
molo (Nm) *Brachystegia spiciformis*
mombo (Ny) *Brachystegia boehmii/ B. longifolia*
mombomba (Ln/ Lv) *Combretum adenogonium*
mompele-mpempe (I) *Hymenocardia acida*
monganga (Lo) *Ziziphus abyssinica*
mongolo (To/ Tk/ Tu) *Swarzia madagascariensis*
mongwe (Tk/ To) *Sclerocarya birrea*
monjolo ((Tk/ To) *Swarzia madagascariensis*
monkey-bread (E) *Piliostigma thonningii*
monkey-bread tree (E) *Adansonia digitata*
monkey flower (E) *Mimulus gracilis*
monkey nut (E) *Arachis hypogaea*
monkey orange (E) *Strychnos innocua*
monkey pod (E) *Senna petersiana*
monkey rope (E) *Cocculus hirsutus*
mono (B/ K/ Lb/ Lo/ Ny/ Tu) *Ricinus communis*
mono-mono (Lo) *Croton megalobotrys*
monoza-nyemba (Ln/ Lv) *Jatropha curcas*
montamfumu (B) *Combretum celastroides/ Combretum molle*
montbrecia (E) *Crocosmia aurea*
monwa (Lo) *Ricinus communis*
mooba-nkonono (To) *Ziziphus abyssinica*
moombwa (I/ To) *Commiphora mollis*
moongola (To) *Swarzia madagascariensis*
moono (B/ Ku) *Ricinus communis*
moonzo (Tk) *Acacia nilotica*
moonzu (Tk) *Croton gratissimus*
mosiya-balozi (Tu) *Acacia seyal*
mosterdboom (Afr) *Salvadora persica*
mountain-spice (E) *Xylopia acutiflora*
moza (I) *Sterculia africana*
mozya (I/ To) *Sterculia africana*

mpabula (Ny) *Gymnosporia senegalensis*

mpaca (Ny) *Phyllanthus muelleranus*

m'paela (Nm) *Adansonia digitata*

mpakasa (M/ Ny) *Combretum adenogonium/ C. molle/ Philenoptera bussei*

mpakulu (Ny) *Gymnosporia senegalensis*

mpala (Ny) *Triumfetta rhomboidea*

mpala-cipopo (Ku) *Laportea aestuans*

mpala-mabwe (Ku) *Jateorrhiza palmata*

mpala-matongwe (Ku) *Erythrocephalum zambesianum*

mpala-mtowasilu (Ku) *Excoecaria bussei/ Holarrhena pubescens*

mpala-mwezamunko (Ku) *Holarrhena pubescens*

mpala-ntanga (Ku) *Chamaecrista falcinella/ Indigofera astragalina/ Senna obtusifolia/ Tephrosia elata*

mpala-sasamwa (Ku) *Evolvulus alsinoides*

mpama (Ny) *Dioscorea bulbifera/ D. dumetorum/ D. spp./ Smilax anceps*

mpama-laza (Ny) *Oncoba spinosa*

mpamba-mvula (Ny) *Rauvolfia caffra*

mpambulu (Ny) *Carissa edulis/ Gymnosporia senegalensis*

mpanda-njovu (Ny) *Cussonia arborea/ Steganotaenea araliacea*

mpangala (M/ Ny) *Dichrostachys cinerea*

mpangala (Ny) *Acacia sieberiana*

mpange (Ku) *Bersama abyssinica/ Craibia affinis*

mpanzi (I/ K/ Le/ Tk) *Brachystegia spiciformis*

mpapa (Ny/ M) *Afzelia quanzensis/ Bauhinia petersiana*

mpapa (Ny/ Tu) *Brachystegia hockii/ B. spiciformis*

mpapa-dende (Tu) *Afzelia quanzensis*

mpasa (K/ Ny) *Bridelia duvigneaudi/ B. micrantha/ Julbernardia globiflora*

mpasu-pasu (Tu) *Hymenocardia acida*

mpata (Ny) *Euclea crispa*

mpatsa-cokolo (Ny) *Senna petersiana/ S. singueana*

mpatugila (Tk) *Hymenocardia acida*

mpavula (Ny) *Gymnosporia senegalensis*

mpazupazu (I) *Hymenocardia acida*

mpazyi (K) *Brachystegia spiciformis*

mpefu (Ny) *Albizia antunesiana/ A. anthelmintica/ Harungana madagascariensis/ Macaranga capensis/ Trema orientalis/*

mpela (Nm) *Adansonia digitata*

mpelele (Ny) *Dalbergia boehmii/ Ochna macrocalyx*

mpelesya (Ny) *Cissus cornifolia/ C. integrifolia/ Rhoicissus tridentata*

mpelu (Ny) *Gymnosporia senegalensis*

mpembu (Ny) *Parinari curatellifolia*

mpempwe (Ny) *Sida acuta/ Hymenocardia acida/ Triumfetta rhomboidea*

mpendo (Ny) *Euphorbia heterophylla*

mpenjere (Ny) *Ziziphus abyssinica*

mpepu (Ny) *Albizia antunesiana/ Trema orientalis*

mpereka (Ny) *Dalbergia boehmii*

mpesa (Ny/ Tu) *Parkia filicoidea*

mpesi (Tk/ To/ Tu) *Mucuna pruriens*

mpeska (Ny) *Cissus cornifolia*
mpeta (B/ Ku) *Afzelia quanzensis (seeds)*
mpetu (Ny) *Boscia angustifolia/ Erythroxylum marginatum/ Hippocratea parvifolia*
mpeuma (Ny) *Ochna macrocalyx*
mpewu (Ny) *Pavetta crassipes*
mpezi (Ku/ Ny) *Mucuna pruriens*
mpfilu (Ny) *Vangueria infausta*
mpicila-nyambo (Ny) *Margarita discoidea*
mpika (Ny) *Phyllanthus muelleranus*
mpika-maungu (Ny) *Antidesma venosum/ Senna petersiana*
mpila-kukuru (Ny) *Rhus longipes*
mpilila (Ny) *Ceiba pentandra*
mpili-pili (B/ I/ Lb/ Le/ M/ To/ Tu) *Capsicum annuum var. frutescens*
mpimbi-nyolo (Ny) *Psychotria zombamontana*
mpinda-nkwanga (Ny) *Dalbergia nitidula/ Dichrostachys cinerea*
mpindimbi (Ny) *Allophylus africanus/ Vitex doniana*
mpindule (Ny) *Cassipourea mollis*
mpingi (Nm) *Ximenia americana*
mpingo (Nm/ Ny) *Dalbergia boehmii/ D. melanoxylon*
mpingu (Nm) *Albizia anthelmintica*
mpingu (Ny) *Hoslundia opposita*
mpinji (Ny) *Terminalia sericea/ Vitex doniana/ Ximenia americana/ X. birrea*
mpinji-pinji (Ny) *Diospyros senensis/ Ximenia caffra*
mpira (Ny) *Calotropis procera/ Euphorbia heterophylla/ Landolphia parvifolia/ Manihot glaziovii/ Saba comorensis*
mpira-nkututu (Ny) *Allophylus africanus*
mpolata (Lo) *Kigelia africana*
mpolo (Ny) *Vangueria infausta*
mpoloni (Ny) *Steganotaenea araliacea*
mpolota (Lo) *Kigelia africana*
mpombo (Lo) *Brachystegia boehmii*
mponbona (Ny) *Securinega virosa/ Flueggea virosa*
mponda (Nm) *Commiphora africana/ C. ugogensis*
mponda-njovu (Ny) *Stereospermum kunthianum*
mponda-nya (M) *Pseudolachnostylis maprounifolia*
mponda-silu (Ny) *Holarrhena pubescens*
mpondesa (Ny) *Antidesma venosum*
mponjela (Ny) *Annona senegalensis*
mponjo (K) *Peltophorum africanum*
mporogo-bongore (Nm) *Albizia amara*
mposa (Ny) *Annona senegalensis*
mpotogola (Ny) *Antidesma venosum*
mpoto-poto (Ny) *Uapaca kirkiana*
mpo-uya (Ny) *Annona senegalensis*
mpovwa (B/ Ku/ Ny) *Annona senegalensis*
mpovya (Ny) *Annona senegalensis*
mpukuso (Ny) *Euclea crispa/ Gardenia subacaulis*
mpuluka (Ny) *Securidaca longipedunculata*

mpuluku-tutu (Ny) *Ehretia obtusifolia/ Vangueria infausta*
mpululu (Ny) *Antidesma venosum*
mpumbu (Ku) *Acacia sieberiana*
mpumbula (Nm) *Calotropis procera*
mpundu (B/ Ku/ Ny) *Parinari curatellifolia*
mpundu (Nm) *Strychnos innocua*
mpunga (Ny/Tu) *Oryza sativa*
mpungabwi (Ny) *Clausena anisata*
mpunga-ziwanda (Ny) *Clausena anisata*
mpungulira (Ny) *Antidesma venosum*
mpuyi (Ny) *Brachystegia spiciformis*
mpuzela-manzi (Ny) *Bridelia cathartica*
mruguyi (Nm) *Balanites aegyptica*
mrunda-runda (Nm) *Cassia abbreviata*
msabuwa (Ny) *Rauvolfia caffra*
msada (Nm) *Vangueria infausta*
msagadzinje (Ny) *Cassytha filiformis*
msagazi (Nm) *Commiphora africana*
msaka-dzinje (Ny) *Cassytha filiformis*
msale (Ny) *Brachystegia bussei/ B. hockii/ B. spiciformis*
msalinja (Tu) *Uapaca nitida*
msamba (Ny) *Brachystegia boehmi/ B. floribunda/ B. glaberrima/ B. longifolia*
msamba-mfu (Ny) *Sterculia quinqueloba*
msamba-mfumu (Ny) *Adenia gummifera/ Afzelia quanzensis/ Sterculia quinqueloba*
msamba-senya (Ny) *Clausena anisata*
msambi (Nm) *Senna singueana*
msambila (Nm) *Senna singueana*
msambirya (Nm) *Senna singueana*
msambi-sambi (Nm) *Senna singueana*
msambitsa-mkanda (Ny) *Harungana madagascariensis*
msanda (Ny) *conyza sp.*
msangalasa (Ny) *Harrisonia abyssinica/ Liquidambar styraciflua*
msanganya (Ny) *Antidesma venosum*
msangasi (Ny) *Allophylus africanus*
msangati (Nm) *Diplorhyncus condylocarpon*
msangu (Ny) *Faidherbia albida*
msangu-sangu (Ny) *Faidherbia albida/ V. amygdalina*
msangwa (Nm) *Stereospermum kunthianum*
msanjipila (Ny) *Maesa lanceolata*
msanjwambeke (Nm) *Crossopteryx febrifuga*
msanku-sanku (Ny) *Albizia glaberrima/ Bathriocline eupatorioides*
msanla-njazi (Ny) *Rourea orientalis*
msanyinja (Tu) *Uapaca nitida*
msanza (Nm) *Crossopteryx febrifuga*
msanza-mbeki (Nm) *Crossopteryx febrifuga*
msapa-tonje (Ny) *Jatropha curcas*
msarasi (Nm) *Grewia villosa*
msasa-mbeke (Nm) *Crossopteryx febrifuga*

msasaula (Ny) *Psychotria zombamontana*
msase (B/ Lb/ Ny/ Tu) *Albizia antunesiana*
msase (Ny) *Albizia anthelmintica/ Burkea africana*
msatsi (Ny) *Ricinus communis*
msatsi-manga (Ny) *Jatropha curcas*
msavala (Nm) *Sterculia africana*
msawa-sawa (Ny) *Allophylus africanus*
msawu (Ny) *Ziziphus abyssinica/ Z. mauritiana/ Z. spp.*
msawu-wathengo (Ny) *Ziziphus mucronata*
msayo (Ny) *Vernonia colorata*
mscera (Ny) *Pittisporum viridiflorum/ Uapaca nitida*
msebe (Ny) *Sclerocarya birrea*
msece (Ny) *Faurea delevoi/ F. speciosa/ Oncoba spinosa/ Passiflora edulis*
msecela (Ny) *Uapaca nitida*
msefu (Ny) *Zanha africana*
msekelia (Nm) *Antidesma venosum*
msekera (Nm) *Antidesma venosum*
msekese (Ny/ Tk/ To) *Piliostigma thonningi*
msele (Ny) *Gloriosa superba*
mselecele (Ny) *Uapaca nitida*
mseme (Ny) *Gloriosa superba*
msenda-luzi (Ny/ Tu) *Brachystegia allenii/ B. boehmii/ B. manga/ B. stipulata*
msenga (Nm) *Margaritaria discoidea*
msengu-lulu (Nm) *Strophanthus eminii*
msenjele (Ny) *Albizia glaberrima/ A. versicolor/ Pennisetum purpureum*
msenya (Ny) *Parkia filicoidea*
msera (Ny) *Antidesma venosum*
msere-cete (Ny) *Flueggea virosa*
msese (Lv/ Ln/ Lv) *Burkea africana*
msese (Ny) *Oncoba spinosa*
mseta-nyani (Ny) *Sterculia quinqueloba*
msewe (Ny/ Tu) *Sclerocarya birrea/ Oncoba spinosa*
msewe (Tu) *Brachystegia spiciformis*
msicisi (Ny) *Trichilia emetica*
msikizi (Ny/ Tu) *Trichilia emetica*
msikwipi (Ny) *Ocimum spp.*
msila-nyama (Ny) *Psorospermum febrifugum*
msimbiti (Ny) *Combretum imberbe/ C. molle/ C. psidioides/ Strychnos potatorum*
msimbiti (To/ Tu) *Ozoroa reticulata*
msindi (Nm) *Diospyros mespiliformis/ Pteleopsis aquilinum*
msinika (Ny) *Antidesma venosum*
msinyani (Ny) *Ocimum canum*
msinyika (Ny/ Tu) *Syzygium cordatum*
msisi (Nm) *Tamarindus indica*
msisisi (Ny) *Cissampelos mucronata/ Cocculus hirsutus*
msitoti (Ny) *Rourea orientalis*
msiwa-kalulu (Lv) *Strophanthus welwitschii*
msoko-lowe (Ny) *Uapaca nitida/ U. sansibarica*

msole (Ny) *Vangueriopsis lanciflora*
msolo (Ny/ Tu) *Pseudolachnostylis maprouneifolia*
msombo (Ny) *Syzygium cordatum/ S. guineense* sbsp. *afromontanum/ S. guineense* sbsp.
  barotsense/ S. guineense sbsp. *huillense/ S. guineense* sbsp. *macrocarpum*
msonde (Ny) *Sorghum spp.*
msondozi (Ny) *Salix mucronata*
msonga (Nm) *Diplorhynchus condylocarpon*
msonga-lukuga (Nm) *Holarrhena pubescens*
msonga-ti (Nm) *Diplorhynchus condylocarpon/ Holarrhena pubescens*
msongoro (Ny) *Strychnos innocua*
msongwa (Nm) *Harrisonia abyssinica*
msoyo (Ny) *Vernonia colorata*
msuce (Ny) *Dodonaea viscosa*
msuka-ana (Ny) *Rotheca myricoides*
msuka-cuma (Ny) *Fadogia ancylantia/ Phyllanthus ovalifolius/ Rourea orientalis/*
  Tacca leontopetaloides
msuka-mfuti (Ny) *Gymnosporia senegalensis*
msuku (Ny) *Uapaca kirkiana*
msuku-mpinini (M) *Uapaca nitida*
msulu (Ny) *Sphenostylis marginata*
msumba-mpungu (Nm) *Ocimum americanum*
msumbuti (Ny) *Brachystegia floribunda/ B. spiciformis*
msumwa (Ny) *Diospyros mespilliformis/ Euclea natalensis*
msunduzi (Ny) *Antidesma venosum*
msunga-wantu (Ny) *Stereospermum kunthianum*
msungu-lula (Nm) *Strophanthus eminii*
msunguru (Nm) *Strophanthus eminii*
msunguti (Nm) *Diplorhynchus condylocarpon*
msungwa (Ny) *Erythrina abyssinica*
msunta (Ny) *Cleome spp./ Cleome gynandra*
msuwa-suwa (Ny) *Harungana madagascariensis*
mswaculu (Ny) *Cleome gynandra*
mswake (Nm) *Salvadora persica*
mswele (Nm) *Hoslundia opposita*
msyagasyi (Nm) *Commiphora africana*
msyala-nkunzi (Ny) *Carissa edulis*
msyelele (Nm) *Hoslundia opposita*
msyinda-mbogo (Nm) *Piliostigma thonningii*
msyisyi (Nm) *Tamarindus indica*
msyowa (Ny) *Piliostigma thonningii*
mtalala (Ny) *Deinbollia xanthocarpa/ Gardenia volkensii/ Lecaniodiscus fraxinifolius/*
  Zanha africana
mtalala-njovu (Ny) *Stereospermum kunthianum*
mtalawanda (Ny) *Allophylus africanus/ Lecaniodiscus fraxinifolius/ Zanha africana*
mtambe (Ny) *Erythrina abyssinica*
mtana-wanjano (Ny) *Euclea natalensis*
mtanda (Ny) *Croton gratissimus*
mtanda-mpira (Ny) *Maesa lanceololata*

mtanda-nyerere (Ny) *Bridelia cathartica*
> **N.B.** There is some confusion between the various spellings of this name. See also mtanda-nyerere, mtanta-nyelele, mtanta-nyerere, mtantha-nyele, ntanda-nyere, ntanda-nyerere, ntanta-nylele, ntanta-yelele, tanta-nyele

mtanga-nga (Ny) *Albizia glaberrima*

mtanga-tanga (Ny) *Albizia anthelmintica/ A. adianthifolia/ A. versicolor/ Indigofera spp./ Parkia filicoidea*

mtangoma (Ny) *Pavetta crassipes*

mtanta (Ny) *Oncoba spinosa*

mtanta-mpete (Ny) *Ochna schweinfurthiana*

mtanta-nyelele (Ny) *Annona senegalensis/ Cassia abbreviata/ Senna petersiana/S. singueana*
> **N.B.** There is some confusion between the various spellings of this name. See also mtanda-nyerere, mtanta-nyelele, mtanta-nyerere, mtantha-nyele, ntanda-nyere, ntanda-nyerere, ntanta-nylele, ntanta-yelele, tanta-nyele

mtanta-nyerere (Ny) *Senna singueana*

mtantha (Ny) *Senna singueana (?)*

mtantha-nyele (Ny) *Phyllanthus ovalifolius*

mtatu (Ny) *Allophylus africanus/ Eriosema burkei/ Rhus longipes/ R. tenuinervis/ Rhus spp.*

mtawa (Ny) *Ficus glumosa/ F. wakefieldii/ Flacourtia indica/ Oncoba spinosa*

mtawe-tawe (Ny) *Senna singueana*

mtebe-lebe (Ny) *Combretum collinum*

mtebe-tebe (Ny) *Steganotaenia araliacea*

mteeyu (Nm) *Securidaca longipedunculata*

mtela (Nm) *Annona senegalensis*

mtelela (Nm) *Stereospermum kunthianum*

mtelezi (Ny) *Stereospermum kunthianum*

mtema (Ny) *Hymenocardia acida/ Flacourtia indica*

mtembe (Ny) *Strychnos spinosa*

mteme (Ku/ Ny) *Strychnos cocculoides/ S. innocua/ S. spinosa*

mtemuwa-syimba (Nm) *Withania somnifera*

mtendasiwa (Ny) *Plumbago zelanica*

mtengene (Ku/ Ny) *Ximenia americana*

mtengere (Ny) *Ximenia americana*

mtengo-wangoma (Ny) *Erythrina abyssinica*

mtere-vere (Ny) *Cayratia gracilis/ Senna petersiana/ S. singueana/ Dalbergiella nyasae*

mteru-teru (Ny) *Hymenocardia acida/ Pavetta crassipes*

mtesa (Ny) *Grewia micrantha/ Trema orientalis*

mteta (Ny) *Peltophorum africanum*

mtete (Ny) *Acacia nilotica/ A. polyacantha & spp./ Faidherbia albida*

mthambe (Ny) *Cissus integrifolia/ C. quadrangularis/ Combretum mossambicense*

mthika (Ny) *Solanum incanum*

mthula (Ny) *Solanum incanum/ S. richardii*

mtinda-mbogo (Nm) *Piliostigma thonningii*

mtindwa-mbogo (Nm) *Piliostigma thonningii*

mtobo (Nm) *Azanza garckeana*

mtombozi (Ny) *Diplorhynchus condylocarpon/ Holarrhena pubescens*

mtomoni (Lv/ Ny) *Diplorhynchus condylocarpon*

mtondwoko (Ny) *Sclerocarya birrea*

mtonga (Ny) *Croton megalobotrys/ Strychnos spinosa*
mtoowa (Ny) *Croton megalobotrys*
mtopa (Ny) *Annona senegalensis*
mtoto (Ny) *Ipomoea batatas/ Uapaca kirkiana/ U. nitida/ U. sansibarica*
mtowa (B/ Ku/ Ny/ Tu) *Diplorhyncus condylocarpon*
mtowa-mtombozi (Ny) *Diplorhyncus condylocarpon*
mtowa-silu (Ku/ Ny) *Holarrhena pubescens*
mtowu (Nm/ Ny) *Azanza garckeana*
mtsatsule (Ny) *Allophylus africanus*
mtse (To) *Stereospermum kunthianum*
mtsiloti (Ny) *Psorospermum febrifugum*
mtsitsila-manda (Ny) *Asparagus africanus*
mtuba-tuba (Ny) *Acacia sieberiana*
mtube-tube (Ny) *Acacia kirkii/ A. sieberiana/ Faidherbia albida*
mtudzu (Ny) *Flacourtia indica*
mtuka-mbako (Tu) *Ozoroa reticulata*
mtuku (Ny) *Imperata cylindrica*
mtuku-mpuku (Ny) *Pavetta schumanniana/ Ozoroa reticulata*
mtukutu (Ny) *Piliostigma thonningii*
mtula (Nm) *Solanum incanum/ S. panduraeforme*
mtulu-tulu (Ny) *Strychnos innocua*
mtumbali (Ny) *Pterocarpus angolensis*
mtumbati (M) *Pterocarpus angolensis*
mtumbu (Ny) *Kirkia accumulata*
mtumburuka (Ny) *Ximenia caffra*
mtumbuzya (Ny) *Flacourtia indica*
mtumbwi (Ny) *Kirkia acuminata*
mtumdulu (Nm) *Dichrostachys cinerea*
mtumu (Ny) *Harungana madagascariensis*
mtumuko (Ny) *Ekebergia benguelensis/ Zanha africana/ Z. golungensis*
mtunda-tunda (Nm) *Sesamum indicum*
mtundira (Ny) *Bridelia cathartica/ Garcinia buchananii*
mtundo (Nm) *Brachystegia spiciformis*
mtundo-molo (Nm) *Brachystegia spiciformis*
mtundu (Ny) *Ficus sycomorus/ F. vallis-choudae/ Salix subserrata/ Ximenia americana/ X. caffra*
mtundulu (Nm) *Dichrostachys cinerea*
mtundu-lukwa (Ny) *Ximenia americana/ X. caffra*
mtundu-waŋono (Ny) *Margarita discoidea*
mtundwa (Nm) *Ximenia americana*
mtunga-njila (Ny) *Diospyros senensis*
mtungba-bala (Ny) *Gymnosporia senegalensis*
mtungulu (Nm) *Senna singueana*
mtungu-jamita (Nm) *Solanum incanum*
mtungu-jamita (Nm) *Solanum incanum*
mtungu-lulu (Nm) *Strophanthus eminii*
mtungwisa (Ny) *Solanum incanum*
mtunguru (Nm) *Pseudolachnostylis maprouneifolia*

mtungusa (Nm) *Solanum incanum*

mtupa (Ny) *Gymnosporia senegalensis*

mtutu (Ku/ Ny) *Croton megalobotrys*

mtutu (Ny) *Cordia abyssinica/ C. pilosissima/Croton macrostachys/ Tephrosia vogelii*

mtutu-muko (Ny) *Bersama abyssinica/ Ekebergia capensis/ Rhoicissus revoilii/ Stereospermum kunthianum/ Xeroderris stuhlmannii*

mtuwe-tuwe (Ku) *Acacia sieberiana*

mtwate (Tu) *Tecomaria capensis*

mtwe-twe (Ny/ Tu) *Acacia sieberiana*

muambwe (To) *Ozoroa reticulata*

muandakati *(N)* Burkea africana

muanga (Lv) *Pericopsis angolensis*

muangi (Nm) *Cleome gynandra*

mubaba (B) *Keetia zanzibarica*

mubaba (Lo) *Piliostigma thonningii*

mubabaju (Nm) *Carica papaya*

mubabala (Lo/ N/ Tk) *Gymnosporia senegalensis/ Gymnospora heterophylla*

mubala (N) *Afzelia quanzensis*

mubalatobe (Lv) *Afzelia quanzensis*

mubale (Nm) *Philenoptera wankieensis/ Strychnos spinosa/ Xeroderris stuhlmann*

mubale (Nm) *Xeroderris stuhlmann*

mubalika (B) *Ricinus communis*

mubalusika (Lo) *Vernonia spp.*

mubambala (I) *Euphorbia tirucalli*

mubamba-ngoma (B/ Ku) *Balanites aegyptiaca*

mubambwa-ngoma (B) *Balanites aegyptiaca/ Flacourtia indica*

mubambwa-ngoma (B) *Flacourtia indica*

mubambwa-ngoma (I/ K) *Carissa edulis*

mubanga (Ln/ Lv) *Dichrostachys cinerea*

mubanga-lala (I) *Cordia goetzei*

mubayi (To) *Cussonia arborea*

mubemu (Lo) *Prorospermum spp./ P. febrifugum*

mubeza-munku (Tu) *Holarrhena pubescens*

mubila (Ln) *Strychnos spinosa*

mubilo (Lo) *Canthium burttii/ Vangueria infausta/ V. randii & spp.*

mubimbi (B/ I/ K/ To) *Rauvolfia birrea/ R. caffra*

mubita (Lo/ Tk/ To) *Boscia angustifolia/ B. salicifolia*

mubombo (I/ K/ Lb/ Ln/ Lo Tk/ To) *Brachystegia boehmi/ B. glaberrima/ B. gossweileri/ B. longifolia/ B. wangermeeana*

mubombo (I/ To) *Lannea discolor*

mubona-ntelemba (I) *Ricinus communis.*

mubongo (B) *Landolphia owariensis*

mubongo (N) *Sclerocarya birrea*

mubu (Lo) *Pseudolachnostylis maprouneifolia*

mubu (Tk) *Vangueria randii/ V. infausta*

mububa (Tk) *Vangueria infausta/ Albizia versicolor*

mububu (Lo) *Combretum collinum/ C. hereroense/ C. zeyheri*

mububu (To) *Albizia versicolor/ Vangueria infausta*

mububwa (Tk) *Lannea discolor*
mubuga (K *Diplorhynchus condylocarpon*
mubuge (K) *Diplorhynchus condylocarpon*
mubukusyu (Lo) *Flacourtia indica*
mubula (I/ Lo/ M/ Nm/ Tk/ To) *Parinari curatellifolia*
mubula (Ln/ To) *Uapaca kirkiana*
mubula-kusya (Lo) *Flacourtia indica*
mubuli (Lo) *Croton megalobotrys*
mubuli (Lo/ N) *Diplorhynchus condylocarpon*
mubulu-kusyu (Ln/ Lo/ N) *Flacourtia indica*
mubulwe-bulwe (Lo/ N) *Burkea africana*
mubulya (Lv) *Diplorrhynchus condylocarpon*
mubumbu (I/ Lb/ Ln/ Lo/ Ln/ N/ To) *Lannea discolor*
mubumbu-culu (Tk) *Lannea discolor*
mubungu-bungu (To) *Sterculia quinqueloba/ Ficus abutifolia*
mubungu-bungu (Tk) *Sterculia quinqueloba/ Albizia tanganyikensis*
mubungwe (Lo) *Senna obtusifolia*
mubuyu (B/ I/ K/ Ku/ Le/ Lo/ Ny/ Tk/ To/ Tu) *Adansonia digitata*
mubuzi (I) *Adansonia digitata*
mubwa-bwa (I/ K/ Lb/ Le/ Lo/ Tk/ To) *Commiphora africana/ C. pyracanthoides & spp.*
mubwa-bwa (To) *Commiphora africana*
mubwali-bwali (To) *Tecomaria capensis*
mubwanga (B) *Flueggea virosa*
muca (Lv/ Ln/ Lv) *Parinari curatellifolia*
mucaca (K/ Lb) *Securidaca longipedunculata*
mucakali (Tk) *Dichrostachys cinerea*
mucakwe (To) *Faidherbia albida*
mucaluwe (Lo) *Ziziphus abyssinica*
mucangwe (To) *Faidherbia albida*
muce (Nm) *Salvadora persica*
muceceni (Tu) *Ziziphus mucronata*
muceceti (Tk) *Ziziphus mucronata*
mucejete (To) *Ziziphus abyssinica*
muceme (I/ To) *Sorghum sp.*
mucenga (Ku/ To/ Tu) *Diospyros mespiliformis*
mucenja (B/ Lb/ Lo/ Ny/ Tu) *Diospyros mespiliformis*
mucenja (B/ Ku) *Diospyros mespiliformis*
mucenje (I/ Lb/ Lo/ Le/ To/ Tu) *Diospyros mespiliformis*
mucenje-mulonga (To) *Diospyros mespiliformis*
mucenje-te (To) *Ziziphus abyssinica*
mucenya-mbulo (B) *Bersama abyssinica*
mucesi (B/ To) *Faidherbia albida*
mucinga-zuba (Ku/ Ny) *Catunaregam obovata*
mucinje (Ln) *Adenia lobata*
mucoka (Ku) *Pavetta crassipes/ Vangueria infausta*
mucoka (To) *Syzygium cordatum*
muconfwa (To) *Ximenia americana*
mucongwe (To) *Albizia anthelmintica*

mucuncu-cuulu (To) *Ehretia obtusifolia*
mudengula (Ln) *Uapaca nitida*
muembembe (B) *Boscia angustifolia*
muenya (B) *Euclea racemosa*
mufifi (B) *Harungana madagascariensis/ Psorospermum spp.*
mufilu (Lb) *Psorospermum febrifugum*
mufinsa (B/ K/ Lb/ Le) *Syzygium cordatum/ S. guineense sbsp.afromontanum/ S. guineense sbsp. barotsense/ S. guineense sbsp. huillense/ S. guineense races*
mufitu-mbwe (B) *Euclea racemosa*
mufofo (Lo) *Nicotiana tabacum/ Oldfieldia dactylophylla*
mufombwa (B/ K/ Ln/ S) *Carissa edulis*
mufufu (To) *Securidaca longipedunculata*
mufufu (Lo) *Combretum collinum*
mufufuma (I/ K/Tk/ To) *Securidaca longipedunculata*
mufufuma (Lb) *Azanza garckeana*
mufuka (B/ K) *Combretum collinum/ C. adenogonium/ C. zeyheri & spp.*
mufula (Lo/ Lv/ N) *Combretum molle/ C. psidioides/ C. zeyheri*
mufulombo-mbo (Lv) *Ximenia caffra*
mufuma-ngoma (Lo) *Jatropha curcas*
mufumbe (B/ Lb/ M) *Bauhinia reticulata/ B. thonningii*
mufumbe-lele (Lv) *Diplorhynchus condylocarpon*
mufumfula (Tk) *Waltheria americana*
mufumpe (I) *Ehretia obtusifolia*
mufumpu (Ln) *Croton leuconeurus/ C. megalobotrys*
mufunda-balu (B) *Flueggea virosa*
mufunda-nzinzi (Ku/ Ny) *Brachystegia boehmii*
mufundu (To) *Pavetta crassipes*
mufunge (Lo) *Lannea edulis/ L. gossweileri*
mufungo (B/ K/ Lb/ Le/ Lo/ N) *Anisophyllea boehmii*
mufungo (Tk) *Gardenia volkensii*
mufungu (K/ Lb/ Le/ Lo/ N) *Anisophyllea boehmii*
mufungu (Lb) *Cassia abbreviata*
mufungu (To) *Xeroderris stuhlmannii*
mufungu-fungu (B/ Lo) *Kigelia africana*
mufungula (K/ Lb/ Le/ Ku/ Tu) *Kigelia africana*
mufungu-nasya (B) *Senna singueana*
mufuno-funo (Ln) *Kigelia africana*
mufupu (Lv) *Flacourtia indica*
mufuti (To) *Steganotaenia araliacea*
mufuwe (Lo) *Carissa edulis*
mufwate (I/ K/ To) *Entada abyssinica*
mufwe-fwe (I) *Acacia polyacantha subsp. campylacantha*
mufwe-fwe (I/ Lo/ Tk) *Acacia polyacantha*
mufwele (I/ Tk/ To) *Acacia polyacantha/ A. erubescens*
mufyelo (Lo) *Croton gratissimus/ C. pseudopulchellus*
mugando (Nm) *Albizia antunesiana/ Burkea africana/ Erythrophleum africanum*
mugende-gende (To) *Ehretia obtusifolia*
mugoga (I/ Le/ Lo/ To) *Acacia nilotica*

mugoga (I/ Le/ Lo/ Tk) *Acacia polyacantha*
mugololo (To) *Albizia glaberrima/ Philenoptera violacea*
mugona-mbwa (Ku/ Ny) *Manilkara mochisia*
mugongo (To) *Sclerocarya birrea*
mugoya (I) *Acacia polyacantha*
mugozya (Tu) *Sterculia africana*
mugu (Nm) *Acacia polyacantha*
mugugu (To) *Croton megalobotrys*
mugungwela (K) *Kigelia africana*
muhaka (Lo) *Uapaca kirkiana*
muhalapoli (Ln) *Smilax anceps*
muhamba (N) *Brachystegia spiciformis/ B. longifolia*
muhani (Lo) *Acacia mellifera*
muhasi (Nm) *Leonotis molissima*
muhlu (Lv/ Ny) *Syzygium cordatum*
muhongo (Lo) *Ehretia obtusifolia*
muhonsia (Nm) *Strychnos innocua*
muhoŋandumba (Ln) *Dalbergia nitidula*
muhota (Lv) *Parinari curatellifolia*
muhota (Lv) *Psorospermum febrifugum*
muhota-hota (Ln) *Psorospermum febrifugum/ Parinari curatellifolia*
muhoto (Ny) *Anisophyllea boehmii/ A. pomifera*
muhubu (I) *Sesbania sesban*
muhuka (N) *Uapaca kirkiana*
muhuku (Lo) *Uapaca kirkiana*
muhulu-hulu (Lo) *Strychnos cocculoides/ S. spinosa*
muhuma (Lv) *Combretum molle*
muhuma (Ln) *Strychnos spinosa/ S. welwitschii*
muhumu (Lo) *Manilkara mochisia*
muhundu (Nm) *Strychnos innocua*
muhungu (I) *Kigelia africana*
muhunu (To) *Xeroderris stuhlmannii*
muhwila (Ln) *Strychnos innocua/ S. spinosa/ Strophanthus welwitschii*
muja-gwe (To) *Faidherbia albida*
muja-minzi (Nm) *Combretum adenogonium*
mujejete (To) *Ziziphus abyssinica*
muka (Ny) *Boscia angustifolia*
mukakabwa (Lo) *Pupalia lappaca*
mukakane (Ln) *Acacia erubescens*
mukakane (Lo) *Acacia sieberiana/ Ziziphus spp.*
mukako (Lo) *Lannea discolor/ Erythrophleum africanum*
mukala (Lo) *Ziziphus mucronata*
mukala (Lv) *Pterocarpus angolensis/ Tricalysia angolensis*
mukala-la (To) *Albizia amara*
mukala-lusaka (To) *Zornia glochidiata*
mukala-ngo (Le) *Capparis tomentosa*
mukalo (Lo) *Ziziphus mucronata/ Z. abyssinica*
mukamba (I/ Lo/ Tk/ To) *Afzelia quanzensis*

mukambila (B) *Craterispermum schweinfurthii*
mukanane (Lo) *Ziziphus abyssinica*
mukanda-cina (Ln) *Rhus longipes*
mukanda-cine (K/ Ln) *Boscia angustifolia*
mukanda-njase (I/ To) *Dalbergia nitidula*
mukanda-ŋombe (I/ Le/ To) *Steganotaenia araliacea*
mukangala (I/ Le/ Lo/ To) *Albizia amara/ A. harveyi*
mukanunkila (Lo) *Croton gratissimus*
mukasala (Le) *Albizia amara*
mukasala (I) *Peltophorum africanum*
mukate (Lo) *Acacia sieberiana*
mukazi-hebe (Lo/ N) *Burkea africana*
mukazutabike (Tk) *Albizia anthelmintica/ A. antunesiana*
mukena (Lo) *Croton gratissimus/ C. pseudopulchellus*
mukena-nene (K) *Antidesma venosum*
mukenjenge (Lv) *Abrus precatorius*
mukenyenge (Lv) *Abrus precatorius*
muketu (Lo) *Steganotaenia araliacea*
muko (Ny) *Croton macrostachys/ C. megalobotrys*
mukoka (I/ K/ Le/ Lo/ To) *Acacia tortilis*
mukoka (I/ K/ Le/ Lo/ Tk/ To) *Acacia polyacantha*
mukoka (I/ Le) *Faidherbia albida*
mukoka (Le) *Clematis brachiata*
mukoka (Lo/ To) *Acacia fleckii/ A. ataxacantha/ A. nilotica*
mukoko (Lo) *Acacia mellifera*
mukoko (Tk) *Acacia polyacantha*
mukokola (Be) *Uapaca kirkiana*
mukole (B/ K/ Ku/ Le/ Ln/ Ny/ Tu) *Azanza garckeana/ Dombeya rotundifolia*
mukole (Lv) *Strychnos spinosa*
mukolo (B) *Smilax anceps*
mukolo (Lv) *Strychnos innocua*
mukolo (Lo) *Strychnos spinosa*
mukolo-bolo (B/ Ln) *Smilax anceps*
mukololo (B) *Gardenia ternifolia*
mukololo (I/ Le) *Albizia glaberrima*
mukolwa (N) *Parinari curatellifolia*
mukolwe (Lo) *Rourea orientalis*
mukoma (Lo) *Oncoba spinosa*
mukomba (I) *Albizia antunesiana/ A. glaberrima*
mukombe (To) *Strophanthus kombe*
mukomboolo (To) *Afzelia quanzensis*
mukome (B/ Lb) *Azanza garckeana/ Strychnos pungens*
mukomfwe (B) *Carissa edulis*
mukona (To) *Acacia nilotica*
mukona (I/ Le/ Lo) *Ziziphus abyssinica*
mukona-ngwe (Lo) *Capparis tomentosa*
mukondo-kondo (B) *Sapium ellipticum*
mukondyo-dyo (Tu) *Flacourtia indica*

mukonkola (K) *Uapaca kirkiana*
mukonkoto (Lo/ Tk) *Dalbergia martinii/ D. nitidula*
mukonkoto (Lo/ To) *Dalbergia nitidula*
mukono-mgwa (Lo) *Annona senegalensis*
mukono-nga (Lo) *Annona senegalensis*
mukono-ngwa (Lo) *Annona senegalensis*
mukonono (I) *Terminalia prunoides/ T. stuhlmannii/ Albizia glaberrima/ Philenoptera violacea/ P. nelsii*
mukonyombwa (To) *Annona senegalensis*
mukorongwe (To) *Capparis tomentosa*
mukosa (Lo/ To) *Sterculia africana/ S. quinqueloba*
mukose (I/ Lo/ To) *Sterculia quinqueloba*
mukoso (B) *Albizia antunesiana/ Burkea africana/*
mukoso (Lv) *Erythrophleum suaveolens*
mukoto-koto (Le/ Lo/ Tk/ To) *Acacia polyacantha*
mukoto-koto (Lo) *Albizia anthelmintica/ Acacia nilotica*
muku (B) *Ficus sur*
mukuba (B/ I/ K/ To) *Gymnosporia senegalensis*
mukubamwe (Lo) *Ocimum spp.*
mukube (Lv) *Abrus precatorius*
mukucumu (Lo) *Diospyros mespiliformis*
mukuka-mpombo (K/ Ln) *Strychnos innocua*
mukuku (Lo) *Uapaca kirkiana*
mukuku-biya (Lo) *Sterculia quinqueloba*
mukuku-buyu (Lo) *Sterculia africana*
mukuku-lama (I) *Combretum psidioides/ C. molle*
mukukuma (B) *Artabotrys brachypetalus/ Azanza garckeana*
mukula (K/ Lv/ Le/ Ln/ Lv/ To) *Pterocarpus angolensis*
mukula (To) *Diospyros mespiliformis*
mukula-kula (K) *Pterocarpus angolensis*
mukuli-kuli (Lo) *Clutia abyssinica*
mukuluka (Ny) *Securidaca longepedunculata*
mukulu-kayubule (B) *Sterospermum kunthianum*
mukulu-kumbe (Ny) *Adansonia digitata*
mukululu (Tk) *Sphenostylis marginata sbsp. erecta*
mukulu-mbisya (B) *Flacourtia indica/ Maytenus ovata*
mukulu-ngufia (Lb) *Flacourtia indica*
mukumbuzu (To) *Oncoba spinosa/ Englerophytum magalismontanum*
mukumbwa-ŋombe (K) *Acacia amythethophylla/ Entada abyssinica*
mukungu (I/ T) *Hibiscus sabdariffa*
mukunka-mpombo (Ln) *Strychnos innocua*
mukunku (Tk) *Pseudolachnostylis maprouneifolia*
mukunku (Lo) *Pseudolachnostylis maprounifolia*
mukunta-mpele (B) *Bridelia cathartica*
mukunthu (Tk) *Pseudolachnostylis maprounifolia*
mukunu (Lo) *Pseudolachnostylis maprounifolia*
mukunya-nsefu (Lo) *Steganotaenia araliacea*
mukunyu (Lo/ To) *Pseudolachnostylis maprounifolia*

mukunyu (B) *Ficus capreifolia/ F. stuhlmannii/ F. sur/ F. sycomorus/ F. wakefieldii & spp.*
mukunza (To) *Combretum collinum*
mukupa-ciwa (B) *Senna petersiana/ S. singueana*
mukupaimbu (Ku) *Lippia woodii/ Ocimum canum*
mukuru-guru (Tu) *Croton macrostachys*
mukusau (Ny) *Ziziphus mucronata*
mukusu (K) *Uapaca kirkiana / U. robynsii*
mukuta-bulongo (I) *Combretum collinum*
mukuta-bulongo (I/ To) *Combretum adenogonium*
mukuta-nkwaji (K) *Asparagus sp.*
mukute (B) *Syzygium cordatum*
mukuti (M/ Ny) *Brachystegia hockii/ B/ spiciformis*
mukuwa-kuwa (Lo) *Gloriosa superba*
mukuyu (B/ K/ Ku/ Lb/ Le/ Ln/ Lo/ N/ To/ Tu) *Ficus glumosa/ F. sur/ F. sycomorus & spp.*
mukuyu-kayubule (B) *Stereospermum kunthianum*
mukuza-ndola (Ku) *Senna singueana*
mukwa (I/ Lo) *Pterocarpus angolensis*
mukwadju (Tu) *Tamarindus indica*
mukwa-mbanziba (I) *Ziziphus abyssinica*
mukwanga (Lo) *Albizia antunesiana/ Faidherbia albida*
mukwanga-banziba (Lo) *Faidherbia albida*
mukwata (Ln/Lo) *Ziziphus abyssinica*
mukweyi-cuulu (Tk) *Ehretia obtusifolia*
mukwikwa (B) *Tricalysia pallens*
mukyengya (K) *Diospyros kirkii/ D. mespiliformis*
mulaka-njovu (Ny) *Stereospermum kunthianum*
mulala (Lo) *Elaeis guineensis/ Raphia farinafera/ Sclerocarya birrea*
mulala-bapalwe (Lo) *Combretum adenogonium*
mulala-lusya (B) *Clausena anisata*
mulalane (Tk) *Helinus integrifolius*
mulala-ntanga (B) *Albizia amara/ A. harveyi/ Entada abyssinica/ Nephrolepis undulata*
mulala-ntete (B) *Albizia amara/ A. harveyi/ Entada abyssinica*
mulalawa (Lo) *Helinus integrifolius*
mulalusya (B) *Allophylus africanus & spp./ Rhus longipes/ R. quartiniana*
mulama (B/ I/ K/ Ku/ Lb/ Le/ Lo/ M/ Ny/ S/ To) *Combretum adenogonium/ C. spp.*
mulama (B/ K/ Nm/ Ny/ To/ Tu) *Combretum molle*
mulamana (Lo/ Tk) *Combretum collinum/ C. adenogonium*
mulamata (N) *Piliostigma thonningii*
mulamata (Ln) *Combretum molle/ C. psidioides*
mulamba (Le/ Tu) *Lannea edulis and spp.*
mulamba-bwato (B) *Antidesma venosum*
mulambe (B/ Ku/ Ny) *Adansonia digitata*
mulandala (Nm) *Combretum collinum*
mulanga-mupati (To) *Combretum molle/ C. psidioides*
mulangu (Lo) *Oncoba spinosa*
mulasa-kubili (Ku) *Catunaregam obovata*
mula-tanga (B) *Entada abyssinica*
mule-baisya (Tk) *Rourea orientalis*

mulebe (B/ M) *Ximenia americana/ X. birrea/ X. caffra*
mulekwa-mbweni (Tk) *Tetradenia riparia*
mulele (B) *Opilia amentalea*
mulele (B/ I/ Ku/ Ny/ To/ Tu) *Sterculia africana/ S. quinqueloba/ S. triphaca*
mulele-nzombo (Ny) *Sterculia quinqueloba*
mulema-nanga (Lo) *Jatropha curcas*
mulemba (Le/ Lo/ N) *Kirkia acuminata*
mulemba-lemba (Lo) *Kirkia acuminata*
mulembe-lembe (Tu) *Cassia abbreviata*
mulembwe (Le) *Hibiscus sabdariffa*
mulembwe (Lo) *Waltheria americana*
mulemu (I) *Protea madiensis*
mulende (K) *Sterculia africana*
mulende (I/ K/ Ln/ To) *Sterculia quinqueloba*
mulenga (Lo) *Crossopteryx febrifuga*
mulengo (Ln/ Lo/ N) *Uapaca nitida/ U. sansibarica & spp.*
mulengula (Ln) *Uapaca nitida/ U. sansibarica*
mulenkanga To) *Margaritaria discoedia*
mulepula-mpapa (B) *Clerodendrum capitatum/ C. tanganyikense*
muleya-mbezo (Tk) *Crossopterix febrifuga*
mulezya (To) *Albizia tanganyicensis/ Sterculia quinqueloba*
muli (Ln/ Lv) *Diplorhynchus condylocarpon*
mulilela (Ln/ Lo/ Tk) *Ozoroa reticulata*
mulilila (I/ Ln/ Lo/ Tk/ To) *Ozoroa reticulata*
mulimbo (Tu) *Diplorhynchus condylocarpon*
mulimpwinini (I/ To) *Cussonia arborea*
mulinde (Lv) *Stereospermum kunthianum*
mulobe-lobe (Le) *Uapaca nitida*
mulokwambula (Lo) *Clutia abyssinica/ Diospyros abyssinica/ D. virgata*
mulolo (Lo) *Lannea edulis*
mulomba (Ny) *Pterocarpus angolensis*
mulombe (I) *Xeroderris stuhlmann*
mulombe (Lb/ Lo/ N/ Ny/ Tk/ Tu) *Pterocarpus angolensis*
mulombwa (B/ Ku/ Ln/ Lo/ Ny/ Tu) *Pterocarpus angolensis*
mulome (Ku) *Pterocarpus angolensis*
mulombe (I) *Xeroderris stuhlmannii*
mulombwa (Lo) *Pterocarpus angolensis*
mulondwe (B) *Combretum molle & sp.*
muloolo (I/ K/ Lb/ Lo/ Lv/ N/ S/ Tk) *Annona senegalensis*
mulosio (Tu) *Psorospermum febrifugum*
mulota (I/To) *Euphorbia tirucalli*
muluja (Nm) *Combretum adenogonium*
muluka (B) *Securidaca longipedunculata*
mulula (Lo) *Sclerocarya birrea*
mulula-lula (Lo) *Flacourtia indica*
mululu (B/ I/ K/ Ku/ Lb/ Le/ Lo/ Ny/ Tk/ To) *Khaya anthotheca*
mululu (K) *Cassia abbreviata*
mululwa-lwa (B) *Anthocleista schweinfurthii*

mululwe (Tk/ To) *Trema orientalis*

mululwe (I/ K/ Le/ Lo/ S/ Tk/ To) *Cassia abbreviata/ Senna singueana*

mulumba-lumba (I) *Phyllocosmus lemaireanus*

mulumbe (N) *Pterocarpus angolensis*

mulumbwe (B/ Lb) *Phyllocosmus lemaireanus*

mulumbwe-lumbwe (B/ K/ Ln/ Lo) *Phyllocosmus lemaireanus*

mulumira (Ny) *Bridelia cathartica*

mulunda (Lo) *Capsicum annuum var. frutescens/ Solanum melongena*

mulundu (I/ Lb/ To/ Tu) *Swarzia madagascariensis*

mulundu (I/ K/ To) *Uapaca nitida*

mulundu-lukwa (Ny) *Ximenia caffra*

mulundu-ngoma (N) *Gymnosporia senegalensis*

mulunga-biba/ wiwa (B) *Salix mucronata*

mulungi (B/ Lb) *Strychnos innocua*

mulungi-kome (B) *Strychnos innocua*

mulungu-culu (Ny) *Zanthoxylum chalybeum/ Z. gilletii*

mulunguti (B/ N/ Ny/ To) *Erythrina abyssinica/ E. excelsa/ E. tomentosa & spp.*

mulungwa (Lo) *Peltophorum africanum*

mulutuluha (Lo) *Ximenia americana/ X. birrea/ X. caffra*

mulutulwa (Lo) *Ximenia caffra*

muluza (Ny) *Senna singueana/ Dalbergiella nyasae/ Pouzolzia mixta*

mulwa (Lo) *Sorghum sp.*

mulwa-bensu (To) *Crossopteryx febrifuga*

mulya (Ln/ Lo) *Diplorhynchus condylocarpon*

mulya (Ln) *Holarrhena pubescens*

mulya-balisyina (I/ To) *Capparis tomentosa*

mulya-lamana (Lo) *Asparagus africanus*

mulya-mbesu (K/ Lo/ Tk/ To) *Crossopteryx febrifuga*

mulya-mbesyi (Ku/ Ny) *Zanha africana*

mulya-mbyo (To) *Crossopteryx febrifuga*

mulya-nduba (B) *Artabotrys monteiroae/ Xylopia katangensis*

mulya-nkanga (To) *Margaritaria discoidea*

mulya-nsyinga (Ln) *Lannea edulis*

mulya-nzovu (To) *Balanites aegyptiaca*

mulya-ŋombe (I) *Entada abyssinica/ Faidherbia albida*

mulya-pwele (I/ To) *Phyllocosmus lemaireanus*

mulya-sefu (Ln/ Lv) *Phyllanthus muelleranus*

mulya-tongo (B) *Tecomaria capensis*

mulya-walisyaka (Ln) *Holarrhena pubescens*

mulyi (Ln) *Diplorhynchus condylocarpon*

mulyozya (I/ To) *Sterculia africana*

mumanga (Lv) *Brachystegia bakeriana/ B. spiciformis*

mumanga (K/ Ln/ To) *Brachystegia hockii/ B. spiciformis*

mumbombwe (To) *Boscia angustifolia*

mumbondo (Ny) *Combretum molle*

mumbovwa (Tk) *Acacia sieberiana*

mumbovwa (Lo) *Acacia gerrardii/ A. luederitzi/ A. robusta*

mumbu (Tk) *Lannea edulis*

mumbu (To) *Lannea schweinfurthii/ Acacia polyacantha*

mumbukusyu (Lo) *Ziziphus mucronata*

mumbulu-mbunye (B) *Lannea edulis*

mumbwe (B) *Psychotria kirkii*

mumbwe-munono (B) *Psychotria kirkii*

mumbwe-ngelenge (Lo) *Vangueria infausta & spp.*

mumfene (Tk) *Acacia polycantha*

mumfumbe (B) *Piliostigma thonningii*

mumfumpu (Lv) *Croton megalobotrys*

mumfumpu (Ln) *Flacourtia indica*

mumina-nzoka (Lo/ Tk) *Cissus cornifolia/ Rhoicissus tridentata*

mumono (B) *Ricinus communis*

mumpangala (B/ Ny) *Dichrostachys cinerea/ Hibiscus diversifolius*

mumpangala (To) *Acacia sieberiana*

mumpanse (I) *Brachystegia spiciformis*

mumpele-fwa (To) *Steganotaenia araliacea*

mumpele-mpemfwa (I/ To) *Steganotaenia araliacea*

mumpele-mpempe (I) *Hymenocardia acida*

mumpempe (I/ To) *Hymenocardia acida*

mumpilipifwa (I) *Steganotaenia araliacea*

mumpola-lundu (Tk) *Erythrophleum suaveolens*

mumundu (Nm) *Strychnos innocua*

mumvunyu (To) *Zanha africana*

mumvuzya-syoti (To) *Combretum adenogonium*

mumwaa (Lo) *Maprounea africana*

mumwa-meme (Lv) *Sapium ellipticum*

mumwaya (Tk) *Maprounia africana*

munanana (Lo) *Sorghum sp.*

munanga (Ku/ Ny) *Croton megalobotrys*

munawe (B) *Ochna pulchra/ O. schweinfurthiana & spp.*

muncimbili (B) *Synsepalum passargei*

mundabwe (Lo/ To) *Hibiscus sabdariffa*

mundale (N) *Swarzia madagascariensis*

mundalilo (I) *Dalbergia boehmii*

mundambi (Lo/ To) *Hibiscus sabdariffa*

mundenda (N) *Azanza garckeana*

mundu-kundu (Lv) *Syzygium cordatum*

mundumbwa-kavaya (Ln) *Combretum collinum*

muneko (Lo/ Tk/ To) *Azanza garckeana*

munembwa (B) *Ozoroa reticulata*

mundendha (N) *Azanza garckeana*

mungai (To) *Croton gratissimus*

munga-lunsi (Lb) *Acacia sieberiana*

munganga (Tk) *Acacia mellifera*

munganga (Tu) *Croton megalobotrys*

mungantube-tube (B/ Ku) *Acacia sieberiana*

munga-nuinci (B) *Acacia sieberiana*

munga-nunsyi (B) *Acacia polyacantha/ Faidherbia albida*

munga-nunsyi (B/ Ku) *Acacia sieberiana*
mungao-musweu (Lo) *Acacia sieberiana*
mungao-munsu (Lo) *Faidherbia albida*
munga-senge (B) *Gymnosporia senegalensis*
munga-siya (I) *Acacia erioloba/ A. tortilis/ Faidherbia albida*
munga-tuba (To) *Acacia seyal/ A. sieberiana*
mungayi (Ku/ Le/ Ny/ To) *Vangueria infausta*
mungele (I/ To) *Peltophorum africanum*
mungolo (Lo/ Tk) *Strychnos innocua*
mungomba (Ny) *Ximenia caffra*
mungonga (Lv) *Acacia seyal*
mungongo (To) *Sclerocarya birrea*
mungo-syia (I/ To) *Faidherbia albida*
mungo-tuba (I/ Le) *Acacia sieberiana*
mungu (Nm) *Erythrina abyssinica*
munguli (Lo) *Kigelia africana*
mungwa-malembe (I) *Ziziphus abyssinica*
mungwena (B) *Acacia sieberiana/ Mimosa pigra*
muni-minga (Lo) *Bridelia cathartica/ B. duvigneaudii & spp.*
munina-mpuku (Ln) *Senna singueana*
muninga (Lb/ Ln) *Arachis hypogaea*
muniya (K) *Xeroderris stuhlmann*
munjindu (Lo) *Erythrina abyssinica*
munjinua (Nm) *Hoslundia opposita*
munkanda (K) *Annona senegalensis*
munkanda-ŋombe (I/ Le) *Steganotaenia araliacea*
munka-ngala (Tk) *Combretum molle*
munka-nsyimba (B) *Sapium ellipticum*
munkila (Tk/ To) *Acacia sieberiana*
munkongi (Lo) *Abrus precatorius*
munkulungu (I) *Ehretia obtusifolia*
munkulu-nkulu (K/ Ln) *Strychnos cocculoides/ S. pungens/ S. spinosa*
munkulwaiba (K) *Strychnos spinosa*
mukuyu-cuulu (I) *Ehretia obtusifolia*
munkwa (Le) *Acacia seyal*
munombe-nombe (E) *Synsepalum brevipes*
munondo (K) *Diospyros mespiliformis*
munondwe *(B) Combretum collinum/ C. mossambicense & spp.*
munongo (To) *Sclerocarya birrea*
munonya-nansi (To) *Syzygium cordatum*
munozya-meenda (To) *Phyllocosmos lemaireanus*
munsangwa (To) *Oncoba spinosa*
munsansa (Lo) *Ampelocissus africanus & spp.*
munselemba (B) *Synsepalum brevipes*
munsoka-nsoka (B) *Cassia abbreviata/ S. singueana*
muntalembe (B/ M) *Diplorhynchus condylocarpon*
muntamba (I/ N/ Tk/ To) *Strychnos spinosa*
muntanoba (Tk/ To) *Strychnos spinosa*

muntengwa (To) *Kigelia africana*
munthu-kutu (Tu) *Piliostigma thonningii*
munto (I/ To) *Diplorhynchus condylocarpon*
muntowa (I/ To) *Diplorhynchus condylocarpon*
muntowo (Tk/ To) *Diplorhynchus condylocarpon*
muntu-fita (B) *Diospyros batocana/ Euclea natalensis*
muntu-kufita (K) *Diospyros batocana/ Euclea natalensis*
muntu-ndulu (To) *Xeroderris stuhlmannii*
muntu-ngwa (Ku/ Ny) *Boscia angustifolia*
munumbi-numbi (B) *Englerophytum magalismontanum/ Synsepalum brevipes*
munumka-nsyimba (B) *Senna singueana*
munundu (I/ Tk/ To) *Uapaca nitida*
mununka (B) *Ocimum americanum*
mununka-calisya (To) *Phyllanthus muelleranus*
mununka-nsyimba (B) *Cassia abbreviata/ Strobilanthopsis linifolia*
munwa-mema (Lo) *Bridelia cathartica/ B. duvigneaudii*
munya-nyamenda (K) *Antidesma venosum*
munyango (Lb) *Ochna schweinfurthiana*
munyanya-manzi (To) *Bridelia cathartica*
munyanya-menda (To) *Bridelia cathartica*
munyawasyi (M) *Harungana madagascariensis*
munyeele (Lo/ Tk/ To) *Peltophorum africanum*
munyele (To) *Entada abyssinica*
munyenye (Nm) *Xeroderris stuhlmannii*
munyenye (To) *Acacia robusta*
munyenyene (To) *Acacia robusta*
munyenyengwe (To) *Acacia robusta/ A. gerrardii*
munyo-ngolwa (Lo) *Carissa edulis*
munyo-nyamanzi (I) *Antidesma venosum*
munyumbu (Tu) *Abrus precatorius*
munzanga (Tu) *Albizia anthelmintica*
munze-neze (Lo) *Oncoba spinosa*
munze-nze (Lo) *Oncoba spinosa*
munzi (Ny) *Hibiscus sabdariffa*
munzoka (Nm) *Cassia abbreviata*
munzumina (Tk) *Sterculia africana*
muŋomba (I/ K/ Lb/ Le/ S/ Tk/ To) *Ximenia caffra*
muŋomba (K/ Lo/ Tk/ To) *Ximenia americana*
muŋomba-gomba (Lo) *Ximenia americana*
muŋomba-ŋomba (N) *Ximenia americana/ X. caffra*
muŋombwa (To) *Ximenia caffra*
muŋonga (Lv) *Acacia xanthophloea*
muŋoongwa (To) *Lannea discolor/ Schinziophyton rautanenii*
mupa (B/ M) *Hallea stipulosa*
mupaapa (I/ To) *Afzelia quanzensis*
mupafu (Lo) *Stereospermum kunthianum*
mupako (Lo) *Diospyros mespiliformis*
mupalapati (B) *Syzigium cordatum*

mupangala (B) *Dichrostachys cinerea/ Hibiscus diversifolius*
mupansi (I) *Brachystegia spiciformis*
mupapa (B/ K/ Ku/ Lb/ M/ Ny/ To/ Tu) *Afzelia quanzensis*
mupapa (Ny/ To) *Brachystegia spiciformis*
mupa-pala (Le) *Securidaca longipedunculata*
mupapi (B/ Ku/ Ny) *Securidaca longipedunculata*
mupapi (Lo) *Capparis tomentosa*
mupasa (B) *Maprounea africana*
mupasupasu (To) *Hymenocardia acida*
mupazupazu (I/ To) *Hymenocardia acida*
mupela (Nm) *Adansonia digitata*
mupelewa (To) *Steganotaenia araliacea*
mupembe (N) *Hymenocardia acida*
mupempe (B/ Ln/ Lo/ M) *Hymenocardia acida*
mupepe (Ln/ Lo/ Lv) *Hymenocardia acida*
mupesya (Ku) *Ampelocissus africanus*
mupetwa-lupe (B) *Phyllanthus ovalifolius/ P. muelleranus*
mupetwalupe-unono (B) *Phyllanthus ovalifolius*
mupheka (Tk) *Steganotaenia araliacaea*
mupiti (Lo) *Abrus precatorius*
mupiti-piti (Lo/ Tk) *Abrus precatorius*
mupolata (Lo) *Kigelia africana*
mupolota (Lo) *Kigelia africana*
mupombo (Lo) *Brachystegia boehmii*
mupopolo (Lv) *Uapaca kirkiana*
mupoposya (To) *Excoecaria bussei/ Sterculia africana*
mupuci (Ln/ Lv/ N) *Brachystegia hockii/ B. spiciformis*
mupula (B/ Ku) *Salix subserrata*
mupula-mpako (B) *Ficus stuhlmanni/ Flacourtia indica*
mupulu-kuswa (B) *Flacourtia indica*
mupumbu (Ku) *Acacia sieberiana*
mupumena (I) *Kirkia acuminata*
mupumu-lenge (Lo) *Vangueria infausta*
mupundu (B/ K/ Ku/ Lb/ Le/ Ln/ Ny/ Tu) *Parinari curatellifolia*
mupundu-kaina (Ln/ Lv) *Ziziphus abyssinica*
mupunga ((B/Lb/M/Ny) *Oryza sativa*
mupungo-pungo (Lo) *Kigelia africana*
mupupu (I/ Ln/ Lv) *Croton megalobotrys*
mupuruputa (Ny) *Gymnosporia senegalensis*
muputi (Lo/ N/ Tk) *Brachystegia spiciformis*
muputu (B/ Lb/ Le/ Ln/ N) *Brachystegia spiciformis*
mupykakulu (B) *Cassipourea mollis/ Pterocarpus angolensis*
muraka (Ny) *Securidaca longepedunculata*
muramo (Ny) *Stereospermum kunthianum/ Combretum adenogonium/ Philenoptera bussei*
murima (Ny) *Combretum adenogonium*
murugara (Nm) *Acacia mellifera*
muruka (Tu) *Securidaca longepedunculata*
muruzya-minzi (Nm) *Combretum collinum*

musafwa (B) *Syzygium cordatum & spp.*
musakala (M) *Zanha africana*
musaka-lala (K/ To) *Oncoba spinosa*
musaka-yaze (Lv) *Dalbergia nitidula (?)/ Albizia anthelmintica*
musalya (Lv) *Pseudolachnostylis maprounifolia*
musamba (B/ I/ K/ Lb/ Le/ Ln/ Ny/ S) *Brachystegia boehmii*
musamba (I/ Le/ Lo/ Tk/ To) *Lannea discolor*
musamba-mfwa (B) *Antidesma venosum/ Cassia abbreviata/ Phyllanthus muelleranus*
musamba-mfwa (B/ Ku) *Senna petersiana/ S. singueana*
musamba-mfwa (K) *Afzelia quanzensis*
musamblya (Nm) *Senna singueana*
musambo (K) *Syzygium cordatum*
musambwe (To) *Ozoroa reticulata*
musambya (B) *Synsepalum brevipes*
musambya (Lo) *Ximenia caffra*
musandalima (To) *Flueggea virosa*
musanga-lemba (Lb) *Zanha africana*
musanga-lwembe (Lo) *Zanha africana*
musangati (B/ M) *Pseudolachnostylis maprouneifolia*
musango (Lv) *Capparis tomentosa*
musangu (B/ Ku/ Ny/ To/ Tu) *Faidherbia albida*
musangula (B) *Bridelia cathartica/ Maesa lanceolata*
musangula (M) *Rhus longipes*
musangu-sangu (Lo) *Salix mucronata*
musangu-sangu (Tk) *Oncoba spinosa*
musansa (To) *Boscia angustifolia*
musansi (To) *Abrus precatorius/ Cussonia arborea*
musanta (I/ K/ To) *Kirkia acuminata*
musasa (B/ I) *Boscia angustifolia/ B. salicifolia*
musase (Tu) *Burkea africana*
musase-mamuna (Tu) *Burkea africana*
musatha (Tk) *Kirkia acuminata*
musato (Le) *Cordia goetzei*
musaviji (U) *Abrus precatorius*
musawu (Lo) *Ziziphus abyssinica*
musaye (B) *Strychnos pungens/ S. spinosa*
musayi (B) *Clerodendrum capitatum/ Strychnos cocculoides/ S. spinosa*
musebe (B/ Ku/ Ny) *Sclerocarya birrea*
museese (Le) *Burkea africana*
musekese (I/ Le/ Lo/ Ny/ Tk/ To/ Tu) *Piliostigma thonningii*
museke-seke (Lo) *Senna occidentalis*
musele (M) *Sclerocarya birrea*
musele-sele (Lo/ N/ Ny/ To) *Dichrostachys cinerea*
musenga-meno (B/ Le/ To) *Ochna pulchra/ Phyllocosmos lemaireanus*
musengwa (Lo) *Diplorhynchus condylocarpon*
musenje (Lo) *Gymnosporia senegalensis/ Gymnospora heterophylla*
musense (Ku/ Ny) *Oncoba spinosa*
musense (Ln) *Acacia polyacantha/ Combretum zeyheri*

musenu (Lo) *Peltophorum africanum*

musenya (To) *Burkea africana/ Faurea rochetiana*

musenze (Le) *Entada abyssinica*

musenzenze *Entada abyssinica*

musepa (B) *Erythrophleum suaveolens/ Parkia filicoidea*

musese (I/ Lo/ Ln/ Lv) *Burkea africana*

musese (Tu) *Faurea rochetiana/ Oncoba spinosa*

musese-wezenzele (Ln) *Burkea africana*

musesye (Ln/ Lo/ To) *Burkea africana*

musewe (B/ Ku/ Ny/ Tu) *Sclerocarya birrea*

musewe (I/ Lo/ Tk/ To) *Brachystegia spiciformis*

musigili (I/ Lo/Tk/ To) *Trichilia emetica*

musiika (I/ To) *Tamarindus indica*

musikili (Lo/ Tk/ To/ Tu) *Trichilia emetica*

musikisyi (B/ K) *Trichilia emetica*

musikizi (B/ Ku/ Ny/ Tu) *Trichilia emetica*

musimbilili (Lb) *Strychnos spinosa*

musipa (Ln) *Erythrophleum suaveolens*

musisisyi (Tu) *Cocculus hirsutus*

musiwe (Tk/ To) *Brachystegia spiciformis*

musoka-nsoka (B/ Ku) *Cassia abbreviata*

musoko-bele (K) *Uapaca nitida/ U. sansibarica*

musoko-lowe (B/ K/ Ku/ Lb/ Le/ Ny) *Uapaca nitida*

musoko-lowe-wafita (B/ Tu) *Uapaca nitida*

musole (K) *Pseudolachnostylis maprounifolia*

musole (Lo) *Antidesma venosum*

musolo (B/ Ny/ Tu) *Pseudolachnostylis maprouneifolia*

musolo-koto (To) *Xeroderris stuhlmannii*

musolo-solo (To/ Tk) *Abrus precatorius*

musoma (B) *Sorghum sp.*

musoma-njaro (Nm) *Harrisonia abyssinica*

musomba (Ln) *Syzygium cordatum*

musombo (B) *Brachystegia longifolia/ B. velutina/ B. wangermeeana/ Julbernardia globiflora/ Maytenus cymosus/ Oncoba spinosa/ Ozoroa reticulata*

musombo (I/ K/ Le/ Ln/ Lv/ To/ Tu) *Syzygium cordatum/ S. guineense sbsp. afromontanum/ S. guineense sspp./ S. owariense*

musombo-lombo (Ln) *Ximenia caffra*

musombu (Ln) *Syzygium cordatum/ S. guineense*

musombwani (To) *Hoslundia opposita*

musompe (I/ K/ To) *Salix mucronata*

musompe (Tu) *Syzygium cordatum/ S. guineense sspp.*

musomvwa (To) *Ximenia americana*

musondozi (Ku/ Ny) *Salix mucronata*

musongo-songo (Lv) *Ximenia caffra/ Englerophytum magalismontanum*

musongwa-songwa (Lo/ Ln) *Ximenia americana*

musoŋa-soŋa (Ln) *Olax obtusifolia/ Ximenia americana*

musosa (To) *Faidherbia albida*

musoso (B) *Protea madiensis*

musosu (Ln) *Boscia angustifolia*
musoyo (I/ Lb) *Boscia angustifolia/ B. salicifolia*
musozi-wabeembela (To) *Phyllocosmus lemaireanus*
musozya-wabasibombe (Tk) *Psychotria kirkii*
mustard tree (E) *Salvadora persica*
musuci (Ku/ Ny) *Salix mucronata*
musugu (Lo) *Uapaca kirkiana*
musuku (B/ I/ K/ Lb/ Lo/ Ny/ To/ Tu) *Uapaca kirkiana*
musuku (B/ Ny) *Uapaka nitida*
musungula (To) *Erythrina abyssinica*
musunsu-culu (To) *Ehretia obtusifolia*
musunta (M) *Cleome gynandra*
mususulu (B?) *Sorghum sp.*
mususwa (B) *Philenoptera violacea/ Rhus longipes*
muswamawa (Le) *Xeroderris stuhlmannii*
muswati (I) *Entada abyssinica*
muswati (I) *Dalbergia boehmii*
muswe (Lo) *Sorghum sp.*
muswe-benga (Lo) *Acacia nilotica/ A. pilispina*
muswesya-namo (B) *Harungana madagascariensis*
musyaka-syala (Lo/ N/ Tk) *Swarzia madagascariensis*
musyale (Lo) *Dichrostachys cinerea*
musyande (Lo) *Peltophorum africanum*
musyasya (Lb) *Pavetta crassipes*
musyeke-sye (Lo) *Piliostigma thonningii*
musyembe (Lo) *Combretum adenogonium*
musyesye (I/ Le/ Lo/ Tk/ To) *Burkea africana*
musyewe (I) *Brachystegia spiciformis*
musyika (Ku/ Ny) *Tamarindus indica*
musyikili (I/ Ku/ Ny) *Trichilia emetica*
musyikisyi (B) *Trichilia emetica*
musyikisyi-culu (B) *Khaya anthotheca*
musyiko (Tk) *Adenia gummifera*
musyikisyi (B) *Trichilia emetica*
musyimbilili (Lb) *Strychnos spinosa*
musyingu (B) *Syzygium cordatum/ S. guineense sbsp. afromontanum*
musyipa (B) *Erythrophleum suaveolens*
musyisyi (B/ Ny) *Tamarindus indica*
musyisyi (Tu) *Cocculus hirsutus*
musyonga-wasyonga (Lo) *Ximenia caffra*
musyongwa-syongwa (Lo/ N) *Ximenia americana/ X. caffra*
musyosya (Lo) *Boscia angustifolia & spp.*
musyuku (I/ K/ Lb/ Tk/ To) *Uapaca kirkiana*
musyukusu (Ln) *Uapaca kirkiana*
mutaba (To) *Kirkia acuminata*
mutaba-nzovu (I) *Sclerocarya birrea*
mutaba-taba (Le) *Kirkia acuminata*
mutabo (Le) *Azanza garckeana*

mutaka (Lo) *Phragmites australis/ P. mauritianus*
mutala (To) *Maclura africana/ Terminalia spp. (spinose)*
mutalala (To) *Lecaniodiscus fraxinifolius/ Zanha africana*
mutalu (Lo) *Grewia villosa*
mutamba (Tk/ To) *Strychnos spinosa*
mutanda-fiwa (B) *Acacia sieberiana*
mutanda-myoba (To) *Asparagus sp.*
mutanda-viwanda (Tu) *Croton gratissimus*
mutanta (Ku) *Oncoba spinosa*
mutanta-nsange (B) *Antidesma vogelianum/ Bridelia micrantha/ Sapium ellipticum/*
    Coptosperma neurophyllum/ Tricalysia congesta/ T. nyassae
mutanta-nyelele (Ku) *Antidesma venosum/ Senna singueana*
mutantwa-sokwe (I) *Ziziphus abyssinica*
mutata (Lv/ Ln/ Lv) *Securidaca longipedunculata*
mutata-mbeba (To) *Dalbergia nitidula*
mutata-mbululu (Ln) *Dalbergia nitidula*
mutebe-tebe (B/ Le/ Tu) *Steganotaenia araliacea*
mutebiti (B) *Steganotaenia araliacea*
muteje (K) *Holarrhena pubescens*
muteju (Nm) *Securidaca longipedunculata*
mutelela-nzovu (Ku) *Cleistochlamys kirkii/ Stereospermum kunthianum*
mutelele (B/ Ku) *Corchorus olitorius*
muteme (I/ Le/ Lo/ To/ Tu) *Strychnos innocua*
mutenda-nkwale (B) *Gymnosporia senegalensis & spp.*
mutendefwa (B) *Maesa lanceolata*
mutengu-musyamba (Lo) *Rourea orientalis*
mutente (K/ Lb/ Lo/ Tk/ To) *Ximenia americana/ X. birrea*
mutese (I/ To) *Stereospermum kunthianum*
mutesi (I/ To) *Stereospermum kunthianum*
mutesu (Lb) *Salix mucronata*
mutesye (I/ To) *Stereospermum kunthianum*
mutesyu (I/ To) *Stereospermum kunthianum*
muteta (Ny) *Peltophorum africanum*
mutete (Lv) *Swarzia madagascariensis*
mutete (Ln/ Lv) *Hibiscus sabdariffa*
mutewe-tewe (B) *Cussonea arborea*
muthamba (Tk) *Strychnos spinosa*
muthithi (Tk) *Erythrina abyssinica*
muthoo (Tu) *Diplorhyncus condylocarpon*
muthunga-phazi (Tk) *Gymnosporia senegalensis*
mutika-lamo (B) *Zanha africana*
mutimba-hula (I/ To) *Flacourtia indica*
mutimba-mvula (I/ To) *Flacourtia indica*
mutimba-vula (Tk) *Entada abyssinica*
mutimbula (Lo) *Flacourtia indica*
mutimbwinini (To) *Cussonia arborea/ C. kirkii*
mutimbya-ndavu (To) *Mucuna pruriens*
mutinza (To) *Brachystegia spiciformis*

mutipilili (I) *Abrus precatorius*
mutipi-tipi (I/ Le) *Abrus precatorius*
mutiti (B/ Le) *Erythrina abyssinica*
mutiti (Lo/ To) *Abrus precatorius*
mutjelele (Lo) *Ziziphus abyssinica*
muto (I/ Le) *Diplorhynchus condylocarpon*
mutoba (I/ To) *Azanza garckeana*
mutobakongwe (K) *Crossopteryx febrifuga*
mutobo (I/ To) *Azanza garckeana/ Dombeya rotundifolia*
mutobo (Tk) *Combretum collinum*
mutobolo (To) *Steganotaenia araliacea*
mutoci (Ln) *Rauvolfia caffra*
mutoka-toka (Ln) *Ilex mitis/ Stereospermum kunthianum*
mutoma-mena (K) *Antidesma venosum*
mutoma-toma (Lo) *Vangueria infausta & V. spp.*
mutomwa (Lo/ Ln/ Lv) *Diospyros mespiliformis*
mutondo-tondo (B/ Ny/ Tu) *Xeroderris stuhlmannii*
mutongwe (B) *Tecomaria capensis*
mutoto (K/ Ln) *Rauvolfia caffra*
mutoto-toto (Tk) *Phyllanthus ovalifolius*
mutowa (Ln) *Diospyros mespiliformis*
mutowa (Ny/ Ku/ To/ Tk/ Tu) *Diplorhynchus condylocarpon*
mutoya (Lo) *Antidesma venosum/ Salix mucronata/ Syzygium cordatum/ S. guineense sbsp.*
    barotsense/S. guineense sbsp.barotsense x afromontanum/ S. owariense
mutoye (Lo) *Syzygium cordatum*
mutsa-wari (Ny) *Catha edulis*
mutu (I) *Brachystegia hockii/ B. spiciformis*
mutu (To) *Strychnos innocua*
mutuba-tuba (Lo/ To) *Acacia sieberiana*
mutuba-tuba (I/ To) *Sterculia quinqueloba*
mutube-tube (I/ Ku/ Ny/ To/ Tu) *Acacia sieberiana*
mutube-tube (Ny) *Faidherbia albida*
mutuhu (Lo) *Leonotis nepetifolia*
mutuka-mbamba (Lo) *Cissus cornifolia/ Rhoicissus tridentata*
mutuku (Lo/ N) *Clutia abyssinica*
mutukumwisi (Ln/ Lv) *Phyllocosmus lemaireanus*
mutukutu (Tu) *Piliostigma thonningii*
mutumbe (B) *Harungana madagascariensis*
mutumbi (Le/ To) *Gardenia resiniflua*
mutumbi-mazi (Tk) *Ficus glumosa*
mutumbo (K) *Diospyros batocana*
mutumbula (I/ Tk/ To) *Flacourtia indica*
mutumbulwa (I/ Lo/ Tk/ To) *Flacourtia indica*
mutumbusya (B) *Flacourtia indica*
mutumbwa (B) *Maprounea africana*
mutumbwisya (B) *Flacourtia indica/ Gymnosporia senegalensis/ Maytenus ovata*
mutumbwisyi (B) *Flacourtia indica*
mutumbwizya – (see mutumbwisya)

mutume (Lo) *Rotheca myricoides*
mutumpu (B) *Trema orientalis*
mutumuko (Ny) *Stereospermum kunthianum*
mutunda (Ln) *Symphonia globulifera*
mutundu (Ln) *Sapium ellipticum*
mutundu (Lo/ Lv) *Aframomum alboviolaceum*
mutundula (To) *Xeroderris stuhlmanni*
mutundulu (To) *Aframomum alboviolaceum* (?)
mutundu-lukwa (Ny) *Ximenia spp.*
mutunduti (I) *Rourea orientalis*
mutunduti (Tk) *Croton sp. cf. polytrichus*
mutundwa (Lv) *Brachystegia spiciformis*
mutunga (Ln) *Strychnos pungens*
mutunga (Lv) *Strychnos spinosa*
mutunga-babala (To) *Catunaregam obovata/ Xeromphis obovata*
mutunga-barbara – (see mutunga-babala)
mutunga-bambala (I) *Canthium glaucum*
mutunga-mbabala (I/ K/ Le/ Lo/ To) *Canthium glaucum/ Xeromphis obovata*
mutunga-mbawala (Tk) *Xeromphis obovata*
mutunga-nkomo (B) *Eriosema psoraleoides*
mutungi (Ln) *Strychnos spinosa*
mutungule (K) *Aframomum alboviolaceum*
mutungulu (B/ Lb/ Ln) *Aframomum alboviolaceum*
mutungulu-wanika (B) *Aframomum angustifolium*
mutungunu-musindwa (Ln) *Aframomum alboviolaceum*
mutungwa (Ku) *Gymnosporia senegalensis*
mutuni (Ln) *Strychnos pungens*
mutunila (Ku) *Cissus cornifolius*
mutunta-mankwaji (Ln) *Hallea stipulosa*
mutuntula (B) *Solanum spp./ S. incanum*
mutuntula (B/ Ku) *Solanum tettense* var. *renschii/ Withania somnifera*
mutuntulwa (Lo) *Solanum spp.*
mutuntulwa (Lo/ To) *Solanum spp./ Withania somnifera*
mututuma (Ln/ Lo) *Albizia antunesiana*
mututume (N) *Albizia antunesiana*
mutuundu (Ln) *Symphonia globulifera*
mutuwa (I/ Tk/ To) *Croton megalobotrys*
mutuwa-tuwa (Lo/ Tk) *Croton megalobotrys*
mutuwi-tuwi (Tu) *Acacia sieberiana*
mutuwila (Le) *Brachystegia utilis*
mutuya (Lo/ N/ Tk) *Brachystegia spiciformis*
mutuyu (Lo) *Brachystegia spiciformis*
mutwa (Lo) *Cynodon dactylon*
mutwa-maila (To) *Terminalia sericea*
mutwa-nakabaya (K) *Schrebera trichoclada*
mutwate (Tu) *Tecomaria capensis*
mutwebu-bacasi (I/ To) *Phyllanthus engleri*
mutwepa (I/ To) *Brachystegia glaucescens/ B. microphylla*

mutyete (Ny) *Acacia spp.*
muugabwa (Tu) *Manilkara mochisia*
muukila (I/ Lo/ To) *Acacia sieberiana*
muula (M/ Ny/ To) *Parinari curatellifolia*
muule (Ny/ Tu) *Parinari curatellifolia*
muulya (Lo) *Diplorhynchus condylocarpon*
muumba (To) *Brachystegia spiciformis/ Julbernardia globiflora*
muumvwa-utuba (Le) *Acacia sieberiana*
muunga (I/ Le/ Lo/ Tk/ To) *Acacia spp./ Faidherbia albida*
muunga (Lo/ Tk/ To) *Acacia amythethophylla/ A. sieberiana/ A. polyacantha*
muunga (To) *Acacia tortilis*
muunga-usiya (I) *Acacia erioloba/ A. gerrardii/ A. tortilis / Faidherbia albida*
muunga-utuba (To) *Acacia sieberiana*
muungu (B) *Acacia polyacantha*
muungula-mabele (To) *Gymnosporia senegalensis*
muunza (I) *Combretum collinum*
muupa (B/ M) *Hallea stipulosa*
muuse (Tk) *Manilkara mochisia*
muuti (N) *Lannea discolor*
muuyu (I/ Le/ Lo/ N/ Tu) *Adansonia digitata*
muva-malopa (M/ Tu) *Pterocarpus angolensis*
muva-mbangona (Tk) *Balanites aegyptica*
muvwa-malowa (I) *Xeroderris stuhlmann*
muvatuba (I/ To) *Acacia sieberiana*
muver-veri (Nm) *Strophanthus eminii*
muveve (To) *Kigelia africana*
muvulama (Ln) *Ximenia americana*
muvulatowo (Lv) *Afzelia quanzensis*
muvule (Lo/ Tk/ To) *Salix mucronata*
muvunda (To) *Gymnosporia senegalensis*
muvungo-vungo (Ln/ Lv/ N) *Kigelia africana*
muvungula (Ku/ Ny/ Tu) *Kigelia africana*
muvungutsi (Ny) *Kigelia africana*
muvunyo-nunyo (Ln) *Kigelia africana*
muvwa-malowa (I) *Xeroderris stuhlmannii*
muwaama (Tk/ To) *Securidaca longipedunculata*
muwako (Lo) *Erythrophleum africanum*
muwale (Ny) *Erythrina abyssinica/ E. caffra*
muwama (To) *Securidaca longipedunculata*
muwambwa-ngoma (Lb) *Gymnosporia senegalensis*
muwana (Ln) *Pericopsis angolensis*
muwanga (Lb) *Burkea africana*
muwanga (Lv) *Pericopsis angolensis*
muwaŋa (Lv/ Ln) *Pericopsis angolensis*
muwasa-nzolu (Le) *Gymnosporia senegalensis*
muwawa (Ku/ Ny/ To/ Tu) *Khaya anthotheca*
muwembebe (B) *Boscia angustifolia*
muwembi (Ku/ Ny/ Tu) *Tamarindus indica*

muwi (To) *Strychnos spinosa*
muwiba-utoka (K) *Acacia sieberiana*
muwinda (Lo) *Securidaca longepedunculata*
muwole (Ny/ Tu) *Adenia gummifera*
muwoliba (Lo) *Gymnosporia senegalensis/ Gymnospora heterophylla*
muwomba (Tk) *Ximenia americana*
muwombe (I/ To) *Acacia nilotica sbsp. kraussiana*
muwombege (To) *Acacia nilotica sbsp. kraussiana*
muwombo (B/ Lb/ Ln/ Lo/ Ny/ Tk/ To/ Tu) *Brachystegia boehmii/ B. glaberrima/ B. gossweileri/ B. longifolia/ B. wangermeeana*
muwongo (I/ K/ Lo/ Tk/ To) *Sclerocarya birrea*
muwongo (To) *Balanites aegyptica*
muwonzo (Lo) *Acacia nilotica sbsp. kraussiana*
muwuwu (Ny) *Syzygium owariense*
muyambwe (Lo) *Ziziphus abyssinica*
muyangula (To) *Swarzia madagascariensis*
muyeele (Le) *Peltophorum africanum*
muyeeye (I) *Dichrostachys cinerea/ Triumfetta spp.*
muyente (Lb) *Ximenia caffra*
muyenzi (Nm) *Brachystegia boehmii*
muyimbe-yimbe (K) *Dalbergia nitidula*
muyimbili (Lo) *Strychnos cocculoides/ S. spinosa*
muyise (Lv) *Securidaca longipedunculata*
muyobo (Le) *Vetiveria nigritana*
muyogo-yogo (Nm) *Vangueria infausta*
muyombo (Lo) *Sclerocarya birrea*
muyombo (Lv) *Kirkia acuminata/ Lannea discolor/ L. schweinfurthii*
muyombo-wazimo (Lv) *Zanha africana*
muyombo-yombo (Ln) *Zanha africana*
muyongolwa (To) *Swarzia madagascariensis*
muyunda (Lo) *Hymenocardia acida*
muyu-ngula (To) *Swarzia madagascariensis*
muyuyu (Lo/ To) *Mucuna pruriens*
muzanda-lima (To) *Flueggea virosa*
muzanga (Ku) *Albizia anthelmintica*
muzatavame (I) *Combretum zeyheri/ C. collinum*
muzau (To) *Ziziphus abyssinica*
muzaviji (Ln) *Abrus precatorius*
muzenze (Ln) *Acacia sieberiana/ Entada abyssinica*
muze-nzenze (To) *Peltophorum africanum*
muzeze (Ln) *Acacia seyal/ A. xanthophloea*
muzeze (Lv) *Acacia polyacantha*
muzezi (Ln) *Capparis tomentosa*
muzi (Ln) *Diplorhynchus condylocarpon*
muzimbi-kolo (Lo) *Strychnos innocua/ S. madagascariensis*
muzimbili (Tu) *Strychnos spinosa*
muzimwa (M/ Tu) *Cleome gynandra*
muzinda-ngulube (Ku) *Ampelocissus africana/ Turbina stenosiphon*

muzinga (Ny) *Ricinus communis/ Schinziophyton rautanenii*
muzombo (B/ Ku) *Oncoba spinosa*
muzombo (M) *Brachystegia boehmii/ B. longifolia/ B. stipulata/ B. wangermeeana*
muzombo (Nm) *Brachystegia spiciformis*
muzuma-ngoma (To) *Albizia tanganyicensis/ Sterculia quinqueloba*
muzumba (B/ Ku/ Ny/ Tu) *Kirkia acuminata*
muzumba-mulyamba (B/ Ku) *Zanha africana*
muzumbu (Ny/ Tu) *Kirkia acuminata*
muzumina (I/ Lo/ Tk/ To) *Kirkia acuminata*
muzuminwa (I/ To) *Kirkia acuminata*
muzunda (I/ Lo/ To) *Acacia sieberiana*
muzunga (Ku/ Ny) *Acacia tortilis*
muzunga (Tu) *Acacia gerrardii*
muzunge (Lo) *Kirkia acuminata*
muzungi (Lo) *Protea madiensis*
muzungula (I/ K/ Le/ Lo/ Tk/ To) *Kigelia africana*
muzwa-malowa (I/ Lo/ To) *Xeroderris stuhlmannii*
muzwa-malowa (Le/ To) *Pterocarpus angolensis*
muzyazya (K) *Cassia abreviata*
muzyimbi-lili (I) *Strychnos spinosa*
muzyu (Ln) *Hallea stipulosa*
mvilu (Ny) *Vangueria infausta/ Vangueriopsis lanciflora*
mvimbe (Ny) *Diospyros abyssinica/ D. mespiliformis/ Psorospermum spp.*
mvukwe (Ny) *Brachystegia allenii/ B. floribunda/ B. manga/ B. spiciformis*
mvula (Ny) *Kigelia africana*
mvule (To) *Salix mucronata*
mvumba-mvula (Ny) *Rauvolfia caffra/ Voacanga spp.*
mvumuti (Ny) *Kigelia africana*
mvunga-njati (Ny) *Flacourtia indica*
mvungula (Ku/ Ny/ Tu) *Kigelia africana*
mvunguti (Ny) *Kigelia africana*
mvungutwa (Ny) *Kigelia africana*
mvungwe (Ny) *Dalbergia nitidula*
mwaabo (To) *Strychnos spinosa*
mwababa (Lo) *Piliostigma thonningii*
mwabo (To) *Strychnos innocua*
mwabvi (Ny) *Erythrophleum suaveolens/ Trichilia emetica*
mwafi (B/ Le) *Erythrophleum suaveolens*
mwage (Nm) *Strychnos spinosa*
mwala (Ln) *Afzelia quanzensis*
mwalabwe (Lo/ To) *Balanites aegyptiaca*
mwale (N/ Ny) *Erythrina abyssinica*
mwalisaka (Ny) *Momordica balsamina*
mwama (To) *Securidaca longipedunculata*
mwamba-ngoma (Nm) *Balanites aegyptica*
mwamfu (Lo) *Polycarpaea corymbosa*
mwamuna-njulo (Ku) *Albizia anthelmintica*
mwana-bere (Ny) *Stereospermum kunthianum*

mwana-nkali (Ny) *Cyphostemma junceum*

mwana-wamfepo (Ny) This name is generic rather than specific, and includes the following species: *Acalypha villicaulis/ Adenia digitata/ A. gummifera/ Allophylus africanus/ Ampelocissus africana/ A. obtusata/ Bulbophyllum sandersonii/ Calyptrochilum christyanum/ Cayratia gracilis/ Cissus cornifolia/ C. faucicola/ C. integrifolia/ C. quadrangularis/ Commelina africana/ Cyphostemma buchananii/ C. crotalarioides/ C. junceum/ C. rhodesiae/ C. subciliatum/ C. zombense/ Cyrtorchis arcuta/ Dolichos kilimanscharicus/ Eriospermum flagelliforme/ Impatiens walleriana/ Jateorrhiza palmata/ Kalanchoe prolifera/ K. verticillata/ Paullina pinnata/ Peucedanum eylesii/ Rhoicissus revoilii/ R. tridentata/ Scadoxus multiflorus/ Sonchus oleraceus/ Stereospermum kunthianum/ Tinospora birrea/ Tridactyle bicaudata/ T. tricuspis.* See note in *Morris*, pp. 40–42.

mwanda (Nm) *Adansonia digitata*

mwanda-bala (Lo) *Xeromphis obovata*

mwanda-kasi (Lo) *Burkea africana*

mwanda-kati (N) *Burkea africana*

mwanda-kazi (Tk) *Burkea africana*

mwanda-koso (Le) *Erythrophleum africanum*

mwande (I/ Ln/ Lo) *Afzelia quanzensis*

mwandu (Nm) *Adansonia digitata*

mwanga (B) *Flacourtia indica*

mwanga (Nm/ Ny) *Pericopsis angolensis*

mwangule (To) *Albizia versicolor*

mwanja (B/ Ku/ Lo/ Lv/ To) *Manihot esculenta*

mwanjane (Ny) *Bridelia duvigneaudii*

mwansa-buso (To) *Burkea africana*

mwansasa-buso (To) *Burkea africana*

mwansusa-buso (To) *Burkea africana*

mwanya (B/ Ku) *Cissus integrifolia*

mwanza-bala (To) *Entandrophragma caudatum*

mwanza-masaka (B/ Ku) *Albizia harveyi/ Pterocarpus sp./ Xeroderris stuhlmannii*

mwasambala (To) *Afzelia quanzensis*

mwavi (B/ Ku) *Antidesma venosum*

mwavi (B/ Ku/ Ny) *Crossopterix febrifuga*

mwavi (Ny) *Erythrophleum suaveolens*

mwavi-kulu (Ku) *Hymenocardia acida*

mwaye (Ny) *Strychnos spinosa*

mwayi (Ny) *Erythrophleum suaveolens*

mweemwe (I/ Le) *Sterculia quinqueloba*

mweeye (I/ Tk/ To) *Dichrostachys cinerea/ Triumfetta spp.*

mwele-wele (Nm) *Azima tetracantha/ Strophanthus eminii*

mwelu (Ku) *Vahlia dichotoma*

mwemba (Ny) *Tamarindus indica*

mwembe (B) *Mangifera indica/ Tamarindus indica*

mwembembe (B) *Boscia angustifolia/ Phyllocosmus lemaireanus*

mwembi (Ku/ Ny/ Tu) *Tamarindus indica*

mwemwe-tuseko (K) *Sterculia africana*

mwenge (B/ K/ Lb/ Lo/ M/ Ny) *Diplorhynchus condylocarpon*

mwenge-busyilu (B) *Holarrhena pubescens*
mwenye (Ny) *Syzygium cordatum*
mwenza (B) *Peltophorum africanum*
mweri-weri (Nm) *Holarrhena pubescens*
mweti (Ny) *Combretum collinum*
mweya (Ln/ Lv) *Combretum collinum*
mweye (I/ Tk/ To) *Dichrostachys cinerea*
mweyema (B) *Phyllanthus muelleranus*
mweyeye (K) *Albizia zygia*
mweza-munko (Ku/ Ny) *Holarrhena pubescens*
mwezengele (Ln) *Peltophorum africanum*
mwezenyele (Ln) *Peltophorum africanum*
mwezya (Nm) *Gymnosporia senegalensis*
mwiba (K) *Acacia erioloba/A. nilotica/ A. sieberiana/ Faidherbia albida*
mwica (Ln) *Parinari curatellifolia*
mwicecele (To) *Phyllanthus reticulatus*
mwicecete (To) *Ziziphus abyssinica*
mwidi (Lv) *Diplorhynchus condylocarpon*
mwihangwe (To) *Albizia anthelmintica*
mwiinya-meenzyi (i/ To) *Phyllocosmos lemaireanus*
mwijimbe (Ln) *Strychnos spinosa*
mwikala-kanga (Lo) *Albizia anthelmintica*
mwikala-nkanga (B/ K/ Le) *Albizia amara/ A. harveyi/ Peltophorum africanum*
mwikala-nkanga (Lo) *Albizia anthelmintica*
mwikala-sosa (B) *Leonotis molissima*
mwikale-senga (B) *Hippocratea indica*
mwila (K) *Entada abyssinica*
mwilatuba (To) *Flueggea virosa*
mwilikano (B) *Adenia gummifera*
mwimba-finoka (B) *Securidaca longipedunculata*
mwimbe (B/ Ny) *Rauvolfia caffra*
mwimbili (Lo) *Strychnos spinosa*
mwinula-mponda (Ny) *Hyptis spicigera*
mwinda (Lo/ Ny/ Tu) *Securidaca longipedunculata*
mwinguti (Ny) *Stereospermum kunthianum*
mwini-munda (Ny) *Cyphostemma junceum/ Tacca leontopetaloides*
mwinula-mponda (B/ Ku/ Ny) *Hyptis spicigera*
mwinzwa-ndimu (Nm) *Margaritaria discoidia*
mwiombo (Ln) *Kirkia acuminata*
mwisi (Lv) *Securidaca longipedunculata*
mwita-ngolwa (K) *Gardenia ternifolia*
mwitwa-ngolwa (K) *Gardenia ternifolia*
mwitwi (Lv) *Strychnos spinosa*
mwiveli (Nm) *Strophanthus eminii*
mwiyenzi (Nm) *Brachystegia boehmii*
mwiyombo (Ln) *Kirkia acuminata/ Lannea antiscorbutica/ L. schweinfurthii & spp.*
mwiyombo (Nm) *Brachystegia boehmii*
mwula (Nm) *Parinari curatellifolia*

myenze (Nm) *Brachystegia boehmii*
myrrh (E) *Commiphora madagascariensis*
myuguyu (Nm) *Balanites aegyptica*
myunga cikobe (B) *Acacia polyacantha*
mzabaza (Ny) *Cassia abbreviata*
mzakaka (Ny) *Zanha africana*
mzakomo (Ny) *Maerua juncea*
mzanga (Ny) *Albizia anthelmintica*
mzaule (Ny) *Allophylus africanus/ A. spp.*
mzayi (Ny) *Strychnos spinosa/ S. cocculoides*
mzaza (Nm) *Protea madiensis*
mzaza (Ny) *Monotes africanus/ M. elegans/ M. katangensis / M. spp./ Tragia benthamii*
mzembe (Ny) *Dalbergia nitidula*
mzenda-nzovu (Ny) *Stereospermum kunthianum*
mziloti (Ny) *Psorospermum febrifugum/ P. spp.*
mzimbili (Ny) *Strychnos cocculoides/ S. innocua/ S. spinosa*
mzimdiwi (Tu) *Brachystegia spiciformis*
mziru (Ny) *Ekebergia benguelensis/ Vangueria infausta*
mzizi (Ny) *Acacia abyssinica/ A. sieberiana/ Cissampelos mucronata*
mzumba (Ny) *Kirkia acuminata*
mzumbu (Ny/ Tu) *Kirkia acuminata*
mzunga (Ny) *Acacia tortilis*
mzungu-newe (Tu) *Acacia seyal*
mzunga-nyewe (Ny) *Acacia gerrardii/ A. robusta*
nacisansa (Lo) *Ampelocissus africana*
nacisungu (B) *Rourea orientalis*
nafangale (Ny) *Dichrostachys cinerea*
naka-bele (To) *Euphorbia hirta*
naka-bumbu (B) *Lannea discolor*
naka-bwazi (Ny) *Securidaca londepedunculata*
naka-fungu (B) *Leonotis mollissima*
naka-kwete (Lb) *Rourea orientalis*
naka-lenge (To) *Dichrostachys cinerea*
naka-meso (Ny) *Euphorbia hirta*
naka-ncete (B) *Bersama abyssinica/ Clerodendrum capitatum*
naka-ngunde (Ny) *Neorauthanenia mitis*
naka-nkhwali (Ny) *Erythrocephalum zambesianum*
naka-nunkhu (Ny) *Lantana trifolia/ Lippia javanica*
naka-sonde (Ny) *Lantana trifolia*
naka-swewe (Ny) *Crotalaria natalitia*
naka-tiko (Ny) *Margaritaria discoidea*
naka-timba (Ny) *Olea europaea*
naka-tobwa (Ny) *Euphorbia hirta*
naka-tuta (Ny) *Euclea natalensis*
naka-umba (B) *Lannea discolor*
naka-uti (Ny) *Steganotaenea araliacea*
naka-walika (Ny) *Croton macrostachys*
naka-yumbe (K) *Burkea africana*

naked-stem carpetweed (E) *Mollugo nudicaulis*
nakumanga (Ku/ Ny) *Asparagus sp.*
nalita jute (E) *Corchorus olitorius*
nalo (Nm) *Acacia kirkii*
nama-bele (Be) *Ozoroa reticulata*
nama-finya (Ny) *Euphorbia heterophylla*
nama-gara (Ny) *Senna singueana*
nama-langa (Ny) *Acacia nilotica*
namalopa (Ny) *Mucuna poggei/ M. pruriens/ Rhynchosia hirta*
namanga (Lo) *Cleome gynandra*
namano-mano (Ny) *Gymnosporia senegalensis*
nama-sakata (Ny) *Trichodesma zeylanicum/ Tridax procumbens*
nama-simba (Ny) *Dalbergia nitidula*
nama-sira (Ny) *Ozoroa reticulata*
nama-tuli (I) *Xeroderris stuhlmannii*
namgoneka (Ny) *Maerua juncea/ Momordica boivinii/ M. foetida*
namonde (Lo) *Ficus sur*
namu-coko (Lo) *Citrullus lanatus*
namu-kondowe (Lb/ Le/ K) *Citrullus lanatus*
namu-kosi (B) *Tecomaria capensis*
namu-lalusya (B) *Rhus ancietae/ R. longipes*
namu-lilo (B) *Fadogia spp./ Psychotria kirkii*
namu-lilo-wakasyika (B) *Psychotria kirkii*
namu-lozi (Ny) *Adenia gummifera*
namu-lomo (Lo) *Grewia villosa*
namu-nsi (B) *Syzygium cordatum*
namu-nunke (I) *Panicum maximum*
namu-nwa (B/I/Lo) *Citrullus lanatus*
namu-tengwa (I/ To) *Kigelia africana*
namu-wale (B) *Ozoroa reticulata*
namu-zungula (I/ To) *Kigelia africana*
namwa-licece (Ny) *Cyphostemma buchananii*
nanda (Nm) *Faidherbia albida*
nankobwe (Ny) *Neorauthanenia mitis*
nanpapwa (Ny) *Cussonia arborea*
nanyole (Ny) *Syzygium cordatum*
napala-pala (Ny) *Bridelia cathartica/ Senna singueana/ Margaritaria discoidea*
napini (Ny) *Combretum molle/ Terminalia sericea/ T. stenostachya*
napose (Ny) *Acalypha villicaulis/ Antidesma venosum/ Brackenridgea zanguebarica/ Diospyros zombensis/ Cryptolepis oblongifolia/ Grewia flavescens/ Heteromophia trifoliata/ H/ arborescens/ Margaretta rosea/ Ochna schweinfurthiana/ Phillippia benguelensis/ Trichodesma zeylanicum*
nard grass (E) *Cimbopogon nardus*
narrow-pod elephant-root (E) *Elephantorrhiza goetzii*
natal indigo (E) *Indigofera arrecta*
natal mahogany (E) *Trichilia emetica*
natal sorrel (E) *Hibiscus sabdariffa*
natal teak (E) *Strychnos henningsii*

nayula (Ny) *Acacia nigrescens*
nazilyango (To) *Cordia goetzei*
ncaca (Ny) *Erythrocephalum zambesianum*
ncenga (Ny) *Diospyros mespiliformis*
ncofu (Ny) *Cayratia gracilis/ Cyphostemma buchananii/ C. subciliatum*
ncowana (Ny) *Oncoba spinosa*
ndale (B) *Diplorhynchus condylocarpon*
ndale (B/ I/ K/ Lb/ Le/ Ny) *Swarzia madagascariensis*
ndambala (Ln) *Hibiscus sabdariffa*
ndawa (Ny) *Flacourtia indica*
ndelele (To) *Dalbergia nitidula*
ndembo (Ln) *Securidaca longipedunculata*
ndemika-ngongo (Ny) *Cyphostemma buchananii/ C. junceum/ Gonatopus boivinii/ Tacca leontopetaloides*
ndiapumbwa (Ny) *Senna singueana*
ndiyu (Ny/ Tu) *Boophane disticha*
ndolola (Ny) *Crotalaria recta*
ndombe (K) *Pterocarpus angolensis*
ndombo (Ny) *Holarrhena pubescens*
ndombozi-cipeta (Ny) *Holarrhena pubescens*
ndombwa (Ny) *Pterocarpus angolensis*
ndozi (Ny) *Adenia gummifera*
ndula-nkwangwa (Ny) *Dalbergia nitidula/ Dichrostachys cinerea*
ndule (Ny) *Euclea racemosa*
ndumba-luango (Tu) *Sorghum sp.*
ndunga (Ny) *Maesopsis eminii*
needle bur (E) *Amaranthus spinosus*
negro coffee (E) *Senna occidentalis*
nembe-nembe (Ny) *Senna petersiana*
nenepa (Ny) *Erythrina abyssinica/ Vigna unguiculata*
nengo-nengo (Nm) *Securidaca longipedunculata*
never-die (E) *Kalanchoe crenata*
ngaci (Ny) *Euphorbia tirucalli*
ngada (Nm) *Albizia anthelmintica*
ngagaga (Ny) *Acacia nilotica*
ngalati (Ny) *Burkea africana*
ngalawa (Ny) *Crotalaria goetzei/ C. natalitia/ C. recta/ Dolichos kilimanscharicus*
ngansa (B/ Lb/ Ny/ Tu) *Brachystegia boehmii/ B. stipulata*
ngata (Nm) *Albizia anthelmintica*
ngawe (I) *Vangueria infausta*
ngaza (Ny) *Albizia anthelmintica*
ngizi (To) *Capparis tomentosa*
ngoka (To) *Acacia tortilis*
ngole (Nm) *Adenia gummifera*
ngolo (Ny) *Swartzia madagascariensis*
ngolyondo (Ny) *Afzelia quanzensis*
ngomba (Ny) *Ximenia americana/ X. caffra*
ngombe-yanina (B) *Adenia gummifera*

ngombo-uti (Ny) *Steganotaenea araliacea*
ngongolo (Ny) *Brachystegia spiciformis*
ngongomwa (Ny) *Afzelia quanzensis*
ngosa (Ny) *Biophytum umbraculum*
ngowe (B/ Ku/ Ny/ Tu) *Acacia polyacantha*
ngu (Nm) *Acacia polyacantha*
ngubalu (Nm) *Crossopteryx febrifuga*
ngula-mabele (To) *Ochna schweinfurthiana*
ngulikila (I) *Cocculus hirsutus*
ngunda-nguluwe (Ny) *Ochna macrocalyx/ O. schweinfurthiana*
ngunga (Nm) *Acacia tortilis*
nguru-gunga (Nm) *Acacia tortilis*
ngwa-langalate (Ny) *Cassia mimosoides*
ngwa-malembe (Le) *Acacia goetzei/ A. nigrescens*
ngwelinte (I/ To) *Phyllanthus sp.*
ngwesi (Tk) *Bolusanthus speciosus*
ngwevula (Ku/ Ny) *Clerodendrum capitatum/ C. spp.*
ngwika (Ny) *Dicoma sessiliflora*
ngwimbe (Ny) *Rauvolfia caffra*
nhlangulane (Ny) *Euclea natalensis*
nightshade (E) *Solanum incanum*
nimbwa (Tk/ To) *Strychnos spinosa*
njale (Ny) *Treculia africana*
njasyi (B/ Ny) *Sorghum sp.*
njefu (To) *Securidaca longepedunculata*
njelele (To) *Dalbergia nitidula*
njiriti (Ny) *Holarrhena pubescens/ Phyllanthus ovalifolius*
njofa (To) *Annona senegalensis*
njoka (Ny) *Albizia antunesiana/ A.glaberrima/ Chenopodium ambrosiodes/*
    Jateorrhiza palmata
njolole (Ku) *Kedrostis sp./ Mukia maderaspatana*
nkadze (Ny) *Euphorbia tirucalli*
nkaka (To) *Momordica balsama*
nkalanga (I/ Tk/ To) *Combretum apiculatum/ C. molle*
nkalanga (Ku) *Euphorbia tirucalli*
nkalangu (To) *Combretum molle*
nkanga (Ny) *Bersama abyssinica*
nkhasi (B/ Ku/ Ny) *Mucuna pruriens*
nkhazye (B/Ku) *Euphorbia tirucalli*
nkholosa (Ny) *Euphorbia hirta*
nkhumba (Ny) *Elephantorrhiza goetzei*
nkoce (Tu) *Sorghum sp.*
nkokona-simba (Ny) *Stereospermum kunthianum*
nkolo-kolo (Ny) *Dalbergia nitidula*
nkololo (B) *Smilax anceps*
nkololwe (B) *Smilax anceps*
nkondo-nkondo (Ku/ Ny) *Flacourtia indica/ Psychotria spp.*
nkonona-simba (Ny) *Stereospermum kunthianum*

nkula-kazyi (K) *Afzelia quanzensis*

nkulo (Ku) *Diospyros batocana/ D. kirkii/ D. mespiliformis*

nkululumbi (Ln) *Cyphostemma buchananii*

nkumba (M) *Sorghum spp.*

nkunga (Ny) *Dolichos kilimanscharicus/ Mucuna spp./ Sphenostylis marginata*

nkunzano (To) *Xenostegia tridentata*

nkuyu (Ny/ Tu) *Balanites aegyptica*

nkwinde (Ku) *Panicum maximum/ P. spp.*

nlonga-ndundu (Ny) *Leonotis mollissima/ L. nepetifolia/ Leucas martinicensis*

nsakambwe (Ny) *Acanthospermum hispidum*

nsala (To) *Strychnos spinosa*

nsalo (Ny) *Dichrostachys cinerea*

nsalu-nsalu (B) *Dichrostachys cinerea*

nsambe (B/ Ku) *Ficus sur*

nsangu (Ny) *Acacia tortilis/ Faidherbia albida*

nsangu (To) *Faidherbia albida*

nsangu-nsangu (Ny) *Acacia tortilis/ Faidherbia albida*

nsangu-nsangu (To) *Faidherbia albida*

nsatsi (Ny) *Ricinus communis*

nsengwa (Ny) *Albizia amara*

nsha- (See nsya-)

nshi- (See nsi-)

nsila (Ny) *Cleome gynandra*

nsitula (Ny) *Allophylus africanus*

nsoma (Nm) *Harrisonia abyssinica*

nsowa (Ln) *Citrullus lanatus*

nsyaba (B/ Ku/ Ny/ Tu) *Arachis hypogaea*

nsyafuta (Ny) *Ricinus communis*

nsyapita (Ny) *Abrus precatorius*

nsyawa (B/ Ny) *Arachis hypogaea*

ntaci (Ny) *Dicoma sessiliflora sbsp. sessiliflora*

ntalula (M) *Acacia polyacantha*

ntanda-nyere (Ny) *Phyllanthus reticulatus*

> **N.B.** There is some confusion between the various spellings of this name. See also mtanda-nyerere, mtanta-nyelele, mtanta-nyerere, mtantha-nyele, ntanda-nyere, ntanda-nyerere, ntanta-nylele, ntanta-yelele, tanta-nyele

ntanda-nyerere (Ny/ Tu) *Aphloia theiformis/ Margaritaria discoidea/ Ormocarpum kirkii/ Phyllanthus ovalifolius/ P. reticulatus/ Rourea orientalis/ Securinega virosa/ Senna petersiana/ S. singueana*

ntanga (B/ Ny/ Tu) *Citrullus lanatus*

ntanga-manze (Tu) *Citrullus lanatus*

ntanta-nyelele (Ny) *Senna singueana*

> **N.B.** There is some confusion between the various spellings of this name. See also mtanda-nyerere, mtanta-nyelele, mtanta-nyerere, mtantha-nyele, ntanda-nyere, ntanda-nyerere, ntanta-nylele, ntanta-yelele, tanta-nyele

ntanta-yelele (Ny) *Phyllanthus reticulatus*

ntapiko (Ny) *Dalbergia boehmii*

ntedza (Ny) *Arachis hypogaea*

ntelwe-lewe (Ny) *Senna singueana*
ntembe-nuko (Ny) *Crotalaria goetzei/ C. natalitia*
nteme (Ny) *Strychnos innocua*
ntemya (Ku) *Gardenia resiniflua/ Strychnos innocua*
ntengele (Ny) *Ximenia americana/ X. caffra*
ntengeni (Tu) *Ximenia americana*
ntente (Tu) *Sorghum sp.*
ntewe-lewe (Ny) *Senna petersiana*
ntewe-tewe (Tu) *Steganotaenia araliacea*
nteyo (Nm) *Securidaca longipedunculata*
ntedza (Ny) *Arachis hypogea*
nthenga (Ny) *Dichrostachys cinerea*
ntheza (Ny) *Diospyros mespilliformis*
nthombolozi (Ku) *Dalbergia boehmii*
ntihenga (?) (Nm) *Dichrostachys cinerea*
ntimba (Ny) *Abrus precatorius*
ntiti (To) *Artabotrys monteiroae (?)*
ntondo-tondo (Ny) *Xeroderris stuhlmannii*
ntowa (Tu) *Diplorrhynchus condylocarpon*
ntowo (M/ Ny) *Azanza garckeana*
ntudza (Ny) *Flacourtia indica*
ntukwinama (Ny) *Clematis brachiata*
ntumbuzya (B/ Ku) *Flacourtia indica*
ntumbwa (Ny) *Kirkia acuminata*
ntunduvya (B/ Ku) *Flacourtia indica*
ntunduwa (Ny) *Kirkia accuminata*
ntunduwere (Ny) *Solanum incanum*
ntungulu (Ln) *Aframomum alboviolaceum*
ntungu-ndwa (Ny) *Kirkia accuminata*
ntungu-ngu (Ny) *Harungana madagascariensis*
ntuza (Ny) *Flacourtia indica*
nulambwe (Ny) *Ozoroa reticulata*
nunka-dale (Ny) *Becium obovatum*
nunkamani (Ny) *Chenopodium ambrosiodes/ Ocimum americanum*
nut grass (E) *Cyperus rotundus*
nvungula (Ny) *Kigelia africana*
nyaka-jongwe (Tu) *Gloriosa superba*
nyaka-mbalilo (M) *Peltophorum africanum*
nyaka-panda (Tu) *Sorghum sp.*
nyaka-zuba (Ku/ Ny) *Helianthus annuus*
nyakoko (Ny/ Tu) *Carissa edulis*
nyalati (Ny) *Burkea africana*
nyalu-mpangala (Ny) *Dichrostachys cinerea*
nyama-kalapi (Ny/ Tu) *Syphonochilus kirkii*
nyama-lokane (Ny) *Gloriosa superba*
nyama-roba (Ny) *Xeroderris stuhlmannii*
nyama-songole (Ny) *Imperata cylindrica*
nyama-toka (Ku) *Euphorbia hirta/ Gisekia africana*

nyambuzi (Ny) *Euclea racemosa*

nyamonga (Ln) *Ozoroa reticulata*

nyamula-katundu (Ny) *Leucas martinicensis*

nyangoloka (Ny) *Steganotaenia araliacea*

nyanise (Lo/ To) *Allium cepa*

nyanyata (Ny) *Entada abyssinica*

nyasuna (Ny) *Coldenia procumbens/ Leonotis nepetifolia/ Polycarpon spp.*

nyele (Ny) *Albizia amara/ Peltophorum africanum*

nyelele (Ny) *Senna singueana (?)*

nyele-nyele (Ny) *Portulaca oleracea*

nyembe (Ln/ Lv) *Jatropha curcas*

nyemu (I/ K/ Le/ Ln/ Lo) *Arachis hypogaea*

nyemu-mafuta (I/ Le) *Arachis hypogaea*

nyengele (M) *Syzygium cordatum*

nyenyese (Lo) *Allium cepa*

nyenyewe (Ny) *Syzygium cordatum*

nyesani (Ny) *Rauvolfia caffra*

nyimbula (Ny) *Manihot esculenta*

nyinu (B) *Clematis spp./ C. villosa/ C.scabiosifolia/ Dicoma anomala*

nyoka (Ny) *Cassia abbreviata*

nyongoloka (Ny) *Steganotaenia araliacea*

nyoswa (Ny) *Acacia tortilis*

nyowe (Ny) *Syzygium cordatum/ S. owariense*

nyumbu (Ku/ Ny) *Abrus precatorius*

nyumbula (Ny) *Manihot esculenta*

nyungwe (Ny) *Acacia polyacantha*

nzombo (B) *Oncoba spinosa*

ŋanza (B/ Ku) *Brachystegia boehmii*

ŋunza (I) *Sterculia quinqueloba*

ŋutate (Ln/ Lv) *Securidaca longepedunculata*

ŋwandu (Nm) *Adansonia digitata*

ŋwica (Nm) *Kigelia africana*

ŋwini-ciyeye (Tk) *Dichrostachys cinerea*

ocenca (Ku/ Tu) *Acacia nilotica*

octopus cabbage-tree (E) *Cussonia arborea*

old man's cap (E) *Polycarpaea corymbosa*

olive (E) *Olea europaea*

ombe (Tk) *Acacia nilotica*

omujete (Lo) *Dichrostachys cinerea*

onion (E) *Allium cepa*

oombe (I) *Burkea africana*

orange-fruited asparagus (E) *Asparagus africanus*

orange-fruit fingerleaf (E) *Vitex buchanani*

orange-milk tree (E) *Harungana madagascariensis*

ordeal tree (E) *Erythrophleum suaveolens*

ouwewenaar (Afr) *Bidens pilosa*

palibe-kanthu (Ny) *Aspilia mossambicensis/ Dicoma anomala/ D. sessiliflora/ Lepidagathis andersoniana*

palma Christi (E) *Ricinus communis*
pamusele (Tu) *Gloriosa superba*
pande (B/ Tu) *Sorghum sp.*
pande-kanondo (B/ Tu) *Sorghum sp.*
papau (Ln/ Lv) *Carica papaya*
papaya (Ny) *Carica papaya*
papopo (To) *Carica papaya*
paraguayan bur (E) *Acanthospermum australis*
parsley tree (E) *Heteromorpha arborescens*
pate (Tu) *Sorghum sp.*
patwe (Ny) *Ochna schweinfurthiana*
patwe (To) *Uapaca nitida*
pawpaw (E) *Carica papaya*
peacock berry (E) *Margaritaria discoedia*
peacock flower (E) *Delonix regia*
peanut (E) *Arachis hypogaea*
peanut senna (E) *Senna didymobotrya*
pemba grass (E) *Stenotaphrum dimidiatum*
perdeskop (Afr) *Boophane disticha*
physic nut (E) *Jatropha curcas*
pigeon berry (E) *Duranta erecta*
pigeonwood (E) *Trema orientalis*
pigweed (E) *Amaranthus spinosus/ Portulaca oleracea*
pili-pili (Nm) *Erythrina abyssinica*
pill-bearing spurge (E) *Euphorbia hirta*
pimento grass (E) *Stenotaphrum dimidiatum*
pink bur (E) *Urena lobata*
pink jacaranga (E) Stereospermum *kunthianum*
pink lady (E) *Dissotis rotundifolia*
pioneer rattlepod (E) *Crotalaria natalita*
pirilango (Ny) *Euphorbia hirta*
piri-piri (B/ Tu) *Capsicum annuum*
piri-piri (Ny) *Tetradenia riparia*
piri-piri (E) *Cyperus articulatus*
pisape (Lo) *Phyllanthus reticulatus*
pitchfork (E) *Bidens pilosa*
popgun tree (E) *Steganotaenia araliaceae*
pod mahogany (E) *Afzelia quanzensis*
poison apple (E) *Solanum incanum*
poison-grub corkwood (E) *Commiphora africana*
porridge-sticks (E) *Craterispermum schweinfurthii*
porselein (Afr) *Portulaca oleracea*
pot marigold (E) *Calendula officinalis*
poubessie (Afr) *Margaritaria discoedia*
powder-bark (E) *Strychnos innocua*
pozo (Ny) *Aloe spp.*
prickly amaranth (E) *Amaranthus spinosus*
prickly-leaved elephant's foot (E) *Elephantopus scaber*

prickly salvia (E) *Pychnostachys urticifolia*
prince of Wales' feathers (E) *Brachystegia boehmii*
pulizangala (Tk) *Boophane disticha*
punda-mbuzi (Tu) *Senna singueana*
punyele (Tu?) *Peltophorum africanum*
pupwe (B) *Hibiscus sabdariffa/ Zanthoxylum chalybeum*
pupwe-culu (B) *Zanthoxylum chalybeum*
purging nut (E) *Jatropha curcas*
purple horn-of-plenty (E) *Datura metel*
purple morning-glory (E) *Ipomoea purpurea*
purple nutsedge (E) *Cyperus rotundus*
purple-top (E) *Hyptis pectinata*
purple-wood flatbean (E) *Dalbergia nitidula*
purshoutplatboontjie (Afr) *Dalbergia nitidula*
puti (Ny) *Brachystegia spiciformis*
pwamalonda (K) *Senna singueana*
quinine tree (E) *Rauvolfia caffra*
rabbit-vine(E) *Teramnus labialis*
railway vine (E) *Ipomoea pes-capra*
raisin-bush (E) *Ozoroa reticulata*
ramshoringbos (Afr) *Hugonia orientalis*
ram's horn (E) *Hugonia orientalis*
rat's ear (E) *Commelina forskaoli*
rattlebox (E) *Crotalaria recta*
red berry (E) *Solanum tettense*
red bird-berry (E) *Psychotria zombamontana*
red bitterberry (E) *Strychnos henningsii*
red hook-berry (E) *Artabotris monteiroae*
red-hot poker coral tree (E) *Erythrina abyssinica*
red jute (E) *Corchorus olitorius*
red mahogany (E) *Khaya anthotheca*
red pepper (E) *Capsicum annuum*
redwing (E) *Pterolobium exosum*
rhenoster (Afr) *Ranunculus multifidus*
ribbed paddle-pod (E) *Reissantia indica*
ribbon grass (E) *Setaria megaphylla*
rice (E) *Oryza sativa*
riceweed (E) *Scoparia dulcis*
rigid star-apple (E) *Diospyros squarrosa*
rooibessie (Afr) *Solanum tettense*
rooibitterbessie (Afr) *Strychnos henningsii*
rooihaakbessie (Afr) *Artabotris monteiroae*
rooipendoring (Afr) *Gymnosporia senegalensis*
rooivlerk (Afr) *Pterolobium exosum*
rooster tree (E) *Calotropis procera*
rose-ginger (E) *Syphonochilus kirkii*
rosella (E) *Hibiscus sabdariffa*
rossberry (E) *Grewia villosa*

rosy periwinkle (E) *Catharanthus roseus*
rough-leaved croton: (E) *Croton menyhartii*
rough-leaved shepherd's tree (E) *Boscia angustifolia*
rubber-hedge euphorbia (E) *Euphorbia tirucalli*
ruku (K/ Lb/ Lo/ N) *Eleusine coracana*
sabola (Ny) *Capsicum annuum var. frutescens*
safsafwilger (Afr) *Salix mucronata*
safsaf willow (E) *Salix mucronata*
saka-saka (Lo) *Aerva lanata*
saka-sinji (Ny) *Cassytha filiformis*
saligna gum (E) *Eucalyptus grandis*
samba (Lo) *Vetiveria nigritana*
samba-nteta (B) *Entada abyssinica*
sambilinga (Ny/ Tu) *Ampelocissus africana & spp.*
saminami (Lo) *Ocimum spp.*
sanda (Ku) *Ipomoea aquatica/ I. pes-tigris*
sand crownberry (E) *Crossopteryx febrifuga*
sandkroonbessie (Afr) *Crossopteryx febrifuga*
sand olive (E) *Dodonea viscosa*
sanga-lumbe (Lo) *Zanha africana*
sanga-lwembe (I/ K/ Tk) *Zanha africana*
sanga-lwendo (Le) *Zanha africana*
sanga-zinje (Ny) *Cassytha filiformis*
sansa (B/ Ku) *Strychnos spinosa*
sansamwa (Tu) *Ranunculus multifidus*
sansapati (Tk) *Ampelocissus africana*
sapole (K/ Ln/ Lv) *Allium cepa*
sarsaparella (E) *Jateorrhiza palmata*
sausage tree (E) *Kigelia africana*
scarlet magic (E) *Emilia coccinea*
scented bells (E) *Rothmannia manganjae*
scrambled egg (E) *Senna singueana*
sea heart (E) *Entada gigas*
seepblinkblaar (Afr) *Rhamnus prinioides*
seeroogli (Afr) *Boophane disticha*
segese (Nm) *Senna occidentalis*
sehoho (Ny) *Acacia nilotica*
sekelbos (Afr) *Dichrostachys cinerea*
sengwi-sengwi (I) *Aloe spp.*
sepe (Lb) *Mucuna pruriens*
septic weed (E) *Senna occidentalis*
sesame (E) *Sesamum indicum*
seseresya (Ny) *Acanthospermum hispidum/ Oxygonum sinuatum*
sessile joyweed (E) *Alternanthera sessilis*
sheep bur (E) *Acanthospermum australis*
shiny-leaf (E) *Rhamnus prinioides*
shoe-flower (E) *Hibiscus rosa-sinensis*
short-pod (E) *Rourea orientalis*

shrub verbena (E) *Lantana trifolia*
siamese cassia (E) *Senna siamea*
sibone-nwisiku (To) *Cassytha filiformis*
sibonyu-wisika (To) *Catha edulis*
sicibula (Tk) *Burkea africana*
sickle bush/ pod (E) *Dichrostachys cinerea/ Senna obtusifolia*
sikabasya (Lo) *Lannea edulis*
sikalilo (Ku/ Ny) *Plumbago zeylanica*
sikalu-lumbu (Tk) *Lannea discolor*
sikalutenta (Lo/ To) *Plumbago zeylanica*
sikamba (Le) *Manihot esculenta*
sikantyo (I)*Triumfetta welwitschii/ Tylophoria sylvatica*
sikomwa (I) *Ziziphus abyssinica*
sikubabe (Lo) *Dioscorea quartiniana*
sikwaluku (Lo) *Eleusine indica & spp./ Oxytenanthera abyssinica*
silenje (Lo) *Imperata cylindrica*
silk cotton (E) *Ceiba pentandra*
silver fern (E) *Pityrogramma calomelanos*
silver vernonia (E) *Vernonia natalensis*
silya-tunga (Ny) *Lannea discolor*
simalembo (Lo) *Gnidia chrysantha*
simeye (Lo) *Antidesma venosum*
simezi (Lo) *Antidesma venosum*
simonyi-monyi (N) *Abrus precatorius*
simple-thorned torchwood (E) *Balanites aegyptica*
simu-zingili (Tk) *Gloriosa superba*
simwakasala (To) *Albizia amara*
simwami (Tk) *Xeroderris stuhlmannii*
sindoke (Lo) *Steganotaenia araliacea*
single greenthorn (E) *Balanites aegyptica*
sinuke (Lo) *Hemizygia bracteosa*
sipewa-musazinoko (I) *Sorgum sp.*
sipumba-ŋombe (Lo) *Entada abyssinica*
sisyanga-syaba (Lo) *Psorospermum febrifugum*
siyaba (Tu) *Arachis hypogaea*
siyungwa (To) *Cleome gynandra*
sjambokpeul (Afr) *Cassia abbreviata*
sjijete (K) *Phragmites australis*
skurweblaarkoorsbessie (Afr) *Croton menyhartii*
skurweblaarwitgat (Afr) *Boscia angustifolia*
skyflower (E) *Duranta erecta*
slangboom (Afr) *Stylosanthes fruticosa*
sleepy morning (E) *Waltheria americana*
slender-leaf rattlebox (E) *Crotalaria ochroleauca*
small bead-bean (E) *Maerua triphylla*
small-fruited potato-bush (E) *Phyllanthus ovalifolius*
smalpeulbasboontjie (Afr) *Elephantorrhiza goetzii*
smelly cat's whiskers (E) *Rotheca myricoides*

smooth-fruit zanha (E) *Zanha golugensis*
smooth-fruited lightning bush (E) *Clutia abyssinica*
smooth tinderwood (E) *Clerodendrum glabrum*
snake apple (E) *Solanum incanum*
snake bean (E) *Swartzia madagascariensis*
snakeweed (E) *Euphorbia hirta*
snot-apple (E) *Azanza garckeana*
snowberry tree (E) *Flueggea virosa*
snuffbox tree (E) *Oncoba spinosa*
snuifkalbassie(Afr) *Oncoba spinosa*
soap creeper (E) *Helinus integrifolius*
sodom apple (E) *Calotropis procera/ Solanum incanum*
sokwe (Tk) *Hypoestes verticillaris*
somali khat (E) *Catha edulis*
sondolien (Afr) *Dodonea viscosa*
songoma (Ny) *Flacourtia indica*
sonkhole-waŋono (Ny) *Dicoma sessiliflora sbsp. sessiliflora*
sopa (Ny) *Lagenaria sphaerica*
sore-eye flower (E) *Boophane disticha*
sourplum (E) *Ximenia caffra*
south-american horseweed (E) *Conyza bonariensis*
soyo (Ny) *Leonotis spp./ Microglossa spp./ Rottboelia exaltata/ Vernonia spp.*
soyo (Tu) *Vernonia spp.*
spanish needled (E) *Bidens pilosa*
spanish shawl (E) *Dissotis rotundifolia*
spanish tea (E) *Chenopodium ambrosoides*
spelonkentee (Afr) *Catha edulis*
spiny amaranth (E) *Amaranthus spinosus*
spiny bur (E) *Acanthospermum australis*
spiny jackal-berry (E) *Diospyros senensis*
spiny monkey-orange (E) *Strychnos spinosa*
springsaadboom (Afr) *Sapium ellipticum*
stamperwood (E) *Ehretia obtusifolia*
star-flowered bitter tea (E) *Vernonia colorata*
stemfruit (E) *Synsepalum brevipes*
stickypod (E) *Senna singueana*
stinkbas (Afr) *Pittosporum viridiflorum*
stinking weed (E) *Senna occidentalis*
stomach-bush (E) *Dicoma anomala*
straw flower (E) *Helichrysum panduratum*
sugarplum (E) *Uapaca kirkiana*
suikerteebossie (Afr) *Pollichia campestris*
sunflower (E) *Helianthus annuus*
sunta (Ny/ Tu) *Cleome gynandra*
sunta-conde (Ku) *Cleome hirta/ Cleome gynandra*
suntha (Ku) *Cleome hirta/ Cleome gynandra*
suurpruim (Afr) *Ximenia caffra*
suwawa (Tu) *Khaya anthotheca*

swamp cabbage (E) *Ipomoea aquatica*
swamp milkweed (E) *Calotropis procera*
swartwaterbossie (Afr) *Plumbago zeylanica*
sweet broom (E) *Scoparia dulcis*
sweethearts (E) *Bidens pilosa*
swinya (Ny) *Gnidia chrysantha*
syacibula (L/ N) *Burkea africana*
syamanya (To) *Clerodendrum capitatum*
syaumbu (Ku) *Lannea discolor*
syava-nkunzi (Ny) *Balanites aegyptica*
syene (Nm) *Ehretia obtusifolia*
sye-wiye (Nm) *Triumfetta rhomboidea*
syimonyi-monyi (Lo) *Abrus precatorius*
syimwaka-sala (K) *Peltophorum africanum*
syonga (To) *Antidesma venosum*
syungwa (To) *Cleome gynandra*
tail-grape (E) *Artabotris monteiroae*
tall morning-glory (E) *Ipomoea purpurea*
tamarind (E) *Tamarindus indica*
tambala (B/ Ku) *Acalypha spp./ Argemone mexicana/ Tricliceras spp.*
tambala (Ny) *Gloriosa superba/ Wormskioldia spp.*
tarberry resin-tree *Ozoroa reticulata*
tassel berry (E) *Antidesma venosum*
tassel flower (E) *Emilia coccinea*
tati (Tk) *Aloe spp.*
tawe-tawe (Ny) *Senna singueana*
tawinda-mazimu (Ny) *Hoslundia opposita*
tea senna (E) *Chamaecrista mimosoides*
teerbessieharpuisboom (Afr) *Ozoroa reticulata*
temya (Ny/ Tu) *Strychnos cocculoides/ S. spinosa*
tena-tena (B/ Ku) *Pistia stratiotes*
tengwa (Ny) *Zornia glochidiata*
tenza (Ny) *Croton macrostachys/ Grewia flavescens/ G. micrantha*
thenda-sipa (Tu) *Plumbago zeylanica*
thicket thorn (E) *Catunaregam obovata*
thorn-apple (E) *Datura metel*
thorny amaranth (E) *Amaranthus spinosus*
thorny mulberry (E) *Maclura africana*
three-flower beggarweed (E) *Desmodium triflorum*
thuysa (Ny) *Crotalaria natalitia*
thyme-leaf spurge (E) *Euphorbia thymifolia*
tick berry (E) *Lantana camara*
tick clover (E) *Desmodium triflorum*
tick tree (E) *Sterculia africana*
tilwena (Ku) *Ocimum spp.*
tinana (Le) *Pavetta crassipes*
tinde (Lo) *Neorautanenia mitis*
tindi-ngoma (B/ Ku) *Corchorus olitorius & spp.*

tipa (Lo) *Flacourtia indica*
tobacco (E) *Nicotiana tabacum*
tokwe (Tu) *Cyperus articulatus*
tomo (Lo) *Ricinus communis*
tondo-tondo (To) *Lannea discolor*
tonje (Ny) *Gossypium barbadense*
toothbrush tree (E) *Salvadora persica*
tossa jute (E) *Corchorus olitorius*
towerneut (Afr) *Maprounea africana*
trailing tibouchina (E) *Dissotis rotundifolia*
traveller's joy (E) *Clematis brachiata*
tree hibiscus (E) *Azanza garckeana*
trellis vine (E) *Pergularia daemia*
triangle tops (E) *Blighia unijugata*
tridax daisy (E) *Tridax procumbens*
tropical chickweed (E) *Drymaria cordata*
true indigo (E) *Idigofera tinctoria*
tsamba (Ny) *Brachystegia utilis/ B. floribunda/ B. spiciformis/ Julbernardia globiflora*
tsoyo (Ny) *Vernonia amygdalina*
tsua (Ny) *Sclerocarya birrea*
tsula (Ny) *Sclerocarya birrea*
tui-tui (Ku/ Ny) *Acacia sieberiana*
tufukusa (I) *Parinari curatellifolia*
tukona (Lo) *Smilax anceps*
tukupu-twambula (Le) *Parinari curatellifolia*
tulama (Tu) *Combretum adenogonium*
tulemba (Lb) *Pterocarpus angolensis*
tumbleweed (E) *Boophane disticha*
tumpo (Lv) *Jatropha curcas*
tungwi (Ny) *Momordica foetida*
tupempe (B/ M) *Hymenocardia acida*
tute (B/ Ku/ Lb/ Le/ Ny) *Manihot esculenta*
twambula (Le) *Parinari curatellifolia*
ubule (B) *Eleusine coracana*
ufukula (Ny) *Lannea edulis*
ugegati (Nm) *Abrus fruticulosus*
ukole (Ku) *Azanza garckeana*
ulula-mweyupe (Nm) *Acacia seyal*
ulunga (Ny) *Abrus precatorius*
ulupapi (Lb/ Le) *Securidaca longipedunculata*
umbanga (I) *Pterocarpus angolensis*
umbomba (I) *Brachystegia boehmii*
umbrella thorn (E) *Acacia tortilis*
umbubo (To) *Vangueria infausta*
umganu (Ny) *Sclerocarya birrea*
umkwakwa (Ny) *Strychnos spinosa*
umnulu (Ny) *Balanites aegyptiaca*
umpumba (Lo) *Acacia polyacantha*

umsasane (Ny) *Acacia tortilis*
umsikili (Lo) *Trichilia emetica*
umthu-mgwa (Ny) *Oncoba spinosa*
umtu-nduka (Ny) *Ximenia caffra*
umufumbe (B) *Bauhinia reticulata*
umulamba-lamba (Lb) *Lannea edulis*
umune-nene (Lb) *Burkea africana*
umuwanga (K) *Burkea africana*
unaha-mbalala (Tu) *Rourea orientalis*
usenga (Nm) *Margaritaria discoidea*
usongwe (Ku) *Acacia hockii*
usyemeli (Nm) *Artemisia afra*
uteka (Ny) *Cymbopogon citratus*
uzambwiya (Ny) *Hoslundia opposita*
uzimwe (Ku) *Acacia hockii/ A. nilotica*
uzimwe (Ny) *Acacia nilotica*
valsassegai (Afr) *Maesa lanceolata*
veldemispel (Afr) *Vangueria infausta*
velvet bushwillow (E) *Combretum molle*
velvet-fruit zanha (E) *Zanha africana*
velvet jackal-coffee (E) *Tricalysia pallens*
velvet leaf (E) *Cissampelos pareira*
vergeet-my-nie boom (Afr) *Duranta erecta*
vick's plant (E) *Plectranthus cylindraceus*
vierk-boom Afr) *Xeroderris stuhlmann*
vikau (Tu) *Manihot esculenta*
vimwa-manze (Tu) *Citrullus lanatus*
vinangwa (Ny/ Tu) *Manihot esculenta*
violet tree (E) *Securidaca longepedunculata*
vipapayi (Ny/ Tu) *Carica papaya*
vlamklimop (Afr) *Combretum microphyllum*
voelsitboom (Afr) *Antidesma venosum*
vumba (Ny) *Lippia javanica*
wabona (Ny) *Ximenia americana/ X. caffra*
walking-stick (E) *Strychnos henningsii*
wanyisi (I) *Allium cepa*
waŋono (Ny) *Senna singueana*
water cabbage (E) *Pistia stratiotes*
water convolvulus (E) *Ipomoea aquatica*
water morning-glory (E) *Ipomoea aquatica*
water-lettuce (E) *Pistia stratiotes*
watermelon (E) *Citrullus lanatus*
water-spinach (E) *Ipomoea aquatica*
waukulu (Ny) *Amaranthus hybridus/ Combretum molle/ Securinega virosa/ Terminalia stenostachya*
waxberry (E) *Pollichia campestris*
west-african sarsparilla (E) *Smilax anceps*
west-indian chickweed (E) *Drymaria cordata*

west-indian woodnettle (E) *Laportea aestuans*
wewenaar (Afr) *Bidens pilosa*
wewenaartjies (Afr) *Bidens pilosa*
wezenzele (Ln/ Lo) *Burkea africana*
whiteball rubber (E) *Landolphia owariensis*
whiteberry bush (E) *Flueggea virosa*
white cape beech (E) *Pittosporum viridiflorum*
white-flowered leadwort (E) *Plumbago zeylanica*
white gardenia (E) *Gardenia ternifolia*
white kirkia (E) *Kirkia acuminata*
white rubber-vine (E) *Landolphia owariensis*
white snow (E) *Drymaria cordata*
white spurflower (E) *Plectranthus laxiflorus*
white stinkwood (E) *Celtis africana*
white thorn (E) *Faidherbia albida*
white weed (E) *Ageratum conyzoides*
wild calabash (E) *Lagenaria sphaerica*
wild coffee (E) *Senna occidentalis*
wild cotton *Xysmalobium undulatum*
wild cucumber (E) *Momordica charantia*
wild custard apple (E) *Annona senegalensis*
wild dagga (E) *Leonotis leonurus*
wildekalbas (Afr) *Lagenaria sphaerica*
wildeknoffel (Afr) *Tulbaghia alliacea*
wilderpietersieliebos (Afr) *Heteromorpha arborescens*
wildesering (Afr) *Burkea africana*
wildevyeboom (Afr) *Ficus sur*
wild garlic (E) *Tulbaghia alliacea*
wild grape (E) *Cyphostemma buchananii/ Lannea edulis*
wild indigo (E) *Tephrosia purpurea*
wild jackfruit (E) *Treculia africana*
wild lucerne (E) *Stylosanthes fruticosa*
wild medlar (E) *Vangueria infausta*
wild rubber (E) *Diplorrhynchus condylocarpon*
wild seringa (E) *Burkea africana*
wild tea (E) *Catha edulis/ Lippia javanica*
wild tomato (E) *Physalis angulata*
wild willow (E) *Salix mucronata*
wind-pod (E) *Xeroderris stuhlmann*
winged bersama (E) *Bersama abyssinica*
winter cassia (E) *Senna singueana*
winter cherry (E) *Physalis angulata/ Withania somnifera*
winter-flowering senna (E) *Senna singueana*
winter thorn (E) *Faidherbia albida*
wiregrass (E) *Eleusine indica*
witbessiesbos (Afr) *Flueggea virosa*
witboekenhout (Afr) *Pittosporum viridiflorum*
withaak (Afr) *Acacia tortilis*

witsering (Afr) *Kirkia acuminata*
witstinkhout (Afr) *Celtis africana*
wollerige kapperbos (Afr) *Capparis tomentosa*
woodland croton (E) *Croton macrostachys*
woolly caper-bush (E) *Capparis tomentosa*
woolly rattlepod (E) *Crotalaria incana*
wormseed (E) *Chenopodium ambrosoides*
worsboom (Afr) *Kigelia africana*
wululu (Ln) *Khaya anthotheca*
wunkomukomu (B/ Ln) *Aporrhiza nitida*
wuzimwa (M/ Tu) *Cleome gynandra*
yagi-yanzovu (Nm) *Cussonia arborea*
yellow commelina (E) *Commelina africana*
yellow miraculous-berry (E) *Synsepalum passargei*
yellow trumpet (E) *Costus spectabilis*
yembe (B/ Ku) *Mangifera indica*
yinanana (Lb) *Vangueria infausta*
zakalanda (To) *Moringa oleifera*
zavuma (Ny) *Keetia zanzibarica*
zeepbas (Afr) *Pittosporum viridiflorum*
zimono (Ln/ Lv) *Ricinus communis*
ziye (Ku) *Brachystegia boehmii*
ziye-nganza (Ku) *Brachystegia boehmii*
zunda (B) *Tecomaria capensis*
zungu-bele (B) *Trichodesma zeylanica*
zungu-mbwa (Ny) *Conyza sp.*

VERNACULAR NAMES: SOURCES

1)   **Astle**, W.L., Phiri, P.S.M., & Prince, S.D. (1997). A dictionary of vernacular - scientific names of plants of the mid-Luangwa valley, Zambia, in *Kirkia* (1997) 16 (2):161 - 203.(Harare: Government Printer)

1a) **Bally**, P.R.O. (1937) Native Poisonous & Medicinal Plants of East Africa , in *Kew Bulletin,* Additional Series,vol.10 (RBG Kew)

1d) **Burkill**,H.M.(1985) *Useful Plants of West Tropical Africa* vol.1 (RBG, Kew)

1e) **Burkill**,H.M. (1994) *Useful Plants of West Tropical Africa* vol. 2 (RBG Kew)

1f) **Burkill**, H.M. (1995) *Useful Plants of West Tropical Africa* vol. 3 (RBG Kew)

1g) **Burkill**, H.M. (1997) *Useful Plants of West Tropical Africa* vol. 4 (RBG, Kew)

1h) **Burkill**, H.M. (2000) *Useful Plants of West Tropical Africa* vol. 5 (RBG, Kew)

1b) **Cunningham**, A.B., & Davis, G.W. (1997) Human Use of Plants, in *Vegetation of Southern Africa*, ed. R.M.Cowling, D.M. Richardson, & S.M.Pierce (Cambridge: CUP)

1ba) **Dale**, I.R., & Greenway, P.J. (1961) *Kenya Trees and Shrubs* (Nairobi: Buchanan's Kenya Estates Limited, in association with Hatchards, London)

1c)   **Derricourt**, R. (1985) *Man on the Kafue* (London: Ethnographia)

2)   **Fanshawe**, D.B., & Clough, C.D. (1967). Poisonous Plants of Zambia. (Forestry Research Bulletin no.1 [revised]) (Lusaka: Govt. Printer)

3)   **Fanshawe**, D.B. (1995) *Checklist of Vernacular names of the Woody Plants of Zambia.* (Lusaka: Govt. Printer)

3a) **Fanshawe**, D.B. (1962, revised 1968) *Fifty common trees of Northern Rhodesia* (Lusaka: Govt. Printer)

3c) **Fanshawe**, D.B. & Mutimushi, J.M. (1965) *A Checklist of plant names in the Bemba languages:* Ministry of Lands & Natural Resources bulletin (Lusaka: Govt. Printer)

3d) **Fanshawe**, D.B. & Mutimushi, J.M. (1971) *A Checklist of plant names in the Lozi languages:* Min.of Rural Devlpt.,Forest Research Bulletin no. 23 (Lusaka: Govt. Printer)

3e) **Fanshawe**, D.B. & Mutimushi, J.M. (1969) *A Checklist of plant names in the Nyanja languages:* Min.of Rural Devlpt., Forest Research Bulletin no. 21 (Lusaka: Govt. Printer)

4)   **Fowler**, D.G. (2000) *Dictionary of Ila Usage 1860 – 1960* Monographs from the International African Institute, 5. (Hamburg: Lit-Verlag)

4a) **Fowler**, D.G. (2002) *The Ila Speaking* Monographs from the International African Institute, 7 (Hamburg: Lit-Verlag}

5)   **Fowler**, D.G. (2002b) Traditional Ila Plant Remedies from Zambia. In *Kirkia* 18 (19): 35-48 (Harare: Government Printer)

5b) **Gelfand**, M., S. Mavi, R.B.Drummond, & B. Ndemera (1985) *The Traditional Medical Practitioner in Zimbabwe* (Gweru: Mambo Press)

6)   **Gilges**, W. (1964) Some African Poison Plants and Medicines of Northern Rhodesia, in *Occasional Papers of the Rhodes-Livingstone Museum,* 1-16 (Livingstone: Rhodes-Livingstone Museum)

7)   **Githens**, T.S. (1948) *Drug Plants of Africa:* African Handbooks no. 8 (3Pennsylvania:University of Pennsylvania Press)

7a) **Haapala**, T., G.P.Mwila, K.Kabamba, & E.Chishimba (1994) *Some introduction to some Medical Plants found in Zambia* (Mufulira: Finnish Volunteer Service Copperbelt Regional Research Station)

7ab) **Haerdi**, F., J. Kerharo, & J.G. Adam (1964) *Afrikanische Heilpflanzen* (Basel: Verlag fur Recht und Gesellschaft AG)

7b) **Honigsbaum**, M. (2001) *The Fever Trail* (London: Macmillan)

7bb) **Kerharo**, J., & Adam, J.G.(1974) *La Pharmacopee Senegalaise Trad.* (Paris:Vigot Freres)

7bc) **Kerharo**, J., & Bouquet, A. (1950) *Plantes medicinales & toxiques de la Cote-d'ivoire-Haute-Volta* (Paris: Vigot Freres)

7c) **Kokwaro**, J.O. (1976) *The Medicinal Plants of East Africa* (Kampala: East Africa Literature Bureau)

7d) **Mccurrach**, J.C. (1960) *Palms of the World* (New York: Harper & Brothers)

8) **Mitchell**, B.L.(1963) *A First List of Plants collected in the Kafue National Park* (Lusaka: Govt. Printer)

9) **Morris**, B. 1996. *Chewa Medical Botany.* Monographs from the International African Institute, 2. (Hamburg: Lit-Verlag)

9a) **Nair**, D.M.N. (1967) *Selected families of Zambian flowering plants* (Lusaka: University of Zambia)

10) **Palgrave**, K.C. (1957) *The Trees of Central Africa* (Salisbury: National Publications Trust, Rhodesia and Nyasaland)

11) **Palgrave**, K.C. (1983) *Trees of Southern Africa* (Cape Town: C. Struik)

11a) **Palgrave**, K.C. (2002) *Trees of Southern Africa* rev. edn. (Cape Town: Struik Publishers)

12) **Sekeli**, P. (1999) Letter to the author (Kitwe: Forestry Department)

13) **Smith**, E.W.,& Dale, A.M.(1920) *The Ila-speaking Peoples of Northern Rhodesia* (London: Macmillan)

14) **Storrs**, A.E.G.(1979) *Know your Trees: some of the common trees found in Zambia* (Ndola: The Forest Department)

15) **Storrs**, A.E.G.(1982) *More about Trees* (Ndola: The Forest Department)

16) **Symon**, S.A. (1959) *African Medicine of the Mankoya District, Northern Rhodesia* (Lusaka: Rhodes-Livingstone Communication no.15)

17) **Torrend**, J. (1931) *An English-vernacular Dictionary of the Bantu-Botatwe Dialects of Northern Rhodesia* (London: Kegan Paul, Trench, Trubner & Co.)

18) **Trapnell**, C.G., and Clothier, J.N. (1957) *The Soils, Vegetation and Agricultural systems of North-Western Rhodesia* (Lusaka: Govt. Printer)

19) **Trapnell**, C.G. (2000) *The Ecological Survey of Zambia: The Traverse Records of C.G.Trapnell, 1932-1943*, vol.1, ed. by P. Smith (London: RBG Kew)

20) **Trapnell**, C.G. (2000) *The Ecological Survey of Zambia: The Traverse Records of C.G.Trapnell, 1932-1943*, vol. 2, ed. by P. Smith. (London: RBG Kew.)

21) **Turner**, V.W. (1964) Lunda medicine and the treatment of disease, in *Occasional Papers of the Rhodes-Livingstone Museum,* nos.1-16 (Livingstone: Rhodes-Livingstone Museum)

22) **Watt**,J.M., and M.G.Breyer-Brandwijk (1962) *The Medicinal and Poisonous Plants of South and East Africa* (Edinburgh and London: E.& S. Livingstone)

23) **White**, F. (1962) *The Forest Flora of Northern Rhodesia* (London: OUP)

The following sources have also been consulted about names:

Bingham, M. 1999. *Preliminary National Checklist of Zambia* (Lusaka: priv. printed).

Phiri, P.S.M. (2005) *A checklist of Zambian vascular plants.* Southern African Botanical Diversity Network Report No.32. (Pretoria: SABONET)

Smith, P.P. 1997. A Preliminary Checklist of the Vascular Plants of the North Luangwa National Park, Zambia. In *Kirkia* 16 (2): 205-245 (Harare: Govt. Printer)

Williams, R.O.(1949) *The useful & ornamental plants in Zanzibar & Pemba* (Zanzibar)

APPENDIX 1: GEOGRAPHICAL SPREAD OF REMEDIES

a) Reported in three continents (12 species):

Adansonia digitata
Ageratum conyzoides
Cardiospermum halicababum
Cissampelos pareira
Evolvulus alsinoides
Leonotis nepetifolia
Ocimum canis
Ricinus communis
Senna occidentalis
Tamarindus indica
Verbena officinalis
Waltheria americana

b) Reported in two continents (19 species):
Aerva lanata
Amaranthus spinosus
Bidens pilosa
Cymbopogon citratus
Crossopteryx febrifuga
Delonix regia
Desmodium gangeticum
Eleusine indica
Flacourtia indica
Helianthus annuus
Lantana camara
Moringa oleifera
Mukia maderaspatana
Ocimum americanum
Paullinia pinnata
Phyllanthus fraternus
Plumbago zeylanica
Psidium guajava
Uraria picta

c) Reported in more than one African country (94 species):
Acacia nilotica
Acacia robusta
Acacia tortilis
Adenia gummifera
Afzelia quanzensis
Albizia zygia
Allium sativum
Ampelocissus africanus
Annona senegalensis

Argemone mexicana
Anthocleista schweinfurthii
Aristolochia albida
Artemisia afra
Balanites aegyptica
Bauhinia reticulata
Blighia unijugata
Brachystegia spiciformis
Bridelia duvigneaudi
Burkea africana
Cannabis sativa
Cardiospermum grandiflorum
Carica papaya
Catunaregam obovata
Centella asiatica
Chenopodium ambrosoides
Clausena anisata
Clematis brachiata
Clerodendrum eriophyllum
Clerodendrum glabrum
Combretum molle
Commelina africana
Corchorus olitorius
Cordia sinensis
Crossopteryx febrifuga
Croton gratissimus
Dicoma anomala
Diospyros mespiliformis
Dodonea viscosa
Entada abyssinica
Faidherbia albida
Flueggea virosa
Gardenia ternifolia
Gymnosporia senegalensis
Hallea stipulosa
Harungana madagascariensis
Heliotropium indicum
Heteromorpha arborescens
Holarrhena pubescens
Hoslundia opposita
Hymenocardia acida
Hyptis pectinata
Jatropha curcas
Kalanchoe crenata
Khaya anthotheca
Kigelia africana
Lannea discolor
Lippia javanica

Mangifera indica
Microglossa pyrifolia
Mikania chenopodifolia
Momordica charantia
Ocimum gratissimum
Olea europea
Oncoba spinosa
Parinari curatellifolia
Pavetta crassipes
Phyllanthus muellerianus
Pittosporum viridiflorum
Platostoma africanum
Psorospermum febrifugum
Pterocarpus angolensis
Salix mucronata
Salix subserrata
Securidaca longepedunculata
Senna petersiana
Senna singueana
Smilax anceps
Steganotaenia araliacea
Strychnos innocua
Strychnos spinosa
Stylosanthes fruticosa
Teclea nobilis
Trichilia emetica
Triumfetta rhomboides
Vangueria infausta
Vernonia amygdalina
Vernonia brachycalyx
Vernonia colorata
Vernonia natalensis
Vernonia species
Ximenia americana
Ximenia caffra
Zanha africana
Zanthoxylum chalybeum

d) Reported in more than one survey in the same country (22 species)
Abrus fruticulosus
Albizia amara
Albizia anthelmintica
Albizia gummifera
Allophylus africanus
Boscia angustifolia
Bridelia ferruginea
Brucea antidysenterica
Cassia abbreviata

Clerodendrum capitatum
Commelina africana
Commelina imberbis
Conyza pyropappa
Crotalaria recta
Croton macrostachys
Dalbergia boehmii
Duranta repens
Entada africana
Erythrophleum suaveolens
Momordica balsamica
Mucuna pruriens
Senna didymobotria

e) The remaining 334 plants are reported in only one survey

APPENDIX 2: PLANTS TESTED AGAINST *Plasmodium falciparum*:

Acacia nilotica *El Tahir et al. (1999a) ; (Kirira et al, 2006)*
Acacia tortilis *Clarkson et al. (2004)*
Acanthospermum hispidum *(Sanon et al. (2003) ; Bero et al. (2009)*
Adansonia digitata *Gessler et al. (1994)*
Aerva lanata *Gessler et al. (1994)*
Amaranthus spinosus *Cai et al. (2003)*
Ampelocissus africanus *Gessler et al.(1994)*
Annona senegalensis *Ajaiyeoba et al. (2005)*
Argemone mexicana *Willcox et al. (2007) ; Graz et al. (2009)*
Artemisia afra *Kassa et al. (1998) ; Kraft et al. (2003)*
Azanza garckeana *Connelly et al. (1996)*
Bidens pilosa *Krettli et al. (2000) ; Andrade-Netto et al. (2004) ; Clarkson et al. (2004)*
Blighia unijugata *Oderinde et al. (2009)*
Brucea antidysenterica *O'Neill et al. (1986)*
Carica papaya *(Bhat et al. 2001; Ngemenya et al. (2004); Ayoola et al. (2008) Titanji et al. (2008)*
Carissa edulis *Kirira et al. (2006)*
Catha edulis *Clarkson et al. (2004)*
Chenopodium ambrosoides *Schwikkard et al. (2002)*
Cissampelos pareira *Gessler et al.(1994); Rukunga et al. (2008)*
Clerodendrum eriophyllum *Muthaura et al. (2007)*
Combretum adenogonium *Maregesi et al. (2010)*
Combretum molle *Schwikkard et al. (2002) ; Gansané et al. (2010)*
Crossopteryx febrifuga *Elufioye et al.(2004)*
Croton gratissimus *Okunade et al. (1987) ; Clarkson et al.(2004) ; Pillay et al. (2008)*
Croton menyhartii *Clarkson et al.(2004)*
Cymbopogon citratus *Obih et al.(1986); Pousset (2004); Clarkson et al. (2004)*
Cyperus articulatus *Rukunga et al. (2008)*

Duranta repens *Nikkon et al. (2006)*
Faidherbia albida *Salawu et al.(2010)*
Ficus sur *Muregi et al. (2003)*
Gymnosporia senegalensis *Gessler et al.(1994); El Tahir et al. (1999b)*
Harrisonia abyssinica *El Tahir et al. 1999b); Kirira et al. (2006); Maregesi et al. (2010)*
Hoslundia opposita *Schwikkard et al. (2002)*
Kigelia africana *Schwikkard et al. (2002)*
Lantana camara *Hakizamungu (1988); Weenen et al.(1990); Nkunya et al.(1991); Sharma et al.(1998); Clarkson et al.(2004); Jonville et al.(2008)*
Leonotis nepetifolia *Clarkson et al.(2004)*
Lippia javanica *Clarkson et al. (2004); Pillay et al. (2008)*
Ludwigia erecta *Muthaura et al. (2007)*
Margaritaria discoedia *Schwikkard et al. (2002)*
Microglossa pyrifolia *Muganga et al. (2010)*
Momordica balsamina *Clarkson et al. (2004) ; Benoit-Vical et al.(2006)*
Momordica charantia *Amorim et al.1991; Ueno et al.(1996); Munoz et al. (2000); Mesia et al. (2007)*
Morinda lucida *Tona et al. (1999); Schwikkard et al. (2002)*
Neorautanenia mitis *Joseph et al.(2004)*
Ocimum americanum *Clarkson et al.(2004); Pillay (2008)*
Pavetta crassipes *Sanon et al. (2003)*
Phyllanthus fraternus *Schwikkard et al. (2002)*
Phyllanthus muellerianus *Zirihi et al. (2005)*
Phyllanthus reticulatus *Omulokoli et al.(1997)*
Physalis angulata *Lusakibanza et al. (2010)*
Pittosporum viridiflorum *Clarkson et al. (2004) ; Muthaura et al. (2007).*
Plumbago zeylanica *Simonsen et al. (2001); Clarkson et al.(2004); Pillay et al.(2008)*
Psidium guajava *Blair (1991); Garcia-Barrega (1992); Nundkumar et al.(1992); Gessler et al. (1994) Pillay et al. (2008)*
Ranunculus multifidus *Clarkson et al. (2004)*
Rhamnus prinioides *Muregi et al. (2003)*
Ricinus communis *Clarkson et al. 2004; Pillay et al. (2008)*
Sclerocarya birraea *Gathirwa et al. (2007)*
Securidaca longepedunculata *Bah et al. (2007)*
Senna abbreviata *Connelly et al.(1996)*
Senna occidentalis *Tona et al. (1999, 2001); Clarkson et al. (2004)*
Senna petersiana *Connelly et al.(1996)*
Sesbania sesban *Maregesi et al. (2010)*
Setaria megaphylla *Clarkson et al.(2004)*
Sida acuta acuta *Banzouzi et al.(2004)*
Strophanthus eminii *Innocent et al. (2009)*
Strychnos spinosa *Bero et al. (2009)*
Trichilia emetica *El Tahir et al. (1999b); Clarkson et al. (2004); Pillay et al.(2008)*
Triumfetta welwitschii *Clarkson et al. (2004)*
Uapaca nitida *Schwikkard et al. (2002)*
Vangueria infausta *Nundkumar et al. (2002)*
Vernonia amygdalina *Hakizamungu et al. (1988) ; Phillipson et al. (1993); Bogale et al. (1996) ; Tona et al. (1999); de Madureira et al. (2002)*

Vernonia brachycalyx *Oketch-Rabah et al. (1999)*
Vernonia natalensis *Kraft et al. (2003); Clarkson et al. (2004)*
Waltheria americana *Jansen et al. (2010)*
Withania somnifera *Kirira et al., (2006)*
Zanthoxylum chalybeum *Gessler et al.(1994); Rukunga et al. (2008); Muganga et al. (2010)*

APPENDIX 3: SENNA OCCIDENTALIS TOXICITY

S. occidentalis, a weed invasive of farmlands, has long been suspected of poisoning stock, but when in 1962 Watt and Breyer-Brandwijk revised their magisterial work on medicinal and poisonous plants the position was still not clear. Trials in Northern Queensland had been ambiguous: *Greshoff, 1913, Webb 1948*. In South Africa, Van der Walt had concluded that the plant was not poisonous to sheep: *Van der Walt 1946; Watt et al.1962.* However, the plant has been extensively tested over recent years, and its toxic properties demonstrated beyond doubt.

For ethnobotany, phytochemistry and pharmacology of *S.occidentalis,*
    see *Yadav et al. (2010)*
For chemical analysis of *S. occidentalis,*
    see *Singh et al. (1985)*
For human deaths associated with *S.occidentalis,*
    see *Vashishta et al. (2007)*
For clinical spectrum and histopathology of *S. occidentalis* poisoning in animals and children, see *Vashishtha et al. (2009)*

Laboratory tests:

cattle, *O'Hara et al. (1970), Barth et al. (1994)*
chickens, *Graziano et al. (1983), Flory et al. (1992), Calore et al. (1997, 1998), Cavaliere et al. (1997), Haraguchi et al. (1998), Silva et al. (2003)*
goats *Suliman et al. (1982, 1986)*
horses *Martin et al. (1981)*
pigs *Colvin et al. (1986)*
rabbits *O'Hara et al. (1974), Tasaka et al. (2000)*
rats *Calore et al. (1999), Calore et al. (2000, 2002), Mariano-Sousa (2009), Abubakar et al. (2010)*

CPSIA information can be obtained at www.ICGtesting.com
Printed in the USA
244363LV00001BA/1/P